HYDRAULIC MOTORS:

TURBINES AND PRESSURE ENGINES

HYDRAULIC MOTORS

TURBINES AND PRESSURE ENGINES

FOR THE USE OF

ENGINEERS, MANUFACTURERS, AND STUDENTS

BY

G. R. BODMER, A.M. INST. C.E.

Fredonia Books
Amsterdam, The Netherlands

Hydraulic Motors:
Turbines and Pressure Engines

by
G. R. Bodmer

ISBN: 1-4101-0191-6

Reprinted from the 1895 edition

Fredonia Books
Amsterdam, The Netherlands
http://www.fredoniabooks.com

In order to make original editions of historical works available to scholars at an economical price, this facsimile of the original edition of 1895 is reproduced from the best available copy and has been digitally enhanced to improve legibility, but the text remains unaltered to retain historical authenticity.

PREFACE TO FIRST EDITION

THE author's attention having been directed to the absence of any modern English work of a sufficiently comprehensive character on the subject of turbines, he was induced to write the present treatise in the hope of to some extent repairing the deficiency. His object has been to give in as elementary a manner as is consistent with accuracy and thoroughness an explanation of the principles underlying the action of turbines and water-pressure engines, and of the application of these principles to the design of such motors, together with descriptions of the most distinctive types, accounts and results of experiments, numerical examples, comparisons of theoretical with practical results, and a brief review of the various methods of water measurement.

Historical matter has been purposely avoided, as well as descriptions of obsolete forms of motors.

The theory of reaction turbines adopted by the author is essentially that of Redtenbacher, modified somewhat in detail on the lines followed by Professor Zeuner in his lectures at continental technical universities.

In applying the theory to the design of turbines, the

author has taken, he believes, a course somewhat different from that generally in vogue, and, in his opinion, better suited for the comparison of different types of turbines, while at the same time it allows the distinctive character of the motor to be determined with accuracy from the very beginning. The leading feature of this method consists in the assumption of the ratio of the guide outflow-area to the bucket outflow-area as a starting-point for the construction of any turbine.

The investigations of the performances and peculiarities of American wheels are, as far as the author is concerned, original, although only the accepted theory has been employed.

The tables of co-efficients of flow, theoretical efficiencies, comparison of theoretical with calculated results, and dimensions of turbines actually made, have been especially compiled and calculated for this work by the author himself.

In writing a technical work on any subject, it is always a matter of some difficulty to know exactly where to draw the line between what should and what should not be included in it, and in any case it will certainly not please all critics; for one there will be too much mathematics, for another too little history, a third will find fault because nothing is said about the actual manufacture in the shop, while a fourth wants to know why all mention of his own particular invention has been omitted!

There is one criticism, however, which in the present case the author thinks it well to meet in advance, as he feels tolerably certain that it will otherwise be made. This is, that mathematics have been too freely employed in the present work. In reply to this, it may be pointed out that in the majority of cases it is shown how, in applying the principles demonstrated to the design of motors, graphical may be substituted for analytical methods. At the same time, nothing beyond elementary analysis is involved in the mathematics employed, and it is essential to completeness that the algebraical expressions for the laws and rules relating to the subject should be given. To those familiar with the use of formulas, they afford the readiest means of stating facts which in the simplest cases can only be verbally defined in very long and elaborate phraseology. Circumstances, too, often arise in which dimensions have to be determined and estimates made for preliminary purposes, under conditions in which a drawing-board and instruments are not available, and then analytical methods come in. The author has himself on more than one occasion made preliminary calculations while travelling on the railway.

In writing the present treatise the author has consulted to a greater or less extent the works or published papers of the following authors: Francis, Lehmann, Meissner, Redtenbacher, Rittinger, Weisbach, Professors Fliegner, Schröter, Thurston, and Unwin, and Mr. Clemens Herschel, and, especially for hydraulic pressure engines, Knoke's work, 'Die Kraftmaschinen des Kleingewerbes.' He is further indebted for much information and for illustrations to the 'Minutes of Proceedings of the

Institution of Civil Engineers,' 'Minutes of Proceedings of the Institution of Mechanical Engineers,' 'Transactions of the American Society of Civil Engineers,' 'Transactions of the American Society of Mechanical Engineers,' 'Engineering,' 'The Engineer,' 'Industries,' 'Civilingenieur,' and 'Zeitschrift des Vereins Deutscher Ingenieure.'

The following gentlemen and firms the author has to thank for information personally supplied : Professor Fliegner of Zürich, Professor Schröter of Munich, Professor Thurston, Blanchod and Co. of Vevey, Bryan Donkin and Co., Mr. W. Günther of Oldham, Mr. John Hastie, Mr. Clemens Herschel of Holyoke, Messrs. Howes and Ewell, the Humphrey Machine Co. of Keene, U.S., Rieter and Co. of Winterthur, Mr. Arthur Rigg, Mr. Theodore Risdon, Mr. Hamilton Smith, Jun., and Mr. Alphonse Steiger representative of Messrs. Escher, Wyss and Co. of Zürich.

In the matter of illustrations of turbines and hydraulic pressure engines Meissner's and Knoke's works have respectively been especially serviceable.

In selecting certain types of motors for description, the author does not thereby imply that such types are superior to others, but as a rule has chosen those about which he was able to obtain the most authentic experimental data and details. Of many firms applied to, only one, Messrs. Rieter and Co., gave the information desired, which has been tabulated in Table B.

The method of measuring the vane angles adopted in this work is different from that generally in vogue, in so

far that the angle formed between the vane and a perpendicular to the direction of rotation is given instead
of the complementary angle. This will be found more
convenient for constructive purposes in radial turbines;
the effect on the formulas is that the sine takes the place
of the cosine, and *vice versâ*. Where, according to the
general practice, an angle is given as say 20°, according
to the author's method—which he has borrowed from
Professor Zeuner—it would be given as $90° - 20° = 70°$.

The author makes no claim to an exhaustive treatment
of the subject, but believes that the matter contained in
his work will be found useful to those for whom it is
intended.

PREFACE TO SECOND EDITION

SINCE the first edition of this work appeared, little of a novel character in the form of hydraulic motors has been introduced to the technical world. In preparing the second edition, therefore, the task of the author has been mainly to carefully revise the original text, where necessary, correct a few slight clerical errors and errors of calculation, and add descriptions of such new installations as appeared especially deserving of notice. Among the latter is the Niagara Falls installation.

In several instances descriptive and theoretical matter has been amplified and supplemented, and numerous diagrams and illustrations have been added.

The new, revised, and supplementary matter will in the main be found under the following headings :—Construction of vane curves; Determination of outflow areas; Girard turbine with draft-tube; "Vortex" turbine; "Pelton" wheel; Turbines at Schaffhausen; Niagara installation; "Purdon-Walters" steam-motor; "Venturi" water-meter; and curves of kinetic energy, pressure difference, and hydraulic losses for various diameter-ratios of radial turbines.

The second part of the work on "Hydraulic Pressure Engines" has not been altered nor enlarged, and is intended merely as a sketch of this branch of hydraulic engineering.

G. R. BODMER.

30, *Walbrook,*
London. E.C.

CONTENTS

CHAP.   PAGE

I.   INTRODUCTION   1

II. ON TURBINES IN GENERAL 24

III. GENERAL THEORY OF REACTION TURBINES ... 47

IV. LOSSES OF ENERGY IN REACTION TURBINES ... 81

V. THE DESIGN OF REACTION TURBINES 99

VI. THE DESIGN OF REACTION TURBINES (*continued*) ... 164

VII. IMPULSE TURBINES 211

VIII. IMPULSE TURBINES (*continued*) 247

IX. SUMMARY OF RULES AND FORMULAS AND NUMERICAL
EXAMPLES 281

X. MEASUREMENT OF THE QUANTITY OF FLOWING WATER 312

XI. DESCRIPTIONS OF AND EXPERIMENTS WITH TURBINES 335

XII. AMERICAN TURBINES 462

XIII. HYDRAULIC PRESSURE ENGINES 500

INDEX 543

HYDRAULIC MOTORS.

CHAPTER I.

INTRODUCTION.

Primitive hydraulic motors.—Real agent in so-called water power.
—Conditions under which water can do work.—Capacity for doing
work or available energy.—Various ways in which available energy
is applicable for performing work.—Different types of motors required.—Usual sources of hydraulic power in Nature ; how made
available, and in what forms.

General principles applicable to hydraulic motors :—
Flow of water in closed channels.—Hydro-dynamic equation.—
Continuity of flow.—Motion, relative and absolute.—Illustrations.

Action of water on curved vanes :—
Deflection of water in turbines.—Force due to deflection.—Avoidance of impact or shock.—Mathematical conditions.—Graphical
construction.—Work done.—Centrifugal action.—Explanation of
increased relative velocity.—Effect of shock.

THE use of water for performing mechanical work dates
back to an extremely early age, some primitive form of
water-wheel having been probably the first hydraulic
motor. In Sicily, what may be considered as a primitive
form of impulse wheel with flat vanes, introduced by the
Saracens, is in use at the present day. The historical
aspect of the subject of water power does not, however,
come within the scope of the present work, which is

B

limited to the principles and practice of the construction
of water motors, and the results obtained with the latter.

In many parts of Europe, and especially in Switzerland
and the mountainous portions of South Germany, the
available water power of the numerous streams and small
rivers is taken advantage of to a large extent for indus-
trial purposes, while in the United States a very considerable proportion of the total power expended in
manufactories, mills, and similar establishments is derived
from hydraulic sources.

In the mountainous districts of Europe, as a rule
the falls are high and the quantities of water small,
while, on the other hand, in America the falls are
moderate or low, and the quantities of water very large,
so that the two regions may be taken as illustrating
extreme phases of the occurrence of water power.
In the United Kingdom there is no doubt that a vast
quantity of water power that might be advantageously
utilized is entirely neglected, but it may be admitted
that in many instances the reasons for this are sufficiently
obvious : cheap coal and choice of locality, and sometimes the necessity for expensive works in order to make
the water power available. When coal becomes dearer,
as it undoubtedly will, more attention must be paid to
the hydraulic resources of the country. Of course the
absence of large rivers or high falls makes it impossible
that a very large proportion of the power required for our
numerous manufactories should be drawn from hydraulic
sources, but making every allowance on this score, there
is yet an extensive field open to the enterprising hydraulic
engineer.

Water power is—perhaps after wind power—the most
natural, and at the same time truly economic source of
energy　Every ton of coal burnt is, as far as our knowledge

goes, irretrievably lost; the process involved in its utilization is *irreversible.* On the other hand, in taking advantage of the energy now uselessly expended on friction, erosion, and other resistances, by our rivers and streams, use is made of a complete cycle already provided by Nature ; the water consumed in producing power is raised again by the sun, falls as rain and replenishes the river or stream from which the supplies are drawn.

In most cases where so-called water power is employed in the performance of mechanical work the real agent is gravity ; the fluid itself being the medium through which the action of gravity is transmitted to the prime motor.

It is quite possible that stones, earth, sand, &c., should be utilized in a similar manner, in each case the motive power being that of gravity.

In order that water may be available for the purpose of doing work, it must be in such a position that it can fall from a higher to a lower level, or must be under pressure produced by some external force, such as that of a weight or spring acting on the surface of the fluid through a piston or plunger.

Under the former conditions its utmost capacity for doing work—potential energy, or energy of position— is the product of the height through which it can fall into the weight of water falling, so that if h denote the available height of the fall, and G the weight of water falling per second, $h \times G$ equals the energy or power available per second.

If the fluid is allowed to fall without resistance under the action of gravity, either free or confined in pipes or vessels, the power available is expended in imparting velocity to the water, and the potential energy or energy of position is converted into kinetic energy or energy of motion, and in this form is available for performing work.

It is, however, not necessary that the potential energy [1] of the water should be first transformed into kinetic energy in order that it may be employed for motive power ; the weight of the fluid can also be allowed to act directly on the prime motor in a manner similar to that in which the weight of a body attached to one end of a rope, passed over a pulley for instance, may be made to raise another body suspended at the opposite end.

A third way of utilizing water power for doing work is by means of its *pressure,* but the difference between this method and the preceding is more apparent than real, and consists chiefly in the form of the motor.

The so-called *pressure* of water is the result of weight or its equivalent, so that in this, as in the preceding case, the mechanical work is really done directly by the weight of the fluid, or in some cases by the pressure of steam or other elastic fluid acting through an accumulator. For practical purposes it may nevertheless be said that there are three ways in which hydraulic power can be applied to the performance of work : (1) as kinetic energy, or through the velocity of the fluid ; (2) by weight, and (3) by pressure.

Each of these three methods requires a different type of motor for its application, denoted respectively as (1) Turbine, (2) Water-wheel, and (3) Water-pressure engine.

The most usual source of hydraulic power in Nature is a river or stream, but to make this available for practical purposes some form of works, such as dams, canals, and aqueducts, is almost invariably necessary. A river-bed has always a certain fall or gradient; for instance, one

[1] In some quarters the term "potential energy" is objected to, but the author considers it less cumbersome and more generally expressive than "energy of position."

foot in every hundred feet of length—a very extreme
case, as a foot to the *mile* would be more nearly what
occurs in England ; to be able to take advantage of this
fall for doing work, the portion of it utilized must be
applied in one or several nearly perpendicular steps.   In
order, for instance, to obtain a fall of six feet from a river
having a gradient of one foot in every hundred, one
plan would be to construct a dam or weir across it of such
a height that the level of the surface of the water on the
up-stream side of the dam is elevated 6 ft. over that of
the water immediately below on the down-stream side ;
the surface of the water above the dam will then be
raised for a length of 600 ft. *at least,* and instead of
having the gradient due to the inclination of the river-

Fig. 1.

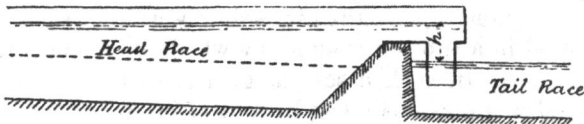

bed, will be *approximately* horizontal. This is plainly
shown in Fig. 1, where the dotted line represents the
original course of the river surface, the full line that
after the construction of the dam.   As a matter of fact,
the effect of a dam on the surface of the river above it
extends beyond the point of intersection of the horizontal full line with the dotted line indicating the
original surface, in some cases nearly twice this distance, and the surface is nowhere truly horizontal.

The head which was previously expended almost
entirely in overcoming the resistance of the river-bed
over a distance of 600 ft. while the water gradually de-
scended 6 ft. down a gentle incline, will now be available

for doing a corresponding amount of useful work by a
sudden drop through the same height as before. The
energy originally wasted in useless friction is accumulated
in the form of "head" immediately behind the dam.
The water flowing over the dam is taken into the
chamber or reservoir containing the motor through
which it passes, generally vertically to the tail-race, or,
as with an overshot wheel, is conducted by a pen-trough
or channel, issuing from the top of the dam, to the point
from which it flows into the buckets. In other cases,
where it is not admissible or possible to construct a dam,
or where only a portion of the water of the river or stream
is required, the necessary quantity is drawn off by a
separate channel, at a sufficient distance above the point
where the power is required to obtain the desired fall.

Occasionally water power is available in the form of a
natural waterfall, and then it is simply necessary to guide
the fluid in a tube or channel by which it is conducted to
the motor. In such cases the greater part of the energy
utilized was previously lost by impact in producing eddies,
heat, and overcoming the resistance of the air.

Sometimes the velocity only of the flowing stream is
employed for working a motor, the "head" in that case
being already converted into kinetic energy.

General Principles applicable to Hydraulic Motors.

Before proceeding to consider in detail the theory of
each kind of hydraulic motor, it will be necessary to give
an outline of the theoretical principles common to all, and
investigate, from a general point of view, some portions of
the phenomena occurring, previous to applying those
principles to the solution of particular problems.

In order to understand the action of hydraulic motors

it is chiefly necessary to have a knowledge of the behaviour of water in pipes and channels, its outflow
through orifices, and its action when flowing over curved
surfaces. It is assumed that the reader is acquainted
with the fundamental principles of hydraulics and me-
chanics, so that it will only be necessary to deal briefly
with those portions of the former most closely connected
with the theory of hydraulic motois.

Flow of Water in Pipes and Closed Channels.

The flow of water under pressure through pipes or
vessels of varying section, always under the assumption
that every part is at all times quite full of the fluid, is
subject to the law of Borda and Carnot. This law is
nothing else than an application of *the principle of the
conservation of energy* to water. Expressed in words it
may be stated generally as follows:

The sum of the potential and kinetic energy of the
fluid flowing through any section of the pipe or vessel is
equal to that sum for any other section plus or minus
any access or loss of energy from external sources.

The potential energy, or energy of position, of, say, a
pound of water at any pressure p is represented by the
product of this pressure and the specific volume or
volume occupied by one pound of water; for any number
of pounds flowing through any section of a pipe, the
potential energy is simply the pressure multiplied by the
total volume. Suppose, for instance, that through some
section of a pipe 62 lbs. of water per second flow at the
pressure of 1000 lbs. per square foot; the total volume of
62 lbs. of water is about one cubic foot, and the potential
energy 1000 foot pounds. It is convenient to express
the pressure of a fluid in the equivalent *head* which
would produce this pressure; as the specific volume is

the inverse value of the specific weight, and since the potential energy of one pound at the pressure p is $p \times$ *specific volume*, it can also be written $\dfrac{p}{\text{specific weight}}$, and this is nothing else than the equivalent head expressed in the unit selected; the equivalent head in feet for a pressure of 1000 lbs. per square foot would be roughly $\dfrac{1000}{62} = 16\cdot1$ ft., and the same figure of course gives the potential energy in foot pounds of one pound of water under the pressure in question.

The kinetic energy of one pound of water due to its velocity c, is $\dfrac{c^2}{2g}$, or since $g = 32\cdot2$, $\dfrac{c^2}{64\cdot4}$.

The sum of the potential and kinetic energy of one pound of water flowing with the velocity c at the pressure p through any section is $S = \dfrac{p}{s} + \dfrac{c^2}{2g}$

where s is the weight of one cubic foot of water (when the foot is the unit employed).

If p_1 and c_1 denote respectively the pressure and velocity at some other section, then, when there is no access or loss of energy from some external source, according to Borda and Carnot's law—

Fig. 2.

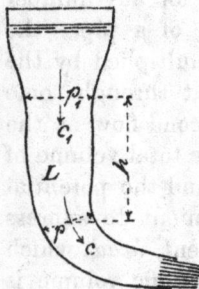

$$\frac{p}{s} + \frac{c^2}{2g} = \frac{p_1}{s} + \frac{c_1^2}{2g}.$$

If the point where the pressure p occurs is lower than that corresponding to p_1, so that there is an increase of head h, then (*vide* Fig. 2):

$$\frac{p}{s} + \frac{c^2}{2g} = \frac{p_1}{s} + \frac{c_1^2}{2g} + h.$$

If, besides this, losses of energy, L, from friction for instance, are incurred between the same points, then

$$\frac{p}{s} + \frac{c^2}{2g} = \frac{p_1}{s} + \frac{c_1^2}{2g} + h - L.$$

This is termed the hydro-dynamic equation, and will be subsequently often employed ; it will be convenient in the sequel to denote by p, p_1, &c., the pressure expressed in equivalent *head* of water, instead of as hitherto the pressure in pounds per unit of surface, so that p takes the place of $\frac{p}{s}$; hence the above equation becomes

$$p + \frac{c^2}{2g} = p_1 + \frac{c_1^2}{2g} + h - L.$$

The condition that every part of the pipes or vessels containing the water to which the above equation applies must be *full,* leads to the equation of continuity of flow, which expresses the fact that an equal quantity of water per unit of time flows through every section of the pipe system, the section being measured at right angles to the direction of flow. If A and A_1 be the sectional areas at any two points, and c and c_1 the corresponding velocities of flow, Q the quantity of water (in cubic feet) per second, then $Q = A\,c = A_1\,c_1$. Owing to this condition, when the dimensions of the pipes or vessels through which the water passes and the velocity at any section are known, the velocity at any other section can be calculated—a change of velocity in any one part affects in the same proportion the motion in every other part.

From the hydro-dynamic equation it is evident that as the velocity increases the pressure decreases, and *vice versâ.* As water flows through a vessel of varying section its velocity must change to satisfy the requirements of continuity of flow and with the velocity the pressure.

There is consequently a continual transformation going on, as the fluid advances, of kinetic into potential and potential into kinetic energy, and provided the changes are gradual, this process is not accompanied by any loss, apart from external causes. Abrupt alterations of velocity of flow due to sudden enlargement or contraction of the inclosing pipes or vessels causes loss, and then the transformation of energy of the one kind into energy of the other cannot take place.

Motion, relative and absolute.

All motion is, strictly speaking, relative, but for the purposes of this work it will be convenient to use the term *absolute motion* or *absolute velocity* in reference to motion or velocity referred to some fixed system on the surface of the earth, and to describe as *relative motion* or *relative velocity* motion or velocity referred to some system in motion relatively to that fixed system.

A passenger sitting in a railway-carriage moving at the rate of forty miles an hour has an absolute velocity referred to the fixed system of the line, of 58·66 ft. per second, but, relative to the moving system of the carriage itself, is at rest. If the same passenger walks across the floor of the carriage at a speed of 3 ft. per second at right angles to the direction in which it is travelling, 3 ft. per second is his relative velocity with reference to the carriage, and his relative direction of motion is at right angles to the longitudinal axis of the carriage ; his absolute velocity is on the other hand compounded of the absolute velocity of 58·66 ft. per second with which the carriage is moving, and the relative velocity of 3 ft. per second, at right angles to the direction of the latter, with which he traverses the floor of the carriage. It is easy to see that this absolute velocity is the resultant as regards value and direction of the absolute velocity of the carriage and the relative velocity of the passenger with respect to the carriage.

As another familiar illustration of a similar character : suppose a passenger to jump out of a train in motion, as it is passing a platform, at right angles to the longitudinal axis of the carriage ; although he jump perfectly straight forwards, he will alight on the platform in an oblique direction with respect to the edge of the latter, assuming that edge to be parallel with the wall of the carriage. The absolute direction and velocity with respect to the platform are given by the resultant of the velocity of the train and the relative velocity with which the passenger

Fig. 3.

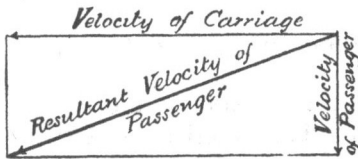

leaves the carriage (*vide* Fig. 3) ; this is equally the case whatever the direction relative to the carriage may be.

Supposing, again, the carriage were moving at only 3 miles an hour, and that the passenger walks in the opposite direction along the floor of the carriage at the same speed, it is clear that *with respect to the line* his body as a whole remains at rest, and therefore his *absolute* velocity is *nil*, although his velocity *relative* to the carriage is 3 miles an hour.

Motion relatively oblique to that of the carriage can always be resolved into two components, one parallel with and the other at right angles to the direction in which the carriage moves, and from this it is easy to see that the absolute velocity of any body moving with a certain velocity relative to another, which again itself has a given absolute velocity, is the resultant both as regards direction

and value of the relative velocity of the one body and the
absolute velocity of the other.

In the sequel this principle will be applied to particles
of water moving over rotating surfaces.   Let *A*, Fig. 4, be

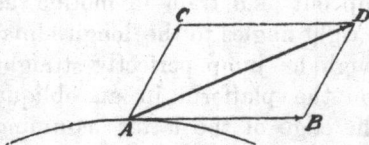

Fig. 4. a particle of water at any
point on a rotating sur-
face which has moment-
arily the absolute velocity
A B tangential to the
circular path of that
point, while the particle
at the same moment has the relative velocity and direction
A C with respect to the rotating surface; then *A D*, the
resultant of *A B* and *A C*, is the absolute velocity and
direction taken by *A*.

The Action of Water on curved Vanes.

The construction of turbines is based on the fact that
when a mass of water (or other material) moving in a
certain direction with given velocity is compelled to
change this direction, force is required to effect the
change.   The intensity of the force necessary is obviously
dependent on the extent to which the mass is deflected
from its original course.   In a turbine a jet of water is
deflected by being brought into contact with a curved
vane, which, preventing further progress in the initial
direction, compels the jet to follow its surface; owing to
the resistance offered by the water to this compulsion, a
reactionary force is exerted on the vane which is employed in driving the turbine-wheel.   When the course
of a particle of water is altered, its motion in certain
directions is retarded or accelerated, and the force exerted
in any of these is measured by the amount of the retarda-

tion or acceleration. Let A B_1 B_2 B_3, &c., Fig. 5, represent portions of a curve, each so short that it may be considered a straight line, and let A C be the initial

Fig. 5.

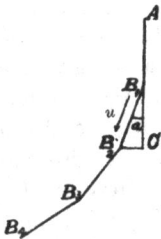

direction of a particle of water compelled to follow this curve; in passing from the element A B_1 to that of B_1 B_2 the particle is deflected through the angle a, and on reaching B_2 has traversed the distance B_2 C relatively to its original course at right angles to the latter; its velocity in this direction, at the point B_1, is proportional to its velocity u in the direction B_1 B_2, and is u sin a; when traversing A B_1 the velocity parallel with C B_2 was *nil*, hence there has been a relative acceleration of velocity at right angles to the original course, and if t be the time required for traversing B_1 B_2, this acceleration will be expressed by $\dfrac{u\ sin\ a}{t}$. By applying this method of reasoning to extremely small portions of a curved vane over which a stream of water is flowing, and extending it to the whole

Fig. 6.

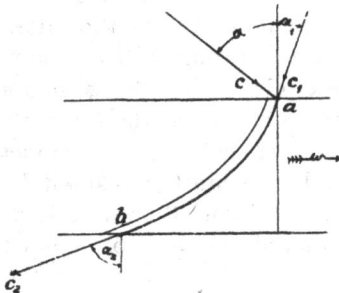

vane, an expression can be found for the total force exerted on the latter by a given quantity of water per second flowing over it.

Let a b, Fig. 6, represent a very narrow curved channel through which a certain mass of water m flows per second. This channel moves with the velocity w, in the direction indicated by the arrow. The water enters at a with the relative velocity c_1, at the angle a_1 formed by the tangent to the curve of the channel at a, with a perpendicular to the direction of motion of the latter; it leaves the channel at the angle a and with the relative velocity c_2.

The *absolute* velocity c—with respect to some fixed point—with which the water is moving at a, and the corresponding angle of direction, bear definite relations to the quantities c_1 w and a_1; resulting from the condition that the water must enter the channel in a direction coinciding with that of the tangent to the curve of the vane at the point of entry a, that is, *without impact* or *shock*.

<table>
<tr><td>Fig. 7.</td><td>Fig. 8</td></tr>
</table>

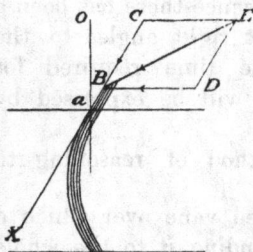 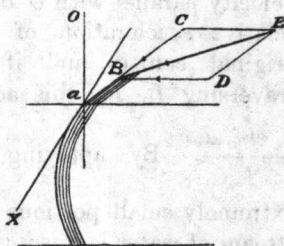

The moment before entering the channel the water has a certain velocity and direction *relative* to the latter, which, as has been explained in dealing with absolute and relative motion, is determined by the absolute velocity of the water in combination with the absolute velocity of the vane. When the water comes in contact with the vane, no immediate change must be caused by the latter in the relative direction of flow if impact is to be avoided, hence the first portion of the vane at a must be parallel with the relative course of the water at the moment of entering the channel. Fig. 7 illustrates the entry of the fluid *without*, and Fig. 8 *with* impact;

in both cases $E\,B$ represents the value and direction of the absolute velocity of flow of the water immediately before coming into contact with the vane, $E\,C$ the absolute velocity and direction of the motion of the vane; $E\,B$ must be the resultant of $E\,C$ and the required *relative* velocity of flow; the latter is found simply by completing the triangle $E\,C\,B$, in which the side $C\,B$ gives its value and direction. The complete parallelogram of velocities is $B\,C\,E\,D$.

In Fig. 7, $a\,X$, the tangent to the vane-curve at a, is a prolongation of $C\,B$, while in Fig. 8 the direction of $C\,B$ does not coincide with $a\,X$, but the course of the fluid on arriving at a is suddenly changed.

From the preceding explanation, the graphical construction necessary to determine in any given case the direction of the first portion of (or tangent to) the vane at a, will be clear, bearing in mind that $B\,E = c$, $B\,C = c_1$, and $C\,E = w$.

Mathematically expressed, the conditions for entry of the water into the channel without impact are the following:—

$(A)...c\ sin\ a = w - c_1\ sin\ a_1,$

$(B)...c\ cos\ a = c_1\ cos\ a_1,$

$(C)...c\ :\ w = cos\ a_1\ :\ sin\ (a + a_1),$

$(D)...c_1\ :\ w = cos\ a\ :\ sin\ (a + a_1).$

The derivation of these will be evident from elementary trigonometrical principles with the assistance of the diagram Fig. 9.

According as the angle a_1 is measured on the same side of the line $O\,Y$ as the angle a_2 (*vide* Fig. 10), or on the opposite side, it is either positive or negative; in the latter case it must not be forgotten that the sine is also negative, consequently formula (A) then becomes

$c\ sin\ a = w + c_1\ sin\ a_1.$

The absolute or *residual* velocity u with which the water leaves the vane, Fig. 10, is the resultant of c_2 and w.

All the forces exerted on the vane or channel by the water may be resolved into two resultant forces, one acting parallel with the direction of w, the other at right angles to it; the first of these may be denoted by X, the second by Y.

Fig. 9.

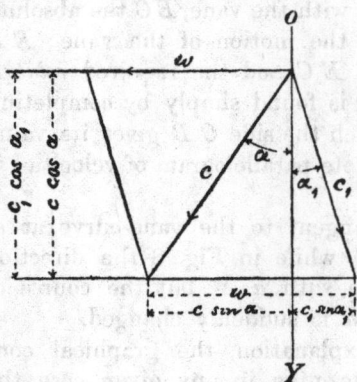

The momentum parallel with X of the water entering with the velocity c, is proportional to the component

Fig. 10.

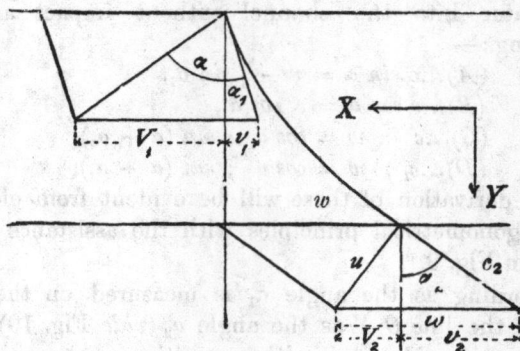

V_1 of the latter, and $V_1 = c \sin a$; on leaving the vane with the residual velocity u, the momentum in the same direction is proportional to the component V_2 of u, where $V_2 = w - c_2 \sin a_2$. Hence the loss of momentum is proportional to $V_1 - V_2$, the actual value being $m (V_1 - V_2)$, and this loss represents the force X exerted on the vane; it is obvious from the diagrams that the difference in the components v_2 and v_1 of c_2 and c_1 parallel with X, representing the increase of momentum *relative* to the moving vane, is identical with $V_1 - V_2$ in absolute value.

Hence $v_2 - v_1 = V_1 - V_2$, or, expressed in terms of the velocities and angles:—

$$c \sin a - (w - c_2 \sin a_2),$$
$$= c_2 \sin a_2 - c_1 \sin a_1,$$
$$X = m \left(c \sin a - (w - c_2 \sin a_2) \right\} \qquad (1$$
$$= m \left(c_2 \sin a_2 - c_1 \sin a_1 \right).$$

The work done by X at the velocity w, per second, is—

$$X w = m w \left(c \sin a - (w - c_2 \sin a_2), \right\}$$
$$= m w \left(c_2 \sin a_2 - c_1 \sin a_1 \right), \qquad \text{or}$$
$$W_m = X w = m w \left(V_1 - V_2 \right) = m w \left(v_2 - v_1 \right).$$

The value of $V_1 - V_2$, it is evident, can very easily be determined graphically without the aid of trigonometrical calculations, by simply measuring V_1 and V_2 (or v_2 and v_1), from the diagram, Fig. 10, already constructed to find c and u.

The equivalence of $V_1 - V_2$ and $v_2 - v_1$ may be thus stated: The *absolute loss* of momentum is equal to the *relative gain* in momentum.

The force Y is found in a similar manner to X, and is simply proportional to the difference of the components of c_2 and c_1 in a direction parallel to Y; hence

$$Y = m \left(c_2 \cos a_2 - c_1 \cos a_1 \right) \qquad (2$$

as the vane is not assumed to have any motion parallel with Y, no work is performed by the latter force; in practice it tends by a downward pressure on the bearings

c

of a turbine to increase the frictional resistance of the shaft.

If, instead of moving in a straight line, the vane or channel rotates about an axis at right angles to the parallel planes in which the filaments of water flow, the expressions for the forces exerted by the water on its passage over the vane are somewhat modified.

Fig. 11.

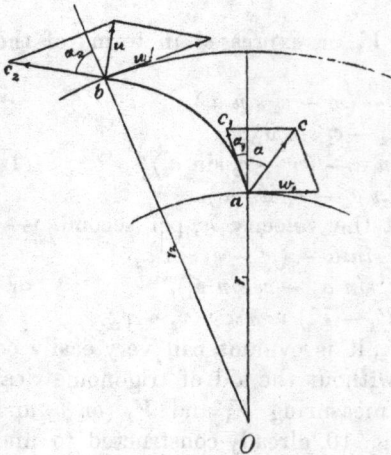

Let a b, Fig. 11, represent a vane or channel rotating round the axis O (perpendicular to the plane of the diagram); the distance or radius O $a = r_1$, and O $b = r_2$,—the water being assumed to enter at a and leave at b; the circumferential velocities at a and b are respectively w_1 and w_2. The angles a, a_1, a_2, are those formed by the direction of the velocities c, c_1 and c_2 respectively with the radii r_1 and r_2.

The work performed by the forces acting tangentially at all points of the vane is in this case expressed by

3) $W_m = m \left[w_2 \, c_2 \, sin \, a_2 - w_1 \, c_1 \, sin \, a_1 - (w_2{}^2 - w_1{}^2) \right]$

The additional quantities, as compared with the corresponding formula for rectilinear motion, are due to the varying velocities of the different points of the vane. Some writers have treated the influence of the varying velocity of rotation at different radii as due to the action of centrifugal force resulting from the rotation of the

vane. This view is misleading, as it gives the impression
that there is a fundamental distinction between the cases
of rectilinear and rotary motion, and that the latter requires the application of principles different from those
which suffice for the former. In fact, precisely the same
reasoning applies to both cases. The water in passing
over a moving vane takes a certain absolute path, the
direction of which at any point x is that of the resultant
of the velocity of flow relative to the vane and the
absolute velocity of the latter. This (absolute) path may
be easily found when the relative velocity of the water
at all points of the vane and the velocity of the latter
are known.[1]

Fig. 12.

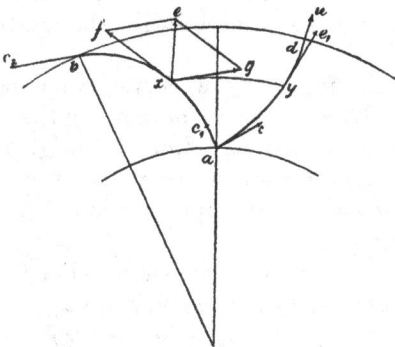

Let a d, Fig. 12,
represent the absolute
path corresponding to
the vane-curve a b.

A particle of fluid
at the point x of the
vane-curve rotating
with the velocity x g,
and with the relative
velocity of flow x f,
has the absolute ve-
locity x e. By the
time a particle en-
tering at a has reached x, this point will have advanced
to y, and therefore y lies in the absolute path of that
particle, which has the absolute velocity y $e_1 = x$ e, the
direction of y e_1 being tangential to the curve of the path,
and obtained by rotating the vane-curve a b, and with
it the constructive parallelogram x f e g, until the point

---

[1] *Vide* Chapter V.

x coincides with y, when the new position of $x\,e$ gives the direction $y\,e_1$.

The force exerted by the water on the vane depends solely on the absolute velocity of the former and *the form* of the path; whether this be the result of a rotary or rectilinear motion is immaterial. The effect of the varying speed of the vane at different points is to modify the curve of the absolute path and the absolute velocity; or if the absolute path be given, then the *relative* course (or vane-curve) is affected from the same cause. A particle of water in passing from a to d over—let it be supposed—a frictionless revolving plane surface, will come into contact as it proceeds with points passing under it with increasing speed, and this of course increases the resultant relative velocity. To consider the so-called effect of the centrifugal action separately, destroys unnecessarily the unity of treatment.

The action of water flowing along a curved vane or channel depends, with given velocities of flow, entirely on the extent to which it is deflected from its original course, and the fact that the motion of all points of the vane is not uniform does not in principle introduce any new element into the problem.

Returning now to the expression previously found for the work done by water flowing over a rotating vane

$$W_m = m \, [w_2\, c_2 \sin a_2 - w_1\, c_1 \sin a_1 - (w_2{}^2 - w_1{}^2)]$$

it may be stated in the form

$$W_m = m \, [\, w_2\, (c_2 \sin a_2 - w_2) - w_1\, (c_1 \sin a_1 - w_1)]$$

in which the quantities $(c_2 \sin a_2 - w_2)$ and $(c_1 \sin a_1 - w_1)$ correspond respectively to V_2 and V_1 as previously defined for the case where the vane has a uniform motion; they can be easily determined graphically in an exactly similar manner, as shown in Fig. 13; w_2 and w_1 however take the place of the uniform velocity w in the expression for W_m.

Hence $W_m = m [w_1 V_1 - w_2 V_2]$,   (3a

where $V_1 = w_1 - c_1 \sin a_1 = c \sin a$,

and $V_2 = w_2 - c_2 \sin a_2 = u \sin \beta$,

β being the angle formed with the radius r_2 by the direction of u.

Fig. 13.

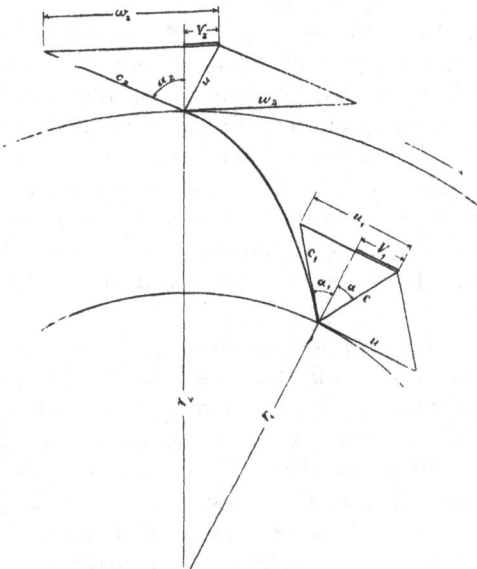

It will be seen that with given velocities the action of the water upon a vane *depends only on the angles of entrance and exit, and not upon the intermediate form of the curve,* assuming the latter to be continuous and the curvature not too abrupt.

By making $w_1 = w_2 = w$ in the formula for W, the latter becomes

$$W_m = m \, w \, (V_1 - V_2)$$

the expression for work done with a uniform velocity w of the vane, as previously demonstrated.

So far it has been assumed that the water comes into contact with the vane without impact, that is, it *glides* along the latter, and the algebraical condition for this [*vide* (A) page 15] is

$$w_1 - c_1 \sin a_1 = c \sin a.$$

$c \sin a$ or V_1 is the component of the absolute velocity of flow in the direction of the motion of the vane, and for a given velocity is independent of the question whether impact takes place or not, the force required to deflect the stream remaining the same; hence the work W_m can always be expressed by $W_m = m \, [w_2 \, (c_2 \sin a_2 - w_2) + w_1 \, c \sin a]$, whether the water enter without shock or not; *only* in the *former* case, however, does $c \sin a$ bear the relationship above given to the speed of rotation w_1 and the relative velocity c_1.

It might be supposed at first sight that as the expression for the work done remains unchanged, it is immaterial whether impact occurs or not, as the work done would in either case be the same; a little consideration will show that this is not so. Impact or sudden retardation of flow causes a loss of energy from which results a diminution of the absolute and relative velocities of flow. Hence under given conditions, although the formula remain the same, different values have to be assigned to the constituents when impact takes place than when the water enters parallel with the guide-vane.

Fig. 14.

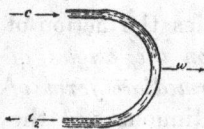

A special case will illustrate the use of the preceding formulas.

Assume that the water enters the vane in a direction parallel with the motion of the latter, and leaves in exactly the opposite sense, so that it is deflected through 360°, the velocity of the vane w being the same at all points. Under these conditions $a = 90°$ and $a^2 = 90°$, consequently $sin\ a = sin\ a_2 = 1$, and the force X exerted on the vane is by (1)—

$$X = m\ (c\ sin\ a + c_2\ sin\ a_2 - w) = m\ (c + c_2 - w);$$

if the vane be held fast, so that $w = o$, and assuming $c = c_2$,

$$X = m\ 2\ c.$$

The relative velocity $c_1 = c - w$, or $c = w + c_1$, when the vane is in motion; assuming $c_1 = c_2$,

$$X = m\ (w + c_1 + c_1 - w) = m\ 2\ c_1,$$

and the work done $W = m\ 2\ c_1\ w$.

CHAPTER II.

ON TURBINES IN GENERAL.

Definition of turbine.—Classification.—Summary of classification and subdivision.—Description of typical turbines.—Reaction turbines.—Impulse turbines.

Definitions.

A TURBINE is a motor for utilizing the energy of water by causing it to flow through curved buckets or channels on which it exerts a reactionary pressure constituting the motive force.

As distinguished from a water-wheel, in the older and narrower sense of the word, a turbine may be defined as a water-wheel in which a motion of the water *relatively to the buckets* is *essential* to its action.

A turbine consists essentially of a ring, or pair of rings, to which are attached curved vanes arranged uniformly round the circumference, revolving on a shaft or spindle to which the ring or pair of rings is connected by a boss and arms, or other suitable means.

Classification.

Turbines may be divided as regards their construction into three classes: radial, axial, and combined or mixed flow.

In *radial* turbines the water in passing through the

wheel flows in a direction at right angles to the axis of rotation, or *approximately radially.*

In *axial* turbines, otherwise called *parallel-flow* turbines, the water flows through in a direction generally *parallel* with the axis of rotation.

In *combined* or *mixed-flow* turbines, both the previously-named systems are combined.

All turbines, as regards the behaviour of the water in them, belong to one of two kinds—that which works with all parts entirely full or *drowned,* or that in which free deviation and air in the buckets is required; these have been named respectively Reaction and Impulse Turbines, and although the terms are not free from objection, it will be convenient to adhere to them as being generally adopted by writers on the subject.

In *reaction* turbines it is essential that there should be continuity of flow of the water in every part of the motor and apparatus connected with it.   Reaction turbines may be used in conjunction with a suction-tube, the nature of which will be subsequently explained.

In *impulse* turbines, the construction is such that the buckets are only partially occupied by the water passing through them, and the atmosphere has free access to the remaining space, so that the inflow always takes place under atmospheric pressure.

In some cases the water only enters into a part of the whole number of buckets at a time; the turbine is then said to have *partial admission.*

Generally both reaction and impulse turbines are provided with guide-vanes from between which the water enters the buckets of the wheel; as the name implies, they cause the water to enter in the desired direction.

The classification and subdivision of types of turbine may be summarized as follows :—

Reaction Turbines.		Impulse Turbines.	
Radial	{ Outward flow. { Inward ,,	Radial	{ Outward flow. { Inward ,,
Axial	{ Downward ,, { Upward ,,	Axial	{ Downward ,, { Upward ,,
Combined	{ Inward ,, up { or { Outward ,, down.	Combined	{ Inward ,, up { or { Outward ,, down.
With or without suction-tubes.		Without suction-tubes.	
With vertical or horizontal axes.		Without total or partial admission ; vertical or horizontal axes.	

Figs. 15, 16, and 17 show in principle respectively the construction of a radial, axial, and combined turbine.

Fig. 15.	Fig. 16.
Elevation	Elevation.

| *Plan* | Plan. |

In Fig. 15, the guide-vanes and wheel of a radial outward-flow turbine are shown. In Fig. 16 both the guide-apparatus and wheel of a Jonval or axial turbine are represented; *B* is the outer casing of the guides, of which *b* is one of the vanes and *e* the stuffing-box through

Fig. 17.[1]

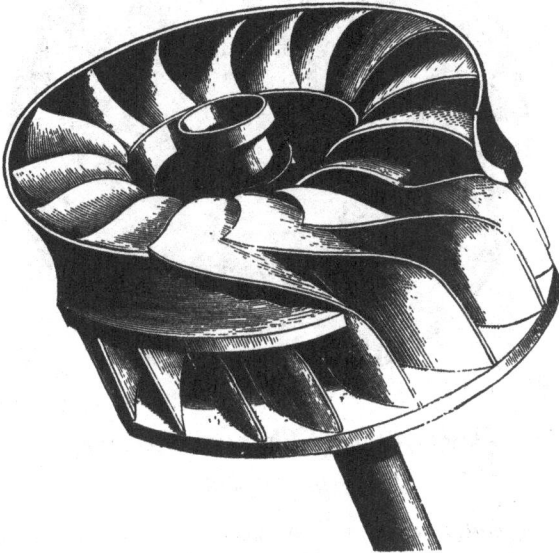

which the shaft passes. *A* is the wheel, and *a* one of the bucket-vanes.

Fig. 17 illustrates the construction of the wheel of a combined or mixed-flow turbine, and Fig. 17A shows a horizontal section through guides and wheel, *B B* being

[1] Figs. 17 and 17A are illustrations of a "Risdon" turbine, for which the author is indebted to the makers.

the guides, and *A A* the wheel-vanes. The guides are
similar to those for a radial inward-flow wheel.

Fig. 17A.

Description of Typical Turbines.
Reaction Turbines.

Radial Outward Flow. As representative of the *radial
outward-flow* class may be taken the *Fourneyron turbine*—
so called, as in other instances, from its inventor; it was
practically the first of the turbines in the modern sense
of the term, although there existed previously primitive
forms of motors which, strictly speaking, were turbines as
regarded the mode of action of the water on them.[1]

---

[1] For the illustrations Figs. 16 and 18, as well as some others,
the author is indebted to Armengaud's work, "Les Moteurs
Hydrauliques."

In Fig. 18 is shown a Fourneyron turbine of an early
type discharging above water. *B* is a tube through

Fig. 18.

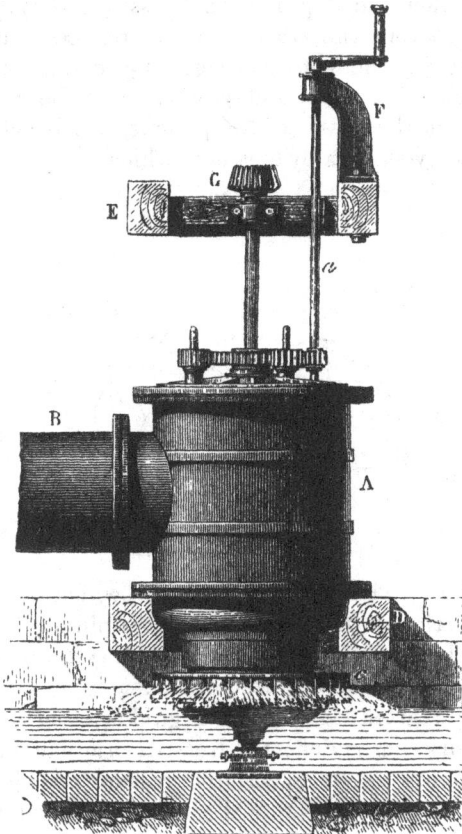

Fourneyron Turbine.

which, under the pressure due to the head, the water
enters the casing *A*; from there it passes through the

guide-passages *p p*, &c., shown in the sectional plan Fig. 19, into the wheel *c* of which *v v*, &c. are the vanes.

The regulation is effected by means of a circular sluice-gate, by which the depth of the passages through which the water enters the wheel can be reduced; motion is transmitted to the gate, through the spindle *a* and an arrangement of a wheel and pinions, from the hand-crank *F*. Through the bosses of the pinions, which act as nuts, pass the screwed rods by means of which the gate is raised

Fig. 19.

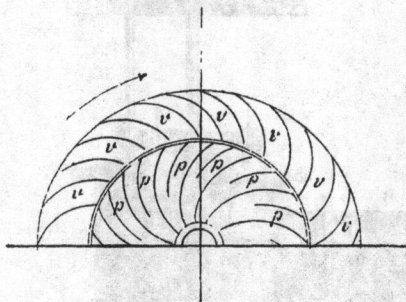

Fourneyron Turbine.

and lowered. The vertical shaft on which the turbine revolves carries at its upper end a bevel pinion, from which the power is transmitted to the machinery to be driven.

The water enters the wheel-buckets *v v*, &c., in the direction imparted to it by the guide-passages *p p*, &c., but after entering is deflected from its original course, and in consequence exerts a certain pressure on the *concave* side of the vanes *v* by which the wheel is driven round.

The guide-vanes and apparatus are stationary, and arranged concentrically with the wheel.

Fig. 20 illustrates another arrangement of a Fourneyron turbine for a low fall.

The wheel *D* is divided by horizontal partitions into three parts, 1, 2, and 3, which are in fact separate turbines connected together, one above the other. The water enters from a wooden reservoir *A* into the guide-wheel *C*, which is furnished with vanes similar to those shown in Fig. 19, and extending from *B* to *C*; thence it flows into the wheel *D*. By means of the sluice-gate *a*, in which

Fig. 20.

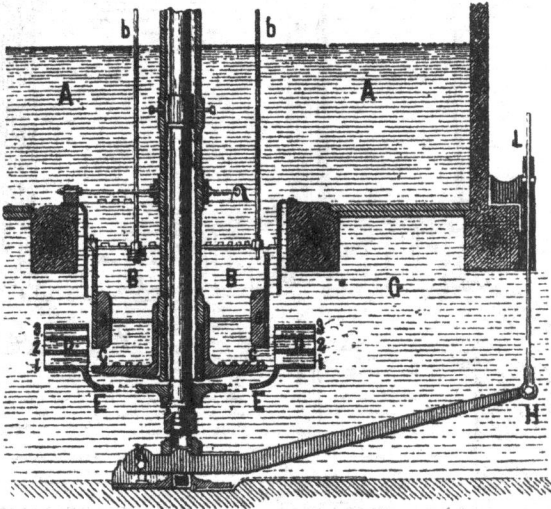

Triple Fourneyron Turbine.

there are slots to admit the guide-vanes, the water can be shut off in succession from the parts 3, 2, and 1 of the wheel. This turbine is shown working drowned; the mechanical construction is essentially the same whether the turbine works in that way, or discharges above water.

When however the discharge takes place above water, there is always a risk of the wheel working as an impulse

turbine owing to the buckets not being filled with water
at starting.

There are, of course, many modifications of the Fourneyron turbine, but they are all essentially of the same
construction, and the same theory applies to all. This type

Fig. 21.

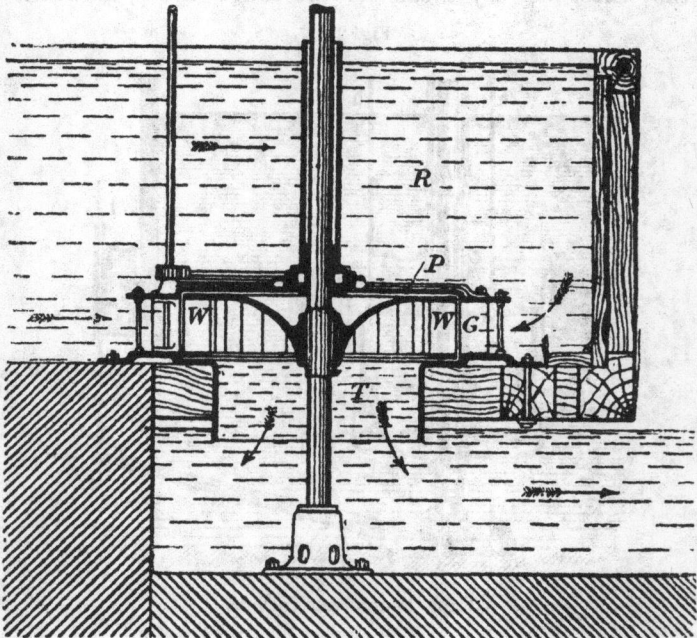

Inward-Flow Turbine.

is chiefly adopted in France, where it originated ; in other
parts of Europe the axial construction is more in favour,
while the Americans prefer inward and mixed-flow turbines.

Fourneyron's wheel is *not* used with a suction-tube.

Radial Inward Flow. Of this class the Vortex Turbine

of Prof. James Thomson and the radial inward-flow wheel of Francis may be taken as types. Fig. 21 shows the general arrangement of one of the latter working with a suction-tube. The water from the reservoir *R* enters the guide-wheel *G* which surrounds the wheel *W*, flowing towards the centre; after passing through the buckets, it is gradually deflected downwards by the curved surface of the top part of the wheel, so that it enters the suction-tube *T* in a vertical direction, and passes out into the tail-race.

The arrangement of the guide and wheel-vanes *P* and *V* is illustrated in Fig. 22. As compared with the Fourneyron turbine, guide apparatus and wheel have changed places, the former being placed concentrically outside the latter.

Fig. 22.

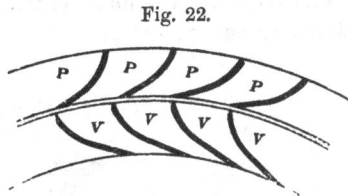

Inward-Flow Turbine.

It is easy to see how the mixed-flow has been developed out of the inward-flow turbine by continuing the vanes into that part of the wheel where the water assumes a vertical direction, so that the radial flow is changed into an axial flow *while* the water is in the buckets instead of *after* it has left them.

In the turbine under consideration the plate *P* sustains the pressure of the column of fluid above the wheel; it is connected at various points with the guide-wheel *G*. The vertical shaft is supported in an ordinary footstep on the floor of the tail-race, and passes through a stuffing-box in *P*.

The outflow from the suction-tube always takes place under water, this being essential to its efficient action with whatever type of turbine it may be employed.

As generally constructed, the suction-tubes of turbines

are of uniform diameter throughout the greater part of their length. Their action, however, is improved, as will be subsequently explained, by a gradual enlargement towards the lower end ; this, however, somewhat increases the cost.

The chief purpose of a suction-tube is to enable the wheel to be placed at a considerable height above the tail-water level, an object which in many cases is for practical reasons very desirable, as it renders the wheel more easily accessible, and a shorter shaft is required than would otherwise be necessary.

Fig. 23.

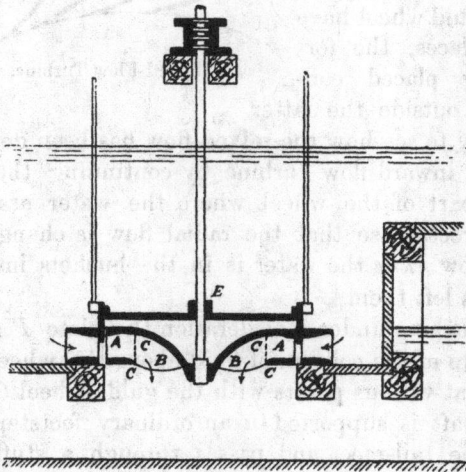

Mixed-Flow Turbine.

Mixed Flow. Fig. 23 shows an inward mixed-flow turbine without suction-tube, in which the regulation

Fig. 24.

Jonval Turbine with Suction-Tube.

is effected by means of a circular sluice-gate passing
between the guide apparatus and wheel, in the clearance
space.

The wheel-vanes extend from A to B, instead of terminating at C, as represented by the dotted lines.

The shaft is suspended from a collar bearing, and passes
through a second bearing at E.

Parallel-Flow or Axial Turbine.   Fig. 24 illustrates an
axial or parallel-flow turbine of the usual construction,
which may be taken as representative of this class.

It is of the type generally known as a Jonval turbine,
from the name of the engineer who first practically introduced it, and is more extensively adopted in Europe than
any other form of reaction wheel.

It consists essentially of three parts: the guide apparatus G, the wheel W, and the suction and off-flow
tube T.

The guide apparatus consists of two concentric casings,
in the annular space between which are secured the guide-
vanes, and a dished or conical cover or plate over the space
within the inner casing, having in the centre a water-tight
bearing or stuffing-box, through which the shaft passes.
This cover is so constructed as to bear the pressure of the
column of water above it, and forms, so to speak, a part
of the bottom of the reservoir from which the turbine
draws its supply.

The shape of the casings and cover should be such
as to offer as little resistance to the flow of the water as
possible.

The wheel or runner consists of two concentric rings
connected by curved vanes which form the buckets, and
secured by arms to a boss keyed to the vertical shaft.   In
some of the older turbines, a single ring only was used for
the wheel, with the vanes projecting from the circumfer-

ence and unconnected with each other at their outer ends.

The vanes are either cast in one with the rest of the wheel, or made of separate sheets of wrought-iron or steel let into the cast-iron rings. They are shown in section in Fig. 25. The water enters vertically between the guide-vanes, is given by them the desired direction in which it enters the wheel, and is then during its passage through the latter deflected by the wheel-vanes, and issues from the passages between them, into the suction - tube or tail-race.

Fig. 25.

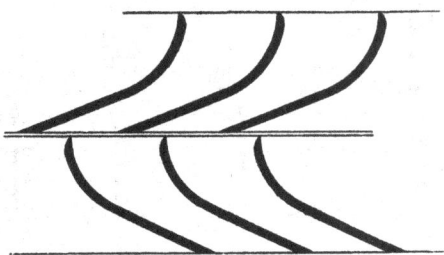

The section shown is on the development of a cylindrical surface of the mean diameter of the wheel.

The suction-tube—or draft-tube—is a large cast-iron pipe, which corresponds in diameter at its upper end with the outer ring of the wheel, where it surrounds the latter with sufficient clearance to insure free motion.

At the lower end is a cylindrical sluice of cast- or wrought-iron, under which the water flows off into the tail-race; the out-flow orifice of the suction-tube must always be submerged to secure its proper action.

The sluice can be raised or lowered from the floor above the turbine by rods actuated by a suitable mechanism.

Fig. 26 represents a Jonval turbine with double buckets, and without suction-tube.

The wheel consists practically of two concentric wheels

which can be shut off from the action of the water by a
series of valves or covers actuated by rods each worked

Fig. 26.

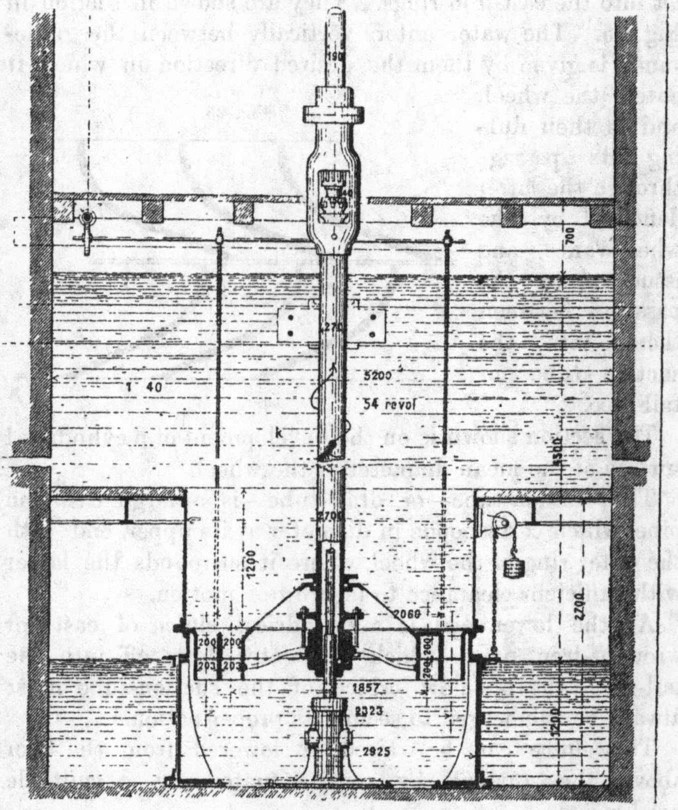

Doubel Jonval Turbine.

from a rack and pinion.   The rest of the construction will
be quite clear after the preceding descriptions.

Impulse Turbines.

Radial Outward Flow. In some cases, as has been already mentioned, Fourneyron's turbines discharged above the tail-water level, and under those circumstances worked as impulse turbines, in which the water leaves the guide passages against atmospheric pressure : generally speaking, however, a Fourneyron wheel is intended to work drowned, and does not come within the system of impulse turbines.

Zuppinger's so-called tangent wheel was one of the first types of radial outward- as well as inward-flow impulse turbines, intended to be used for high falls, for which the dimensions of reaction turbines, obliged to work full of water, would become impracticably small. An illustration of this class of motor is given in Fig. 27, with inward flow.

The water is conducted from the tank or reservoir A through the pipe B into the nozzle or mouthpiece C, which is furnished at the outlet with guide-vanes if sufficiently wide.

From the nozzle the water enters the wheel F on the outer circumference, and flows through the buckets in a radial direction, leaving at the inner circumference.

It will be seen that the quantity of water used is only sufficient to act upon a few vanes at a time; hence the tangent wheel is a turbine with only *partial admission.* The regulation is effected by means of the slide D actuated by a rack and pinion E. The tangent wheel has generally been made for inward flow, but many with outward flow have been constructed.

The French engineers Callon and Girard commenced in 1856 to design impulse turbines for all possible conditions,

Fig. 27.

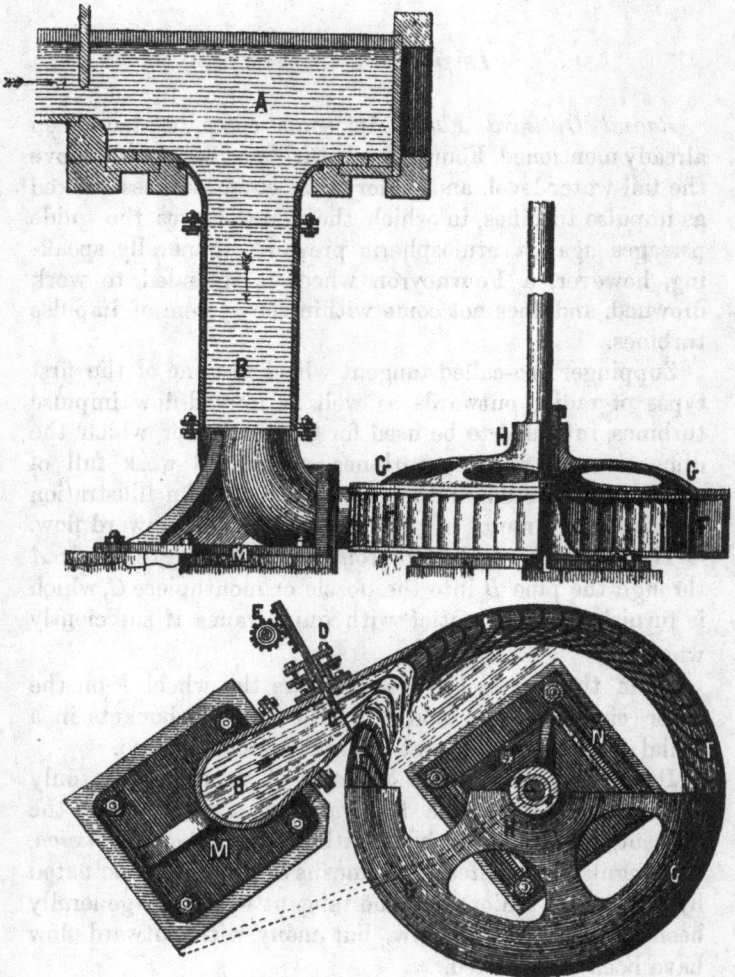

Tangent Wheel.

high and low falls, large and small quantities of water.
These turbines were made both axial and radial, with
horizontal, vertical, and inclined axes. The impulse turbine

Fig. 28.

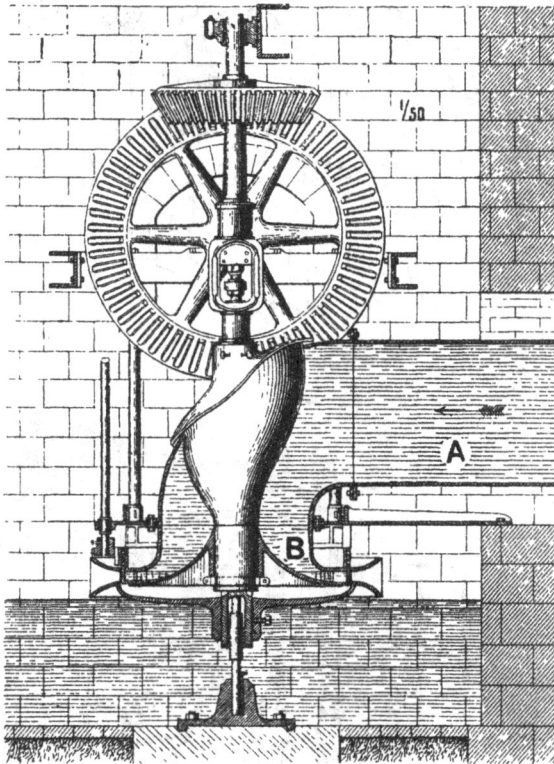

Outward-Flow Impulse Turbine.

has in consequence become associated with the name of
Girard, and every variety of impulse turbine now goes by
the name of a "Girard" turbine. One of the improve-

Fig. 29.

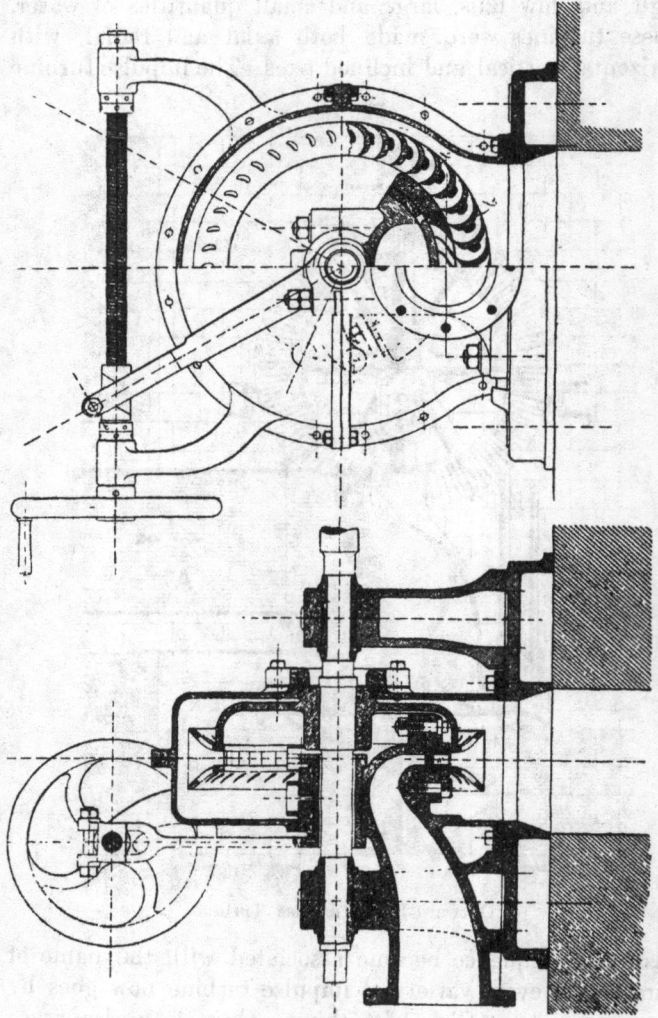

Outward-Flow Impulse Turbine with Horizontal Shaft.

ments introduced by Girard consisted in the ventilation
of the buckets for the purpose of insuring atmospheric
pressure in the latter, and preventing the water from
filling them and working as in a reaction wheel.

Radial Outward-Flow Girard Turbine. Fig. 28 shows a
turbine of this class, in which the water is conducted to
the guide apparatus *B* through the tube *A*, entering the
wheel at the inner circumference. It will be seen that
the cross-section of the wheel-casing is bell-mouthed, the
buckets being wider at the outlet than at the inlet ; the
object of this construction will be subsequently explained
—nearly all turbines of the Girard type are made in this
way. The bearings for the shaft are so arranged as to be
both above water. This is effected by using a hollow
shaft having in its upper part a kind of open frame or
lantern in which the pivot is fixed which carries the
weight of the motor; this pivot bears on a fixed spindle
carried up through the hollow shaft and supported in a
socket on the bed of the tail-race.

In Fig. 29 is shown a radial outward-flow Girard turbine with a horizontal axis. The admission of water through
the pipe and guide passages is only partial; the water
leaves the wheel-buckets in a vertical absolute direction.
Turbines of this pattern have recently been made in
considerable numbers for very high falls.

Axial-Flow Girard Turbine. Fig. 30 illustrates an axial
impulse turbine. In general arrangement it resembles a
Jonval turbine, but has ventilated buckets, and the latter
are widened at the lower end, where the outflow takes
place. The regulation of this type of wheel is effected by
various means, to be subsequently described ; that shown
in the illustration consists of a series of vertical slides,
by which the guide passages are closed. These slides
are raised or lowered from the floor above the turbine by

Fig. 30.

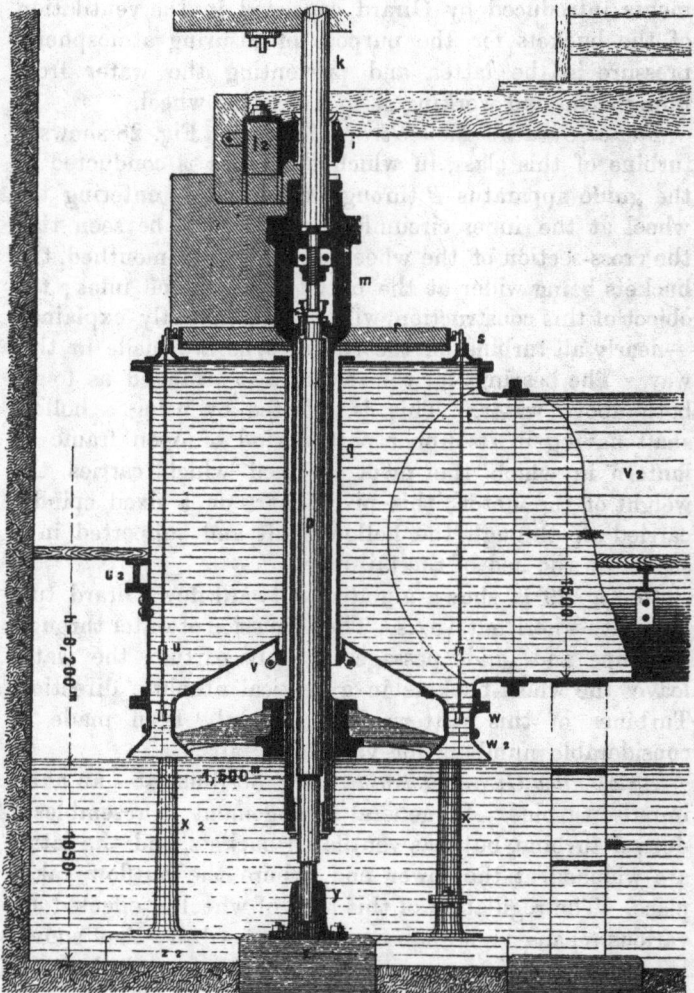

Axial Impulse Turbine.

vertical rods actuated from a kind of horizontal cam ; each rod works three slides.

The cross-section of the wheel-casing and buckets and form of the vanes is shown in Figs. 31 and 31A, with the apertures for ventilation, a being the guide-vanes, b the wheel-vanes, and c the ventilating apertures.

Fig. 31.

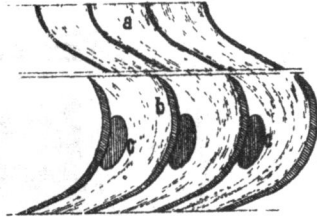

Referring again to the illustration Fig. 30, w is the guide apparatus, w_2 the wheel, x x_2 cast-iron columns, four in all, supporting the guide-casing; v is a cover, connected with the guide apparatus, which supports the column of water in the chamber above the wheel; y a socket carrying the fixed spindle p passing up through a hollow cast-iron shaft terminating at the top in the lantern m, which is connected with a bearing spindle running on the upper end of p; k is the vertical wrought-iron driving shaft coupled to the hollow cast-iron shaft; i a bearing for k, bolted to a horizontal cast-iron girder l_2; r the horizontal cam by which the rods t u, attached to the regulating slides of the guide apparatus, are raised or lowered; v_2 the conducting tube by which the water is conveyed to the turbine; u_2 a girder supporting the floor of the turbine house; z and z_2 foundations for the socket y and the columns x and x_2. The dimensions shown are metric.

There is of course a vast number of varieties of turbines, but the preceding examples are sufficient to give the reader a fair idea of the leading features of various types, and it does not come within the scope of this work to enter at length into details of construction.

Several kinds of turbines have, moreover, merely an historic interest, their coustruction having become obsolete. It

Fig. 31A.

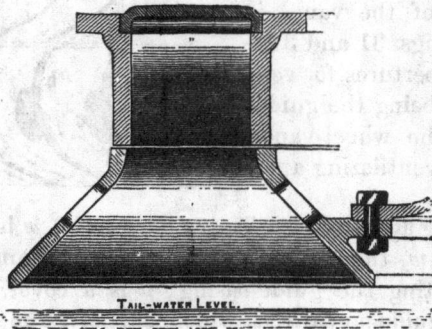

TAIL-WATER LEVEL.

should be mentioned that some turbines have been made without guide-vanes, in which case the water is generally assumed to have on entering an absolute direction at right angles to that of the motion of the wheel-vanes.

The relative merits of reaction and impulse turbines, the methods of regulation, construction of the vanes, &c., will be discussed in another part of this work; but before dealing with the theory of turbines, the author has considered it necessary to give the foregoing description of the essential construction of different systems.

CHAPTER III.

GENERAL THEORY OF REACTION TURBINES.

Essential parts of reaction turbine with suction-tube.—Comparison with system of pipes.—Available energy.—Nature of resistances to flow of water.—Fundamental relations.—General equations for work and efficiency.—Theory of suction-tube.—Pressures at orifices.—Available energy for wheel.—Formula.—Pressure-differences at orifices.—Limit of height for wheel in suction-tube.—Further explanation of action.

Useful work.—Expression for useful work.—Application to turbines of action of water on curved vanes.—Arrangement of vanes. —Formulas for total work done for radial, axial, and mixed-flow wheels.—Velocity of flow.—Interdependence of velocity of flow in all parts of turbine.—Differences between reaction and impulse turbines.—Method of determining velocity of flow.—Formulas for velocity of flow.—Numerical example.—Dependence of velocity of flow on speed of rotation.—Effect of absence of suction-tube.—Effect of losses on velocity of flow.—Complete formula.—Range of velocity of flow compared with that due to head.—Experimental results showing relation between velocity of flow and speed of rotation.

Notation employed.

c_o = velocity of flow in supply or conducting-pipe.

c = absolute velocity of water leaving guide passages = " velocity of flow."

c_1 = relative velocity of water entering wheel-buckets.

c_2 = relative velocity of water leaving wheel-buckets.

c_3 = velocity of water in suction-tube.

c_4 = velocity of water leaving suction-tube.

w_1 = velocity of rotation of wheel at inflow.

w_2 = velocity of rotation of wheel at outflow.

a = angle of outflow from guides.

a_1 = relative angle of inflow into wheel.

a_2 = relative angle of outflow into wheel.

A = effective outflow area of guide passages.

A_1 = effective inflow area of wheel-buckets.

A_2 = effective outflow area of wheel-buckets.

A_3 = sectional area of flow for suction-tube.

A_4 = effective outflow area for suction-tube.

r_1 = radius of wheel at inflow.

r_2 = radius of wheel at outflow.

E_u = available energy per unit weight of water.

ϵ = relative efficiency.

W_u = useful work done per unit weight of water per second.

W = total useful work per second.

L_u = losses per unit weight of water by hydraulic friction, &c.

ζ_1 ζ_2, &c. = experimental co-efficients for frictional losses, &c., corresponding with velocities of flow, c, c_1, &c.

p_1 = pressure of water (in "head") on leaving guide passages.

p_2 = pressure of water on leaving wheel-buckets.

a = atmospheric pressure.

g = acceleration due to gravity = 32·1889 at sea-level, London, taken for practical purposes as 32·2.

h = available head.

FOR the purpose of investigation it will be convenient to take an inward-flow (Francis) turbine as a type of the reaction turbine. Fig. 32 represents the general construction of one of these, which, with the aid of the illustrations previously given, will be clear without detailed description. The water enters the guide passages from a conducting channel or tube of the length l_o and sectional area A_o, in which it has the velocity c_o; it passes through the guide passages into the turbine wheel, and from the latter with the velocity c_3 through a suction-tube of the length l_3, sectional area A_3 and diameter d_3, out under a cylindrical sluice. The area of the outflow orifice below the sluice is A_4, the corresponding velocity of outflow c_4.

Fig. 32.

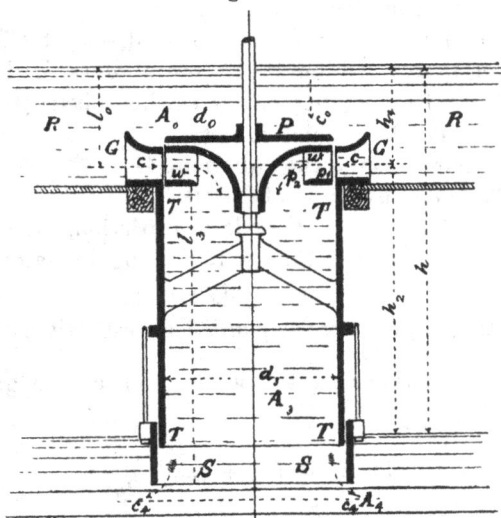

Inward-Flow Turbine.

The total available head of water is h; the depth of the middle of guide passages below the upper water level h_1; that of the tail-water level from the former h_2; hence $h_1 + h_2 = h$.

The whole arrangement forms, so to speak, *one system of pipes*, to which apply the general principles of hydrodynamics for the relations between pressure and velocity with continuity of flow.

E

In passing through the system the water meets with various resistances, to overcome which it has to do work, and leaves the apparatus with a certain residual velocity which has been denoted by c_4.

The available energy for each pound of water passing through the system under the action of the pressure due to the head h is $E_u = h$; this is expended in overcoming resistances and producing velocity.

Now the resistances are of two kinds—that offered by the wheel, in overcoming which useful work is done, and that due to friction, impact, and other obstructions, the energy expended on which is lost.

Hence there results the equation :—

Available energy = Useful work,
+ Work lost on friction, &c.,
+ Energy remaining in water (due to velocity c_4).

Denoting these three quantities respectively by W_u, L_u, and $\dfrac{c_4{}^2}{2g}$ —where g is the acceleration due to gravity— there follows :—

$$(1) \quad E_u = W_u + L_u + \frac{c_4{}^2}{2g}$$

and

$$W = E_u - \left(L_u + \frac{c_4{}^2}{2g} \right).$$

Assuming for a moment that there are no losses by friction, impact, &c., and therefore that $L_u = O$, then

$$(1a) \quad W_u = E_u - \frac{c_4{}^2}{2g}$$

and the only portion of the available energy not expended in doing useful work is that carried off in the kinetic form by the outflowing water.

It is evident from this that the less the value of c_4 the greater will be W_u, the work done by the turbine, and consequently the greater the efficiency of the latter; for the extreme case that $W_u = O$, *i. e.* no work is done,

$$E_u = \frac{c_4^2}{2g} = h \text{ or } c_4 = \sqrt{2gh},$$

the whole available energy is converted into kinetic energy, and the velocity c_4 is that due to the whole head.

The efficiency of the turbine is the ratio of the useful work done to the available energy or $\frac{W_u}{E_u}$, substituting the foregoing values from (1)

$$(2) \text{ Efficiency } \epsilon = \frac{W_u}{W_u + L_u + \frac{c_4^2}{2g}} = \frac{W_u}{h}$$

The preceding formulas apply both to reaction and impulse turbines; in the case of the latter, and of reaction turbines without a suction-tube, c_4 denotes the absolute velocity with which the water leaves the wheel.

To reaction turbines the principle of continuity and the hydro-dynamic equation apply.

The conditions of continuity are, *vide* p. 9,

$$A c = A_1 c_1 = A_2 c_2 = Q,$$

where A A_1 A_2 represent the sectional areas of the pipes and passages through which the water flows at different points, measured in planes at right angles to the direction of motion; c c_1 c_2 the velocities of flow at these points— similar indices denoting corresponding quantities—and Q the quantity of water passing through the system per second.

The hydro-dynamic equation has the form, *vide* p. 8,

$$p + \frac{c^2}{2g} = p_1 + \frac{c_1^2}{2g} + h_1,$$

where p and p_1 denote the pressures expressed as *heads* of water (in feet) at any two points, c and c_1 respectively the corresponding velocities, and h_1 the difference in level of the two points, the increase or diminution of head.

Theory of Suction-Tube.

It has been already stated that within certain limits the wheel may be placed at any height within the suction-tube; it is now necessary to explain the reasons for this fact.

The water leaving the guide passages and entering the wheel-buckets has a certain absolute velocity c representing a corresponding amount of kinetic energy available for doing work on the turbine. At the mouth of the guide passages, in an imaginary cylindrical surface dividing the guide apparatus from the turbine, there exists a pressure p_1; at the orifices of the wheel-buckets where the water leaves the latter a pressure p_2, which may be assumed to be less than p_1. It is evident that the difference in the pressures p_1 and p_2 is also available for doing work on the turbine, by increasing the velocity of the water during its passage through the buckets.

Hence the *total available energy* for the *wheel* is that due to the velocity of entrance c, *plus* that due to the difference of pressure $p_1 - p_2$ or

(3) Available energy $E_1 = \dfrac{c^2}{2g} + p_1 - p_2$ for the unit of weight.

If the water leaves the buckets at a lower level than that at which it enters, as in a mixed or axial-flow turbine, then, h_o being the difference of level,

$$E_1 = \frac{c^2}{2g} + p_1 - p_2 + h_o.$$

Now with reference to Fig. 32 it will be clear that

$$h_1 - p_1 + a = \frac{c^2}{2g} + \text{Loss of energy } L_1;\ L_1 \text{ being the}$$

loss of energy incurred by the water up to the moment of its exit from the guide passages, and a the atmospheric pressure on the upper surface of the water. Similarly

$$(3a)\ \frac{u^2}{2g} + p_2 + h_2 - a = \frac{c_4^2}{2g} + L_2,$$

where L_2 is the loss of energy sustained by the water from the moment of leaving the wheel-buckets to its exit from the suction-tube, and u the absolute velocity with which the water leaves the wheel.

From a combination of the two equations the value of the difference of pressure is obtained—

$$(4)\ p_1 - p_2 = (h_1 + h_2) - \frac{c^2}{2g} - \frac{c_4^2}{2g} - L_1 - L_2 + \frac{u^2}{2g}.$$

Now $h_1 + h_2 = h$, and owing to the condition of continuity all the different velocities may be expressed in terms of any one of them—c for instance—and the ratios of the corresponding sectional areas $A\ A_4$, &c. Further, the losses $L_1\ L_2$ may be stated in the form

$\zeta_1 \dfrac{c^2}{2g}$ and $\zeta_2 \dfrac{c^2}{2g}$, where ζ_1 and ζ_2 are factors made up of the ratios of the sectional areas and certain experimental co-efficients; they may be combined with the equivalents

for $\dfrac{c_4^2}{2g}$ and $\dfrac{u^2}{2g}$ and all expressed in the form

$$(5)\ \zeta^1 \frac{c^2}{2g} = \left(\frac{c_4^2}{2g} - \frac{u^2}{2g} + L_1 + L_2\right).$$

Substituting this in equation (4) there follows—

$$(6) \quad p_1 - p_2 = h - (1 + \zeta^1) \frac{c^2}{2g},$$

and this again in equation (3) —

$$(7) \quad E_1 = h - \zeta^1 \frac{c^2}{2g};$$

that is to say, the energy available for doing work in the wheel is that due to the *total head h, less* the various losses incurred previous to entering and after leaving the wheel-buckets, and is quite independent of the level at which the wheel is placed in the suction-tube, within the limits already mentioned.

The velocity of flow c from the guide passages is—as will be subsequently shown—determined quite independently of the position of the wheel relative to the tail-water level. Hence it is clear from equation (6) that the pressure difference $p_1 - p_2$ is also unaffected by the height of the wheel in the suction-tube.

Apart from mathematical demonstration, this phenomenon is explained by the statement that, as the wheel is placed higher or lower, the pressures where the water enters and where it leaves the buckets are respectively lowered or raised by the same amount, so that *the difference* remains constant.

It must be borne in mind that each of these pressures p_1 and p_2 depends not only on the depth of the points where it occurs below the surface of the water, but also on the velocity of flow at those points.

It remains to be shown what is the limit of height at which the wheel may be placed in the suction-tube above the tail-water surface.

By transposing the equation (3*a*) the following expression for p_2 is obtained—

$$p_2 = a - h_2 + \frac{c_4{}^2 - u^2}{2g} + L_2;$$

if no flow were taking place, or the velocity of flow were
so slow as to be negligible, c_4 and u would both be *nil*,
and therefore also the sum of the losses L_2; in that case

$$p_2 = a - h_2.$$

The smallest possible value of p_2 is of course *nil*, and
would occur when

$$a = h_2,$$

that is to say, if there were no flow the maximum possible
height of the wheel above the tail-water would be equal
to the atmospheric column; the atmospheric pressure
exactly balancing the head h_2. If under these conditions
flow commences, it is clear that the pressure of the
atmosphere is no longer capable of supporting the same
height of column as before, since to the action of the
head h_2 is now added the momentum of the water issuing
from the tube.

The following considerations will serve to render the
action of the suction-tube in a turbine more clear :—

Let $X\ Y$, Fig. 33, be a tube in which a column of
water of the height $B\ D = h$ is maintained; at the
lower end B there exists atmospheric pressure a (expressed in head of water), opposed to which is the head
h, *plus* the atmospheric pressure a acting on the upper
surface at D; the atmospheric pressure at the lower end
balances that on the upper surface, so that the fluid flows
out under the pressure due to the head h. Suppose the
tube continued and bent up as shown in Fig. 34, and
that at the end C, which is closed, there is a vacuum,
while instead of the atmospheric pressure at B there is
substituted a column of water $B\ C = a$.

The column $B\ C$ balances an equal column $B\ A$ in the
other arm of the tube, so that $B\ A = a$; therefore the
excess of pressure at A tending to produce a flow of the
water is $A\ D = h_1$ *plus* the atmospheric pressure at D,

but this is nothing else than the total head h, since $A\,B$ = atmospheric pressure and $A\,B + A\,D = a + h_1$ = h. The *effective* head, therefore, at A, 34 feet above B, is the same as at the latter point. A corresponds to the extreme limit of height at which a turbine may be placed in the suction-tube. If any other point below

Fig. 33.   Fig. 34.

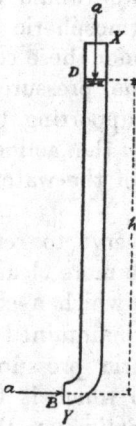

A, A_1 for instance, be taken, the column $B\,A_1$ may be considered as balanced by the equal column $B\,C_1$ in the other arm of the tube; the effective pressure at A_1 is consequently that due to the column $A_1\,D$ *plus* atmospheric pressure, or $A_1\,A + A\,D + a$ *less* that equivalent to $C\,C_1$; $C\,C_1$ however equals $A\,A_1$, so that $A_1\,A + A\,D + a - C\,C_1 = A\,D + a = h$ as before.

Hence for any point below A the effective head for producing flow is the same as at B. In practice the height $A\,B$ is very considerably less than a.

Useful Work.

Under *"useful work"* is to be understood the work done by the force, resulting from the deflection by the vanes of the streams of water entering the wheel-buckets, turning the wheel at a uniform speed, and overcoming an equivalent resistance. In the "useful work" is included that required to over-

Fig. 35.

come the friction of the bearings of the turbine shaft and of the surface of the wheel against the surrounding water; it is therefore the equivalent of the indicated work of a steam-engine.

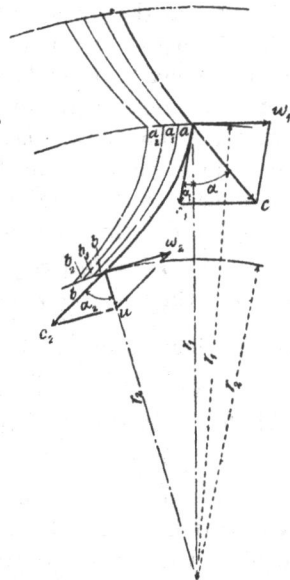

The "effective work" is the work actually given off by the turbine shaft, as measured for instance by a brake; it equals the "useful work" *less* that required to overcome friction.

It has been shown (*vide* p. 22) that the work performed on a curved vane by the action of a jet of water passing along it, when the vane has a rotary motion about an axis at right angles to the planes in which successive parallel filaments of the jet move—or approximately at right angles to a plane parallel with the resultant course of the jet—is expressed by

$$W_m = m \left[w_2 \left(c_2 \sin a_2 - w_2 \right) + w_1 c \sin a \right],$$

the notation being as already explained and indicated by the diagram Fig. 35.

Strictly speaking this applies only to a very shallow stream or jet moving along the vane; but provided the depth be not too great, the formula may be used without appreciable error for a stream of a relatively considerable depth, which can be considered as made up of a series of very shallow streams $a\,b$, $a_1\,b_1$, &c., Fig. 35, superimposed, and all forming the same angle at the points of entry and exit respectively with the radii—drawn from the axis of rotation—at those points.

In an actual turbine vanes are arranged round the circumference of the wheel at equal distances, each in the same position relative to the corresponding radii, and the formula given above may be applied to express the total work done on the wheel, provided that, instead of the small mass m the total mass M of all the water passing through the turbine per second be substituted; then.

$$(8)\quad W = M\,[w_2\,(c_2\,\sin a_2 - w_2) + w_1\,c\,\sin a].$$

If G be the *weight* of water flowing through the turbine per second, then $M = \dfrac{G}{g}$, and if further Q be the corresponding *quantity* in cubic feet, $G = Q\,s$, where s is the weight of a cubic foot. It is clear from the conditions of continuity,

$$(9)\quad Q = A\,c = A_1\,c_1 = A_2\,c_2 = A_x\,c_x,$$

that all the velocities of flow at various points may be expressed in terms of any one of them and of the ratio between the respective sectional areas; it will be convenient to express all these velocities in terms of c, the absolute velocity with which the water leaves the guide passages; hence—

$$(10)\quad c_2 = \frac{A}{A}\,c\,;\ c_1 = \frac{A}{A_1}\,c\,;\ c_x = \frac{A}{A_x}\,c_x.$$

The formula (8) may be written, *vide* p. 21,

$$(8a) \quad W = (w_1 \, V_1 - w_2 \, V_2) \, M;$$

it applies to *radial turbines* with either inward or outward flow.

In axial turbines a very narrow stream of water following the course of a vane on a cylindrical surface at a certain radius from the axis, advances with an equal velocity of rotation at all points, and for the work done by the water the formula applies—

$$W_m = m \, w \, (c_2 \sin a_2 - w + c \sin a).$$

The actual stream of water flowing over the vane of an axial turbine may be considered as made up of a large number of such very narrow streams, the distance of which from the axis, and consequently also their velocity of rotation, varies. It can be proved that the combined work of all these streams acting at different distances from the axis is represented by simply substituting for w in the above formula the *mean* velocity of rotation, that is, the velocity of the middle of each vane at the mean radial distance r from the axis; then

$$(11) \quad W = M \, w \, (c_2 \sin a_2 - w + c \sin a), \text{ or}$$
$$(11a) \quad W = M \, w \, (V_1 - V_2).$$

The formula (11) results out of (8) if for both w_1 and w_2 the same value w be used

$$w_1 = w_2 = w,$$

and hence the expression W for radial turbines includes axial turbines as a special case.

There still remains the mixed-flow turbine to be considered. Generally speaking, in this the water enters radially, flowing inwards, and leaves axially, and the vanes have consequently a double curvature. Assuming, as before, that the total stream flowing over any vane is made up of a large number of very narrow streams or filaments, each one of these enters *radially* at the radius r_1, *vide* Fig. 36, and leaves axially at a radius which varies from r_2 for the filament furthest from the axis, to $r^1{}_2$ for that nearest to the axis. To each of these filaments or very narrow streams may be applied the formula for the work done in radial-flow turbines, so that the work done by the outermost filament will be

Fig. 36.

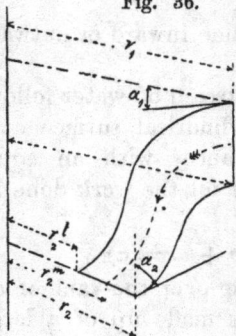

$$W_1 = m\,(w_1\,V_1 - w_2\,V_2),$$

and that performed by the innermost filament

$$W_2 = m\,(w_1\,V_1 - w_{21}\,V_2),$$

where w_2 and w_{21} are the velocities of rotation corresponding to r_2 and $r^1{}_2$ at the inner and outer extremities of the lower edge of the vane; the sum of the work done by these and all the intermediate filaments is expressed by

$$(12)\quad W = M\left(w_1\,c\,\sin a + c_2\,\sin a_2\,\frac{w_2 + w_{21}}{2} - \frac{w_2{}^2 + w_{21}{}^2 + w_2\,w_{21}}{3}\right).$$

It will be seen that when $w^2 = w_{21}$ this formula becomes identical with (8); hence it is applicable with the necessary special assumptions to all three classes of turbines, the radial, axial, and mixed-flow.

It is obvious that

$$w_1 = \frac{r_1}{r_2}\,w_2 = \frac{r_1}{r_2{}^1}\,w_{21},$$

and consequently W can be expressed in terms of one of the velocities w_1, w_2, or w_{21} only.

Generally speaking, therefore, W may be written in the form

$$(13) \quad W = (B \, w_1 \, c - C \, w_1^2) \, M,$$

bearing in mind equations (10) ; so that the work done on the wheel with given dimensions depends both on the velocity of flow c and the velocity of rotation w_1 (or angular velocity ω).

The angle a in mixed-flow turbines is that formed by the direction of flow at any point where the water enters the buckets with a *radial* line to that point; the angle a_2 is that formed by the relative direction of outflow at any point with a line through that point *parallel* with the axis, or, more generally expressed, with a line perpendicular to the momentary direction of rotation in the plane defined by the latter and the relative direction of outflow. This more universal definition is necessary, because in many turbines the plane in question is not parallel with the axis of the wheel.

In a reaction turbine the water as it traverses the vanes of the wheel completely fills the space between any two of them, and when— as generally happens— the sectional area of the passage diminishes towards the outflow orifice, the relative velocity is increased from c_1 on entering to c_2 on leaving, in the inverse ratio of the sectional areas, and as a necessary consequence the pressure at A, Fig. 37, is in excess of that at B. This does not, however, modify in any way the method of calculating the work due

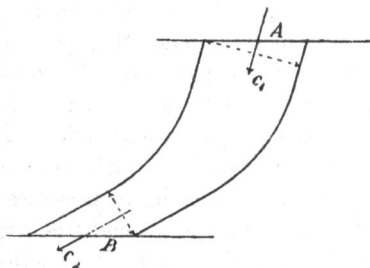

Fig. 37.

to the deflection of the stream by the vanes, which, whether the fluid be under pressure or not, is expressed solely in terms of the velocities of flow. The effect of an excess of pressure at A makes itself felt in a corresponding acceleration of the relative velocity from c_1 to c_2, or when the pressure at B is greater than at A, by a retardation of the flow. As the proportions between c_1 and c_2 can be determined solely from the ratio of the sectional areas of the passages at A and B, the calculation is much simpler than if the pressures were introduced, and as a matter of fact the latter can only be expressed through the velocity of flow.

In applying the hydro-dynamic equation to the flow and pressure of water in turbines, care must be taken to compare *relative velocities with relative velocities only*, and, similarly, *absolute with absolute velocities*, otherwise the deductions will be wrong.

Velocity of Flow.

It has been seen that the velocity of flow at any one point in a reaction turbine determines that at every other point, assuming that the machine is working as it should do. In dealing with this part of the subject, it will be convenient for purposes of comparison and otherwise to consider the velocity c, with which the water leaves the guide passages, as the standard velocity.

If—as in impulse turbines—the water left the guide passages at atmospheric pressure and entered a space to which the atmosphere had free access, the velocity of flow c would depend—apart from friction and similar influences —solely on the head of water above the point of outflow, and would be quite independent of what subsequently became of the fluid.

The pressure, however, at which the water in a reaction turbine leaves the guide passages is not atmospheric pressure, but, as has been seen, a greater or less pressure p_1, and this in its turn is a variable quantity which cannot be independently determined.

Owing to the continuous connection between the fluid in every part of the apparatus both before, during, and after its passage through the wheel, any resistance or facility offered to the flow of the water in any one part reacts on that in every other part. Hence, if after leaving the guide passages the water has to do work in turning the wheel, the velocity of flow c from the former will not be the same as when no work is being done.

The velocity of flow is determined from equation (1) (p. 50).

$$E_u = W_u + L_u + \frac{c_4{}^2}{2g}$$

Both W_u and L_u are functions of the velocity of flow c and the velocity of rotation w_1 or w_2, while E_u is proportional to the head only. Hence from equation (1) c can be calculated in terms of h and w_1 (or w_2).

In making this calculation, the losses of energy by shock as the water enters the wheel must be taken into account, as they very materially affect the result. The detailed calculation is very lengthy, and is never employed in practice; it will therefore be sufficient to state the result, which is as follows:—

$$(14) \quad c = \left(\frac{Bw_2}{1 + \zeta} + \sqrt{\frac{2gh}{1 + \zeta} + \frac{C + B^2}{(1 + \zeta)^2} \, w_2{}^2} \right) \frac{A_2}{A}$$

in which

$$B = \frac{r_1}{r_2} \frac{A_2}{A_1} \sin a_1$$

$$C = (1 + \zeta) \left[1 - \left(\frac{r_1}{r} \right)^2 \right]$$

$\zeta = a$ co-efficient depending on various experimental co-efficients and on the dimensions of the turbine, including guides and supply-pipe.

The value of ζ can be experimentally determined by holding the wheel fast, and allowing the water to flow through under the pressure of the head. In that case the velocity of rotation $w_2 = O$, and hence

$$c = \frac{A_2}{A} \sqrt{\frac{2gh}{1 + \zeta}}.$$

$\frac{A_2}{A}\sqrt{\frac{1}{1 + \zeta}}$ is then the ratio of the actual velocity of flow to the velocity due to the head. Assuming the formula to be correct, ζ should be constant for all speeds of a given turbine.

It is probable that the above formula for c does not give in every instance values which coincide with practical results in detail, since the assumptions involved regarding the loss by shock are not correct for extreme values of the velocity of rotation. The losses by shock, as given by the formulas employed in the calculation—which will be subsequently considered—are too great for speeds considerably above or below the best speed.

In spite of this, the formula in question is useful for showing the general tendencies of an increase or decrease of speed with turbines of various systems.

In the case of a Jonval turbine, in which $r_1 = r_2$, the term $C = O$, since $1 - \left(\frac{r_1}{r_2}\right)^2 = O$. If it be further assumed, that $a_1 = O$ — as is very usual — then $sin\ a_1 = O$ and therefore also $B = O$.

Under these conditions

$$c = \frac{A_2}{A} \sqrt{\frac{2gh}{1 + \zeta}};$$

that is to say, the velocity of flow is *constant* for all speeds of the turbine. This is quite in accordance with experience, as will be subsequently proved.

On the other hand, if in a Jonval turbine the value of a_1 is *negative*, then the first term in the formula for c, viz. $\dfrac{B\,w_2}{1+\zeta}$ also has a negative value, and consequently the velocity of flow diminishes as the speed increases.

For a radial inward-flow turbine in which $\dfrac{r_1}{r_2} > 1$, C is *negative*, and hence the flow decreases as the speed increases. The greater the ratio, $\dfrac{r_1}{r_2}$, the more rapid is this decrease of the velocity of flow. This peculiarity of the inward-flow turbine has a certain practical importance, as if carried far enough it tends to make the motor self-regulating. Let it be supposed that when such a turbine is running at its best speed, a portion of the load is suddenly removed, the wheel will begin to run faster, and the reactionary force exerted by the water on the vanes will in the first place diminish, owing to the increase in the velocity of rotation.

From equation

(8) $W = M\left[w_2\left(c_2\,\sin\,a_2 - w_2\right) + w_1\,c\,\sin\,a\right]$

it is evident that this diminution commences as soon as $w_2 > c_2\,\sin\,a_2$. For normally designed turbines $w_2 = c_2$ $\sin\,a_2$ corresponds with the best speed.

If simultaneously with the increase in w_2 (and w_1) the velocity of flow is reduced, so that c and c_2 diminish, then the decrease in the reactionary force will be more rapid than otherwise, and the more rapid this decrease, the lower is the speed at which effort and load again balance each other.

F

A self-regulating action of this kind is promoted by making the ratio $\dfrac{r_1}{r_2}$ as great as practicable; it will also be intensified by giving the angle a_1 a negative value.

In radial outward-flow turbines $\dfrac{r_1}{r_2}$ has a positive value, and it is obvious from the formula that an increase in the speed is accompanied by an increase in the velocity of flow.

Fig. 39 illustrates the relations between speed and velocity of flow for the three systems of turbine. Theoretically, the velocity of flow is affected by the presence or otherwise of a suction-tube with the turbine.

If there is a suction-pipe, the velocity of discharge

Fig. 38.

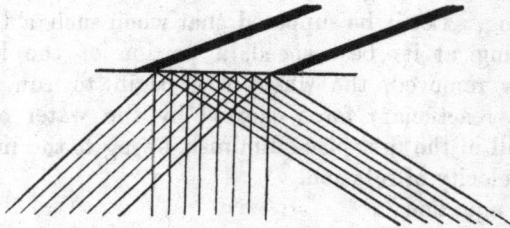

c_4 from the latter bears a fixed ratio to the velocity of flow c, which ratio is unaffected by variations of speed.

When, however, no suction-pipe is employed, and the wheel discharges direct into the tail-race, the velocity of discharge u from the wheel varies with the speed. As has previously been shown, it is the resultant of the relative velocity of outflow c_2 and the velocity of rotation w_2. It is clear that (*vide* Fig. 38) the section of the stream of water leaving each bucket must vary in area according to the direction in which it issues; in other words, the sectional area of discharge will change with different values of c_2 and w_2, but as the variation of c_2—proportional to c—is slight compared with that of w_2, this area will depend chiefly on w_2. Practically, the effect of the suction-tube on the velocity of flow is of little importance.

Fig. 39.

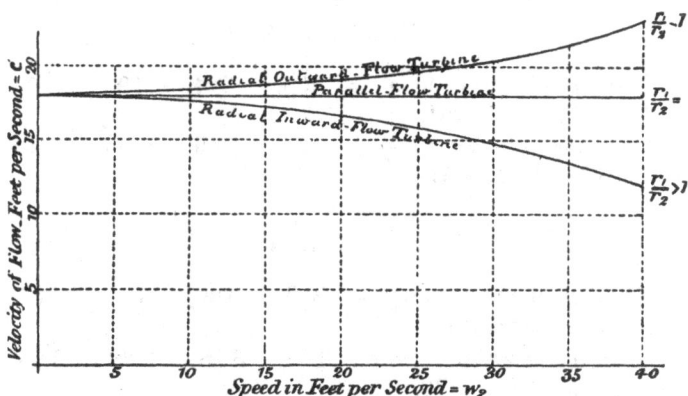

The following figures, taken from the results obtained by experiment with several Jonval turbines by Mr. P. Rittinger of Vienna in 1862, illustrate the effect of the speed on the velocity of flow as indicated by the quantity of water used per second :—

Turbine No. 1.

Number of revolutions = 160, 147, 140, 122, 110, 99, 93, 80.

Quantity of water = 4·98, 5·25, 5·40, 5·74, 5·92, 6·12, 6·25, 6·43.

Turbine No. 2.

Number of revolutions = 242,   201,   177,   150
135,   127,   102,   84.
Quantity of water   =   5·25,   5·15,   5·11,   5·06
5·04, 5·02, 4·99, 4·97.

Turbine No. 3.

Number of revolutions = 268,   224,   198,   168
151,   142,   113,   91.
Quantity of water   =   6·97,   6·66,   6·55,   6·43
6·38, 6·35, 6·24, 6·19.

Turbine No. 4.

Number of revolutions = 289,   186,   171,   164,
156,   141,   126,   112.
Quantity of water   =   3·06,   3·78,   3·89,   3·94,
4·00, 4·05, 4·17, 4·30.

Turbine No. 5.

Number of revolutions = 289,   184,   176,   167,
159,   150,   143,   115.
Quantity of water   =   3·03,   3·60,   3·66,   3·71,
3·77, 3·82, 3·95, 4·08.

Turbine No. 6.

Number of revolutions = 586,   540,   493,   400
354,   308,   285,   216.
Quantity of water   =   2·41,   2·48,   2·55,   2·69
2·75. 2·82. 2·86. 2·96.

Explanation of Effect of Speed on Velocity of Flow.

It will be seen from these figures that in the case of turbines 1, 4, 5, and 6, the quantity of water, and with it the velocity of flow, decreases as the speed increases,

Fig. 40.

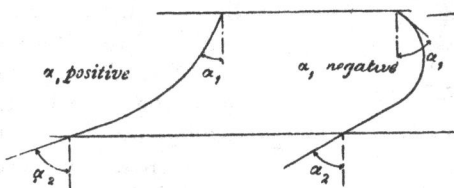

while for turbines 2 and 3 the contrary is the case. The reason for this difference is to be found in the fact that in the former the relative angle of entrance a_1 is negative as compared with a_2, while in the latter it is positive, *vide* Fig. 40. When a_1 is negative, *sin* a_1 also becomes negative, and consequently the co-efficient $B = \dfrac{A_2}{A_1}$ *sin* a, for a Jonval turbine, has a *minus* sign, and $\dfrac{B}{1 + \zeta} w$ must be deducted from the quantity under the root sign; it increases with w more rapidly than the latter, and c consequently decreases simultaneously.

For turbines 2 and 3, a_1 is positive, and therefore $\dfrac{B}{1 + \zeta} w$ also positive, so that c becomes greater with an increase of speed.

According therefore, to the proportions of the buckets, the velocity of flow may be either increased or retarded by the motion of the wheel, while in radial turbines the ratio of the radius at which the water enters to that at

Fig. 41.

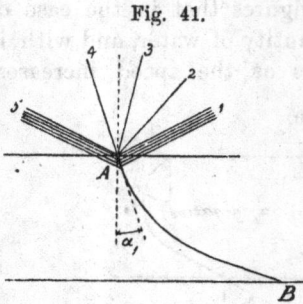

which it leaves exerts an additional influence.

The effect of the inclination of the vanes where the water enters the buckets on the velocity of flow may be explained as follows :—

Let A B, Fig. 41, represent the wheel-vanes of a turbine for which the angle a_1 is positive. When the motor is held fast, the direction relative to the vanes of the stream entering the buckets is A 1, and is in that particular case the same as the absolute direction of flow with which the water leaves the guide passages. The fluid impinges against the surface of the vane at an obtuse angle, and owing to the shock a loss of energy takes place, and the velocity of flow is retarded. When the wheel moves the relative direction in which the water impinges on the vanes is changed and becomes less obtuse ; at a certain low speed let this direction be represented by A 2, and at higher speeds by A 3 and A 4. As the angle of impact becomes less obtuse, the effect of the shock in retarding the flow is reduced, until finally some speed is attained at which the entrance of the water into the buckets occurs without shock ; A 4 represents the relative direction of flow under these conditions.

Up to this point the *shock* has diminished, and consequently the velocity of flow increased. Suppose the speed still further accelerated, so that the relative direction of inflow is A 5. Now it is evident that the direction of impact has changed, and that the shock acts on the back of the vanes in a sense opposed to the motion of the latter; the upper portion of the back convex vane surface tends to *force* the water through the buckets, acting in fact *as a pump*, while in the lower part the water still exerts a pressure on the concave surface, and drives the wheel. The turbine is therefore working simultaneously as a pump and motor, a certain proportion of the power developed in the latter capacity being absorbed in the former. If the increase of the speed be continued, a point is at length reached where the work done as motor is exactly balanced by that expended as pump, and no useful effect is produced. Beyond this the turbine works simply as a pump, and requires power from some external source to drive it.

With a negative angle a_1 the action of the vanes on the water is the reverse of that just described. From a condition of rest up to the speed at which the water enters without shock, there is a diminution of the resistance—from impact—offered to the entry of the water which tends to increase the velocity of flow, but this is generally more than counteracted by the tendency of the velocity of flow, apart from shock, to diminish. When the speed for inflow without shock is exceeded, the impact, as before, is on the back of the vanes; but now, in consequence of their being inclined in the opposite sense, an upward push is exerted on the inflowing water by the convex surfaces of the *negatively* inclined parts, which increases the resistance offered to the flow. As a result of the action just explained, in turbines with a negative value of the angle of inflow a_1 the velocity of flow is greatest when the wheel is held fast, and *decreases* as the speed increases. Theory on this point is fully confirmed by experiment, as a careful inspection of the various tabulated data relating to trials with turbines will show.

Efficiency and Best Speed of Turbines.

By equation (2) the general expression for the efficiency of a turbine is $\epsilon = \dfrac{W_u}{h}$, or $\epsilon = \dfrac{W_u}{W_u + L + \dfrac{c_4{}^2}{2g}}$, or, including under L all losses, whether from friction, impact, or velocity of off-flow, $\epsilon = \dfrac{W_u}{W_u + L}$.

Substituting for W_u its value from equation (8), there follows :—

$$(15) \quad \epsilon = \frac{w_2 (c_2 \sin a_2 - w_2) + w_1 c \sin a}{gh}$$

It is evident that if c and c_2 in the above expression are stated in terms of w_2 from (14), the efficiency E will then be given in terms of the latter only as $w_1 = w_2 \dfrac{r_2}{r_1}$; it is then possible to find a value of w_2 for which the efficiency E is a maximum. The calculation of this value is somewhat complicated, requiring the use of the differential calculus, and it is unnecessary to give it here.

It is clear, however, from the following considerations, that there must be a certain speed of rotation at which the turbine is more efficient than at any other.

The force exerted on a vane by the water flowing over it depends, with a given velocity of the latter, on the extent to which the stream is deflected from its original course; it will therefore be greatest when the vane is at rest, or when the wheel is held fast, because as soon as the latter moves in the opposite direction to that in which the stream is deflected, the deflection must become less. The work done is the product of the force exerted on the vanes and the velocity with which they move. When the wheel is held fast there is no motion, and consequently no work done, although the force acting on the vanes under these conditions is a *maximum;* on the other hand, when a certain speed of the vanes is attained, this force becomes *nil,* the wheel revolves—supposing there is no friction—without doing work. Between these two extremes the efficiency has risen from zero to some maximum value and fallen again to zero, and the speed at which this maximum efficiency occurs is the *best speed,* at which the wheel ought to be worked in order to obtain the highest results. Assuming the velocity of flow c to be constant for all speeds of the turbine, it can be proved mathematically that the *best speed* is half that at which the wheel would run without a load. The efficiency by formula (15) is—

$$\epsilon = \frac{w_2 \left(c_2 \sin a_2 + \frac{r_1}{r_2} c \sin a \right) - w_2{}^2}{gh}$$

When there is no load $E = 0$, and therefore—as gh has a finite value—

$$w_2 \left(c_2 \sin a_2 + \frac{r_1}{r_2} c \sin a \right) - w_2{}^2 = 0,$$

whence the speed (velocity of rotation) without a load

$$w_2 = c_2 \sin a_2 + \frac{r_1}{r_2} c \sin a;$$

the best speed is one-half of this quantity,

$$w_b = \tfrac{1}{2} \left(c_2 \sin a_2 + \frac{r_1}{r_2} c \sin a \right).$$

As a matter of fact the velocity of flow is not constant for all speeds of the motor, but in some cases varies considerably. For axial turbines, in which $a_1 = 0$, the above formula gives correct results, since the velocity of flow in such motors is constant.

The best speed is generally greater than the value above given, but in designing turbines a somewhat different and simpler method of calculating the best speed is followed, which will be explained when the construction of turbines is under consideration.

It is based on the obvious truth that, apart from hydraulic resistances, the less the unutilized energy the greater the efficiency of the motor, consequently the absolute velocity of off-flow from the wheel or suction-tube should be as small as possible. Unless, however, combined with a suitable construction in other respects, the fulfilment of this condition is in practice not sufficient to insure maximum efficiency, as the losses from impact or shock alone on the water entering the wheel-buckets may exceed the whole of the loss due to the residual velocity, and neutralize any advantage gained by reducing this to a minimum. It follows that, simultaneously with the reduction of the residual velocity to the lowest practicable value, the entry of the water into the wheel should be effected without impact, that is, the water must enter parallel with the direction of the extreme end of the wheel-vane.

The efficiency of reaction turbines is reduced by all methods of regulation when working otherwise than at full power. A common method of regulating turbines with suction-tube is by means of a sluice varying the off-flow orifice. By partially closing the sluice the quantity of water flowing through the turbine is reduced, but at the same time the efficiency is reduced in much greater proportion. The effect of diminishing the outflow orifice on the quantity of water passing through the wheel is not by any means proportional to the area of the orifice, as might be supposed, while the work done decreases as the square of the velocity of flow.

Regulation by means of a throttle-valve and by a sluice in the head-race are equally bad.

The method of regulation now generally adopted, of closing a portion of the guide passages, is much less detrimental to the efficiency than any of those above mentioned; this part of the subject will be more fully investigated in dealing with the construction and design of turbines.

The foregoing formulæ and considerations apply to reaction turbines under perfectly general conditions, no assumptions tending to simplify the demonstration having been made. The water enters the wheel, except at one particular speed, at an angle relative to the latter which does not correspond with the inclination of the guide-vanes at the point of entry, and leaves it with an absolute velocity which varies in amount and direction with the speed of the wheel.

Let it be supposed that the wheel is at first held fast by imposing on it a load greater than the action of the water on the vanes is able to overcome, and that then the load is reduced in stages, so as to allow the turbine to revolve at successively increasing speeds. As the speed increases, the following results will take place in the case of an inward-flow turbine:—

(1) The force exerted by the water on the vanes will *decrease;*

(2) The velocity of flow of the water passing through the turbine will first *increase* and afterwards *decrease,* and with it the quantity of water and available total energy;

(3) The difference between the pressure at the entrance and that at the outlet of the wheel-buckets will *decrease,* and

(4) The work done will *increase* up to a certain point and afterwards *decrease;*

(5) The efficiency will *increase* up to a certain point and afterwards *decrease.*

The point of maximum efficiency does not exactly coincide with that of maximum work, for the reason that the available energy varies with the speed, owing to a greater or less quantity of water flowing through the wheel.

Fig. 42.

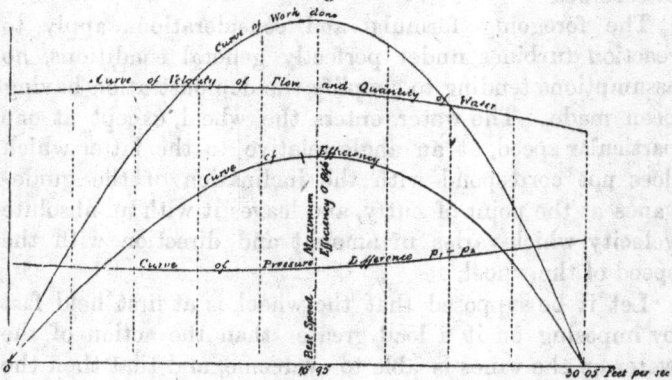

The relations existing between the various quantities are illustrated graphically by Fig. 42, which shows the curves representing velocity of flow c and other quantities proportional with it (Q and E), the difference of pressure $p_1 - p_2$, the work done W, and the efficiency E.

The values of these are plotted as ordinates to corresponding values of w_2 as abscissæ; they have been calculated for an inward-flow turbine. It will be seen by an inspection of the diagram, that while the increase and decrease of the velocity of flow c is very gradual, the efficiency rises and falls very rapidly, so that a comparatively slight deviation from the best speed materially reduces the efficiency. This fact has been amply confirmed by experiment.

For radial outward-flow and axial turbines the course of the various curves above referred to is somewhat different, and in the following particulars :—

(1) The velocity of flow increases continuously with the speed, or remains constant.

(2) The difference between the pressure at the entrance and that at the outlet of the wheel-buckets decreases continuously as the speed increases, or remains constant. With respect to this latter value, it has been found better that it should be positive, that is, p_2 less than p_1, although this tends to cause loss by forcing water out through the space between the guide and wheel-buckets. It may be negative from the commencement, if the ratios which A, the sectional area of outflow from the guide passages, bears to the areas of other portions of the apparatus, be small.

The following results obtained with the experimental turbines of Rittinger, previously mentioned, p. 67, illustrate the connection between speed and efficiency :[1]—

Turbine No. 1.

Number of revolutions = 160, 147, 140, 122, 110, **99**, 93, 80.

Efficiency = 0, 0·326, 0·455, 0·626, 0·659, **0·673**, 0·667, 0·631.

Turbine No. 2.

Number of revolutions = 242, 201, 177, 150, **135**, 127, 102, 84.

[1] The thicker figures denote the best results.

Efficiency   =   0,   0·452, 0·601, 0·687,
0·697, 0·695, 0·650, 0·586.

Turbine No. 3.

Number of revolutions = 268,   224,   198,   168,
151, 142,   113,   91.

Efficiency   =   0,   0·405, 0·543, 0·622,
0·632, 0·630, 0·587, 0·519.

Turbine No. 4.

Number of revolutions = 289,   186,   171,   **164**
156,   141,   126,   112.

Efficiency   =   0,   0·693, 0·711, **0·715**,
0·711, 0·705, 0·674.

Turbine No. 5.

Number of revolutions = 289,   184,   176,   167,
159, 150,   143,   115.

Efficiency   =   0,   0·623, 0·636, 0·641,
0·645, 0·639, 0·618, 0·575.

Turbine No. 6.

Number of revolutions = 586,   540,   493,   400,
354,   **308,** 285,   216.

Efficiency   =   0,   0·231, 0·410, 0·633,
0·687, **0·700**, 0·694, 0·626.

As a further illustration of the same fact, the following
figures giving the results of experiments with a Jonval
turbine, carried out by Mr. G. Meissner, and quoted in
his work on turbines, may be studied :—

Jonval Turbine.

Number of revolutions = 159,   103,   94,   **82**,
71, 65, 53, 41, 31, 15, 0.

Efficiency   =   0,   0·692, 0·735, **0·743**,
0·704, 0·688, 0·605, 0·525, 0·440, 0·257,   0.

In this instance the experiments extended to the whole range of efficiency from zero when the wheel was held fast, to zero when it was running without a load. It will be noted that in every case the *best* speed is greater than one-half of that corresponding to the un-loaded condition, but it must be borne in mind that in practice it is impossible for the wheel to run absolutely without resistance, as there is always a certain amount of friction of the bearings and between the surface of the wheel and the water to be overcome, so that the *hydraulic* efficiency at the maximum actual number of revolutions would be, not 0, but 3 to 5 per cent.

Referring once more to the general expression for the efficiency $\epsilon = \dfrac{W_u}{W_u + L}$, it is easy to see that for a given value of L, the sum of all the losses of energy of the water during its passage through the turbine, the greater W_w, the work done, the greater the efficiency E.

If it were possible to reduce L in the same proportion as W when the latter diminished, then the efficiency would remain the same for all values of W. In fact, however, not only does L *not* decrease, but it actually increases as W becomes less, until, when the turbine is held fast, or runs without a load (the two extreme cases in which no work is done), the whole available energy is absorbed by friction and other resistances, or carried off in the kinetic form.

This is shown by the equation

$$h = W_u + L_1 + \frac{c_4^2}{2g};$$

when in this $W_u = O,$

$$h = L_1 + \frac{c_4^2}{2g}.$$

In fact, for given dimensions, c_4 the velocity of off-flow does not differ very greatly whether the wheel be running at its best speed or standing still; in the latter case the available energy is swallowed up chiefly by friction, impact, and similar resistances within the apparatus.

CHAPTER IV.

LOSSES OF ENERGY IN REACTION TURBINES.

Enumeration of losses, character, and mathematical expression.—Losses from pipe friction.—Examples.—Loss in guide passages.—Co-efficients.—Loss from shock.—Example.—Loss by leakage.—Loss in wheel-buckets.—Co-efficients.—Loss from resistance of sluice.—Loss from residual velocity of flow.—Numerical values.—Summaries of losses.—General summary of facts relating to reaction turbines.

Losses of Energy in Reaction Turbines.

IN considering the losses of energy to which the water on its passage through the turbine and its adjuncts is subject, it will be best to follow the course of the fluid from the reservoir or conducting pipe over the wheel to its off-flow under the sluice at the bottom of the suction-pipe, where one is used.

The first loss incurred is that due to friction in the pipe or casing conducting the water to the turbine. If the diameter of the supply pipe be denoted by d_o, the length corresponding to this diameter by l_o, the velocity of flow by c_o, then the loss of energy is expressed by

$$L_o = \zeta, \frac{l_o}{d_o} \frac{c_o^2}{2g},$$

or, in terms of c, the velocity of outflow from the guide passages,

$$L_o = \zeta_r \frac{l_o}{d_o} \left(\frac{A}{A_o}\right)^2 c^2 \qquad (a)$$

where ζ_r is an experimental co-efficient.

The next loss takes place in entering, passing through, and leaving the guide passages, and is made up, (1) of the resistance offered by the upper edges of the guide-vanes, causing a sudden deflection of portions of the stream and an abrupt contraction, (2) of the friction between the water and the surface of the guide passages, (3) of the loss on the water leaving the orifices of the guide passages owing to internal friction and friction of the vane edges, (4) of the loss resulting from the curved form of the passages.

It is usual to include all these in one expression of the form

$$L_1 = \zeta_1 \frac{c^2}{2g}. \qquad (b)$$

The loss due to friction in the guide passages is, owing to the shortness and comparatively great area of the latter, extremely small.

When the water leaves the guide passages and enters the wheel-buckets, it does not—as already stated—under general conditions, take the direction of the tangent to the vane curve at the point of entry, relative to the wheel, and in consequence impinges against the vane instead of moving parallel with it, the result being a sudden change of velocity causing a loss of energy.

If the relative velocity of flow *after* entering the wheel be c_1, this loss of energy is represented by

$$L_2 = \frac{1}{2g} \left[(c \cos a - c_1 \cos a_1)^2 + (c \sin a - w_1 + c_1 \sin a_1)^2 \right].$$

$$(c)$$

A reference to the diagram, Fig. 43, will show that the quantities in the brackets are the differences between the components of the absolute velocities of flow c and c_a before and after entering the wheel-buckets, parallel with and at right angles to the axis of the wheel. The losses of energy in each of these directions is represented by

Fig. 43.

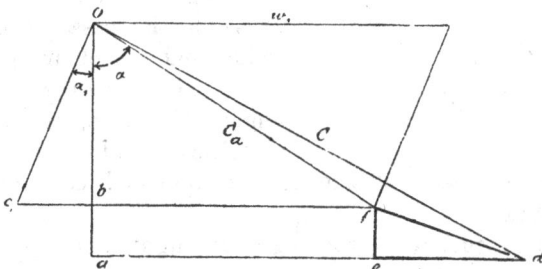

the squares of those differences; parallel with the axis, or radially, according to the type of the wheel, the velocity has been suddenly reduced from $O a$ to $O b$, the difference $b a = c \cos a - c_1 \cos a_1$; at right angles to the axis, the reduction is from $a d$ to $a e$, the difference

$$c d = c \sin a - (w_1 - c_1 \sin a_1);$$

hence $L_2 = \dfrac{1}{2g} \left[(a b)^2 + (d e)^2 \right] = \dfrac{1}{2g} (f d)^2,$

the quantity within the brackets being easily determined by the graphic method, as evident from Fig. 43. In general the difference $c \cos a - c_1 \cos a_1$ will be small, as, owing to the condition of continuity, when the clear distances between the guide-vanes and wheel-vanes measured in the direction of rotation are equal, $c \cos a = c_1 \cos a_1$, except in so far as there is a difference in the width of guide and wheel vanes, or in their thickness, or both. This will be clear on referring to the diagram Fig. 44,

Fig. 44.

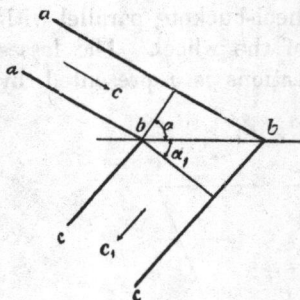

where $a\ b$ and $b\ c$ represent the terminations of the two guide-vanes and two wheel-vanes respectively of equal pitch t, measured on the corresponding circumference. The depths and consequently areas of the passages of the guide and wheel are proportional to the cosines of the angles a and a_1, while the velocities c and c_1 must be in the inverse ratio of the former, provided continuity is maintained.

When a turbine is working as it ought to do, the water entering the wheel-buckets without impact, the source of loss under consideration disappears.

The next loss to be determined is that due to leakage through the clearance space between the guide-casing and wheel; it depends theoretically on the pressure difference inside and outside the clearance space. As the author does not consider that any reliance is to be placed on a detailed calculation of the loss from the leakage in question, he has followed Redtenbacher's method of including it with the other losses incurred during the passage of the water through and out of the wheel-buckets; these are due to similar causes to those sustained in the guide passages, and all are included in an experimental co-efficient, and expressed in the form

$$L_3 = \zeta_2 \; \frac{c_2^2}{2g} = \zeta_2 \left(\frac{A}{A_2}\right)^2 \frac{c^2}{2g}. \qquad (d)$$

The absolute velocity u with which the water leaves the wheel-buckets, as a rule, is not the same either in value or direction as that with which its flow in the suction-tube, immediately below the wheel, is continued. In consequence of this sudden change—in every case from a higher to a lower velocity—there is a loss of energy of the same kind as that previously denoted by L_2.

The water leaves the wheel-buckets with an *absolute* velocity u, which, as previously stated, is the resultant of the *relative* velocity c_2 of outflow and the speed of rotation of the bucket at the point in question. It may be assumed that on its passage downwards through the suction-tube the water takes a spiral course, this latter and the corresponding velocity immediately on leaving the wheel-buckets being represented in the diagram, Fig. 45, by $O\,x$, while the absolute resultant velocity of outflow u is denoted by $O\,d$; the vertical component of $O\,x$ is $O\,b$, that of $O\,d$ is $O\,a$; the sudden change in velocity, in axial or radial direction, is therefore $O\,a - O\,b$, and the square of this is proportional to the loss of energy incurred; similarly, in a circumferential direction the change of velocity is $a\,d - b\,x = d\,y$, and the square of this corresponds to the loss of energy. The latter is small, as the circumferential component of velocity $a\,d$ probably remains nearly unchanged, so that it will be sufficient to consider only the axial (or radial) loss.

Fig. 45.

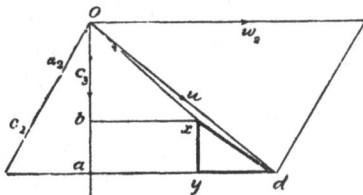

If c_3 be the velocity in a radial or axial direction

immediately after the water leaves the wheel, then the loss of energy in question,

$$L_4 = \frac{1}{2g}(c_2 \cos a_2 - c_3)^2 = \frac{1}{2g}[a\,b]^2, \qquad (e)$$

where, as is clear from the diagram, $O\,a = c_2 \cos a_2$ and $O\,b = c_3$.

It is assumed that c_3 is also the velocity in the suction-tube immediately below the wheel, or, in the case of an inward-flow radial turbine, it comes to the same thing if it be supposed that the absolute velocity of outflow remains unchanged after leaving the buckets until the course of the water has been deflected from a horizontal to a vertical direction, or, more strictly speaking, from a direction at right angles to the axis of rotation to one parallel with the axis of the suction-tube (*vide* Fig. 46).

Fig. 46.

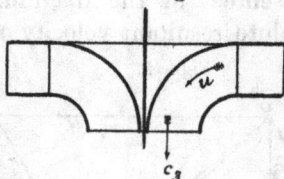

In passing through the suction-tube the water meets with resistance from the friction of the sides; the loss due to this cause is expressed in the same way as that already considered under (*a*) and denoted by L_0; hence its value is

$$L_5 = \zeta_r \frac{l_3}{d_3}\frac{c_3{}^3}{2g} = \zeta_r \frac{l_3}{d_3}\left(\frac{A}{A_3}\right)^2\frac{c^2}{2g}. \qquad (f)$$

In finally leaving the suction-tube under a sluice the water encounters a resistance offered to its passage by the edge of the sluice; the amount of this depends very materially on the form of the sluice, whether the edge is sharp or well rounded. The loss incurred from this cause may be expressed by

$$L_6 = \zeta_4 \frac{c_4{}^2}{2g} = \zeta_4 \left(\frac{A}{A_4}\right)^2\frac{c^2}{2g}. \qquad (g)$$

The last loss to be considered is that which is represented by the energy contained in the water when it leaves the suction-tube; this is obviously

$$L_7 = \frac{c_4^2}{2g} = \left(\frac{A}{A_4}\right)^2 \frac{c^2}{2g}. \qquad (h)$$

It is convenient to combine L_6 and L_7 together in the form

$$(L_6 + L_7) = (1 + \zeta_4) \left(\frac{A}{A_4}\right)^2 \frac{c^2}{2g}.$$

Having enumerated the various losses and briefly explained their character, it will be necessary to examine somewhat more closely into their nature and relative proportions.

Loss in Supply-pipe. The first loss L_0, due to friction in the supply-pipe or channel, is, in many cases, for moderate falls very small; for a Jonval or inward-flow turbine working with a suction-tube and placed tolerably high above the tail-water level, the water above the wheel is contained in a tank or reservoir of such dimensions that the loss from friction in it is quite inappreciable. On the other hand, the loss L_5 from friction in the suction-tube of course disappears where no suction-tube is employed, as in the majority of cases.

The co-efficient ζ_r will be the same in both instances where the material used for the tubes is similar, and for practical purposes with ordinary velocities the value $\zeta_r = 0.03$ may be taken.

In reality ζ_r varies with the velocity of flow, as shown by the following table :—

For $c_3 = 0.82 \quad 1.64 \quad 3.28 \quad 6.56 \quad 16.40$ feet.

    $\zeta_r = 0.0350 \quad 0.0289 \quad 0.0246 \quad 0.0216 \quad 0.0189.$

The velocity in the parts of the turbine in question

will be within the above values, under ordinary circumstances.

In cases where the diameter of the suction-tube varies the ratio $\dfrac{l}{d}$ will be different at various points; the velocity of flow will of course also vary. This may be taken into account, where the tube is conical, by introducing the *mean diameter* and *mean velocity* of flow into the calculation.

As an example taken from practice, suppose the head of water to be 15 feet, the length of the suction-tube 10 feet, the diameter 8 feet, and the velocity of flow 3 feet per second. According to the above table the co-efficient ζ_r suitable to this case would be, say, 0·025, hence the loss of energy

$$L_5 = 0·025 \ \frac{10}{8} \frac{3_2}{64} = 0·0312 \times 0·109$$

$$L_5 = 0·0034 \text{ feet,}$$

expressed as *head* of water; this is so small a proportion of the total available head of 15 feet as to be practically negligible.

With a tube of twice the above length, 20 feet, and double the residual velocity of flow, 6 feet, the value of L_5 would be 8 times as great as the preceding, but even then would only amount to 0·0272 foot, and if the head were, say, 24 feet (almost as small as possible with 20 feet of suction-tube), the percentage of loss would be only 0·11 per cent.

Loss in Guide Passages. The resistance offered to the water in flowing through the guide passages may be represented by a single co-efficient, the value of which does not vary much for the usual construction of this part of a turbine. The lower portions of the guide-vanes are generally made parallel with each other in axial turbines for a considerable length above the orifices, so that there is no contraction; also the passages are so formed that in flowing through them the stream of water undergoes no sudden change of section. The conditions under which the outflow takes place are therefore peculiarly favourable. Under these circumstances, if the surface of the vanes were smooth and polished, the co-efficient ζ_1 (in the expression (*b*) for L_1) would be from 0·042 to 0·025, or otherwise expressed, the velocity of flow would range from 0·96 to 0·975 of the value it would have were there no resistance. In fact, the surfaces of the vanes are comparatively rough, so that a somewhat higher co-efficient must be used, say, 0·053, corresponding to a ratio of the actual to theoretical velocity of 0·95.

A very usual value for the velocity of flow c in a reaction turbine of the Jonval type is $c = 0·7 \sqrt{2gh}$, corresponding to about half the actual head h. Taking this velocity to be 0·95 of that which would occur were there no resistance in the guide passages, the ratio of the energy due to the former velocity to that due to the latter would be as $(0·95)^2 : 1$, or 0·90, corresponding to a loss of 10 per cent.; but as the velocity of flow only represents one-half of the available head, the actual loss as compared with the head would be 5 per cent.; with polished surfaces it might be 3 to 4 per cent.

Weisbach found by experiment the value of ζ_1 to be from 0·05 to 0·10, Hänel gives $\zeta_1 = 0·10$ to 0·20, and Francis 0·20 and over; for practical purposes $0·125 = \frac{1}{8}$ may be taken as an average value, but is rather high for a good turbine. In impulse turbines ζ_1 has been found to range from 0·11 to 0·17, and there is no reason why it should be different for reaction wheels.

Loss due to Impact. This loss L_2, it has been already stated, does not occur to any appreciable extent in a turbine working at the best speed with properly constructed vanes, in which the water enters the wheel-buckets parallel with the surface of the vanes. There will, it is true, even under the latter conditions, be a certain loss of this kind, owing to two causes : (1) the slight excess of width of the wheel-buckets over that of the guide passages, and (2) the leakage through the clearance space between guide passages and wheel ; in consequence of these, the velocity relative to the wheel is suddenly retarded on entering the latter. This source of loss will however be included under the losses in the wheel-buckets for practical calculations. For an example of the calculation of the loss L_2, the following data, taken from an actually constructed turbine, may be used : $a = 73° 30'$, $a_1 = 90°$, $sin\ a = 0.96$, $cos\ a = 0.284$, $c = 20$ feet per second, $w_1 = 29.2$ feet per second, the proper speed being 20 feet; allowing for an excess of width in the entrance to the wheel-bucket and leakage through the clearance space, c_1 may be taken at 5.34 feet instead of $5.68 = c\ cos\ a$; ($cos\ a_1 = cos\ 0° = 1$).

Hence, using formula (*c*) (approximately)

$$L_2 = \frac{1}{64}\,(0.1156 + 100) = \frac{100.1156}{64} = 1.564 \text{ foot}$$

expressed as head.

The total head for the turbine in question is 14·75 feet; the loss therefore amounts to 10·6 per cent., and is due to the wheel running at a speed much greater than the best.

It will be found that for $w_1 = 19.2$ the second quantity in the brackets becomes *nil*.

There is good reason to believe that the loss from

shock as determined by the preceding method is in excess of that actually taking place, especially when the deviation from the best speed is considerable. Taking an extreme case, when a stream of water impinges at right angles against the surface of a vane, according to the formula given the *whole* of the energy represented by the velocity of the stream would be lost; this in fact does not happen, as a portion of the water itself forms a guide for the rest of the stream, interposing between the latter and the vane, and causing a deflection of the fluid over a *curved surface* instead of an abrupt stoppage. It should also be mentioned here that the theoretical conditions for avoidance of impact are not always those corresponding to the minimum loss. This will be further considered in the sequel.

Loss by Leakage through Clearance Space. It is obvious that to avoid friction between the wheel and the guide apparatus there must be a certain amount of clearance, and the excess of pressure inside the turbine at the points where the water leaves the guide passages over that outside will force it out through the annular space between the wheel and guide passages. If the pressure inside should be less than outside—which is possible —water would be forced into the wheel through this space; in practice this is rarely the case, and it will be assumed, for the purpose of investigation, that the pressure p_1 inside is greater than p_2 the pressure outside. The quantity of water which will leak out is obviously the product of the velocity of flow through the clearance space and the area of the latter. The velocity is that due to the head $p_1 - p_2$ multiplied with a co-efficient of discharge, which may be taken as 0·7.

The clearance for turbines of moderate dimensions cannot, for the present purpose, be safely taken as less than

$\frac{1}{8}$ inch, as there is always a tendency for it to increase with the wear of the footstep. If D be the diameter of the wheel where the water enters, then the quantity leaking through the clearance space must be

$$q = 0.70 \sqrt{(p_1 - p_2)\, 2g} \times 2 \, D \, \Pi \, t$$

where t is the clearance.

It has been shown that $p_1 - p_2 = h - (1 + \zeta^1) \dfrac{c^2}{2g}$;

hence $q = 0.70 \sqrt{2g \left(h - (1 + \zeta^1)\, \dfrac{c^2}{2g} \right)} \times 2 \, D \, \Pi \, t.$

The loss of energy from leakage is clearly proportional to q, and the percentage of loss increases as the square root of the head h, and is inversely proportional to the total quantity of water flowing through the turbine. It is impossible to determine exactly what occurs at the clearance, since the conditions are very complicated; as a portion of the water leaks out before reaching the wheel-buckets, the conditions of continuity *in the guide and wheel passages* are not strictly fulfilled, and in consequence some interference with the assumed theoretical state of affairs occurs, which must react on the pressures p_1 and p_2, and modify the leakage in a manner which it is impossible to calculate.

Loss in Wheel. This includes, in addition to the loss due to friction and the curved form of the passages, that resulting from the obstruction offered by the edges of the wheel-vanes to the free flow of the water.

Its effect is to cause a more or less sudden contraction of the stream entering the buckets. If the edges of the buckets were flat with sharp corners, the loss would be proportional to the square of the difference between the absolute velocities of flow before and after entering the buckets.

In practice, the edges of the wheel-vanes at the inflow are tapered off, as shown in Fig. 47, and the loss due to the obstruction caused by them is slight.

The loss in the wheel-buckets is similar in character to the loss incurred in the guide passages, and its ratio or percentage to the total head is about the same.

For practical purposes it would be a too laborious process to attempt to calculate separately all the losses of energy sustained by the fluid from the moment it leaves the guide passages to that at which it issues from

Fig. 47.

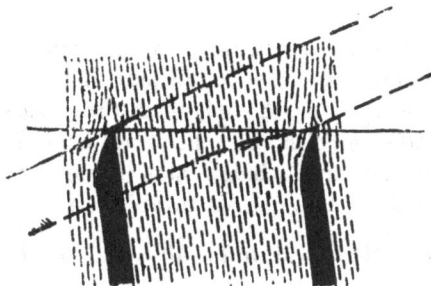

the wheel-buckets, and the uncertainty of the phenomena actually occurring is too great to give the results much value even if this were done. As already indicated, all the losses occurring during the period referred to, which for convenience will be termed *loss in the wheel*, may be combined and represented by a single co-efficient ζ_2.

Meissner is of opinion that the sum of the losses in the guide passages and in the wheel, expressed in percentage of the total head, are about the same for all reaction wheels, as the causes which tend to increase the resistance in the one diminish it in the other, and *vice versâ*. A high velocity in the guide passages is accompanied by a relatively low velocity in the wheel-buckets, while a high rate of flow in the latter corresponds to a relatively low rate in the former.

According to Hänel $\zeta_2 = 0.10$ to 0.20.

Experiments made by Professor Fliegner tend to show that ζ_2, contrary to what might be expected, decreases as the excess of pressure $p_1 - p_2$ increases; but a sufficient number of results are not available to be of any practical use.

Loss from Resistance of Sluice.

This depends almost entirely on the shape of the lower edge of the sluice, and the co-efficient in formula (*g*) may range from 4 to 1, as the edges of the sluice are sharp or well rounded off; there is, however, little doubt that it may be reduced much below this value by giving the lower end of the sluice-gate a suitable form.

If the co-efficient 1 be applicable, the loss $L_6 = \zeta_4 \dfrac{c_4{}^2}{2g}$ is equal to that due to the residual velocity c_4, and the latter may be kept within small limits by giving the sluice-gate a large opening.

Loss from Residual Velocity of Off-flow.

The velocity with which the water leaves a turbine, whether direct from the wheel-buckets or under the sluice of a suction-tube, ranges in practice, as a rule, from $\dfrac{1}{5}$ to $\dfrac{1}{3.5}$ of the velocity due to the total head $(\sqrt{2gh})$; for high falls it may be lower, but for large turbines and moderate falls $\frac{1}{5}\sqrt{2gh}$ may be considered the lowest limit.

If u be the absolute velocity of off-flow, the corresponding loss is expressed by $\frac{u^2}{2g}$, and for $u = \frac{1}{5}\sqrt{2gh}$ will be $\left(\frac{1}{5}\right)^2 = \frac{1}{25}$ of the whole available head, or 4 per cent.; for $u = \frac{1}{4}\sqrt{2gh}$ the loss would be $\frac{1}{16}$, or 6·25 per cent.

One of the losses enumerated disappears when a turbine works at its proper speed, viz. L_2 the loss from impact; others occur only with a suction-tube, viz. L_5 and L_6.

Meissner gives the following summary of the usual losses in a Jonval turbine :—

(1) Sum of losses by friction and impact in guide apparatus and wheel	$10\frac{1}{2}$ to 14 %	
(2) Loss due to residual velocity .	6 „	6 „
(3) Loss by leakage through clearance space.	$4\frac{1}{2}$ „	$4\frac{1}{2}$ „
(4) Friction in air and water and of bearings	2 „	$3\frac{1}{2}$ „
Total	23 „	28 „

A careful estimate of the average losses in a turbine, made by Mr. Lehmann from an analysis of experiments on thirty-six turbines varying from 1 to 500 h.p., is the following :—

Loss per cent. due to	Axial-flow Turbine.	Outward-flow Turbine.	Inward-flow Turbine.
Hydraulic resistances	12	14	10
Unutilized energy	3	7	6
Shaft friction	3	2	2
Total	18	23	18
Efficiency	0·82	0·77	0·82

In estimating the *hydraulic* efficiency, which has hitherto been understood by the term *efficiency*, the shaft friction is not taken into account; but in practice, unless otherwise stated, the expression denotes the ratio of the work as measured on a brake—exclusive of shaft friction—to the available energy.

Mr. Lehmann's estimate of the efficiency is for ordinary wheels, and for the purpose of design rather too high.

Summary.

(1) Every part of the apparatus forming a reaction turbine must be full of water, and consequently there must be continuity of flow, the whole arrangement being comparable to a system of pipes of varying section.

(2) The available energy is the product of the total head and the weight of water flowing through the apparatus per unit of time (second or minute).

(3) The useful work done is the available energy less the various losses arising from hydraulic resistances (friction, impact, &c.), leakage, and the unutilized energy due to the remaining velocity of the water on leaving the apparatus.

(4) The efficiency is the ratio of the useful work done to the available energy, or—what comes to the same thing —the ratio of the useful work to the sum of the useful work and the various losses.

(5) The use of a suction-tube with a reaction turbine enables the wheel to be placed within a certain limit at any height above the tail-water level without diminishing its efficiency; the theoretical limit of this height at which the outflow orifices of the wheel may be placed, is that of a column of water equivalent to the pressure of the atmosphere, 33·9 feet, but in practice it is considerably lower, and varies according to the proportions of the suction-tube.

(6) The work done in turning any turbine wheel depends for a given speed on the change in the momentum in the direction of rotation, produced by the deflecting action of the wheel-vanes on the water passing through them.

Apart from losses, the work done is equal to the difference between the energy of the water entering and that leaving the wheel.

(7) The available energy of the water entering a reaction wheel is due partly to its velocity and partly to its excess of pressure over the atmosphere. This excess may be negative; its amount depends on the proportions of the turbine under any given conditions as to head and height of the wheel in the suction-tube.

(8) The difference between the pressure of the water entering and that leaving the wheel is independent of the position of the latter in the suction-tube.

(9) The velocity of flow in any one part of a reaction turbine determines that in all other parts. It is convenient for purposes of comparison to deal with the velocity with which the water leaves the guide passages. This velocity is primarily dependent on the head of water above the guide passage orifices and the excess of pressure in the latter, but it can be shown that it is unaffected by the position of the wheel in the suction-tube.

H

(10) The velocity of flow is determined for a given turbine by the total head of water and the speed of the wheel.

According to the construction of the wheel, the manner in which the velocity of flow varies with the speed is different.

For outward-flow turbines in which the angle a_1 is positive or 0°, as is usually the case, the velocity of flow increases continuously with the speed. For a radial inward-flow turbine the velocity of flow at first increases, and subsequently decreases as the speed becomes greater, or decreases continuously.

For an axial turbine, according to the angle a_1, the velocity of flow increases, remains constant, or decreases with an increase of speed.

(11) The velocity of flow from the guide passages may be either greater or less than the velocity due to the head $V = \sqrt{2gh}$, according to the proportions between the various parts of the turbine, by which the pressure p_1, against which the water issues, is affected.

This velocity decreases as the ratio of the effective sectional area of the guide passages to that of the outflow orifices of the wheel increases.

(12) The efficiency of a reaction turbine of given dimensions varies with the speed, and is greatest for a velocity of rotation somewhat in excess of *half* that at which the wheel would run without a load.

The extent to which the sluice of the suction-tube is opened affects the efficiency very materially; by partly closing the sluice the efficiency is diminished; other methods of regulation, by closing some of the guide passages for instance, have a similar effect.

CHAPTER V.

THE DESIGN OF REACTION TURBINES.

Simplifying assumptions.—Mathematical conditions.—Relations between pressures and velocities of flow.—Velocity of flow from guide passages.—Another method of calculation.—Efficiency, various formulas.—Connection between various dimensions of turbines.— Effective and measured areas of passages —Methods of determining areas.—Radii of wheels.—Width and depth of buckets.—Number and thickness of vanes.—Summary of formulas.—Application to axial turbines.—Usual data for designing turbines.—Influence of relative angle of outflow on efficiency.—General rules regarding suction-tube.—Alternative data.—Vanes of turbines.—Usual construction for axial turbines : guide-vanes ; wheel-vanes.—Back-vanes. — General remarks. — Vanes of inward-flow wheels. — Absolute path of water.

HITHERTO, in dealing with the theory of reaction turbines, the construction of the latter has been taken as a given quantity, and their actions investigated under perfectly general conditions, that is, at various speeds and with no special assumptions as to the mode of construction.

In applying the theory to the design of turbines, the problem to be solved is : Under given conditions, what must be the proportions and speed of a turbine constructed so that it may work to the greatest advantage, that is, with the maximum efficiency ?

It has been shown that when all the dimensions of an existing turbine are known, it is possible to calculate, for a particular fall or head of water, the speed of the wheel

at which the greatest efficiency is realized. The various
quantities entering into the calculation are, however, so
numerous, and their relations so involved, that it is not
practicable to apply the formula used to the determination of the best proportions for a new turbine, and for
the latter purpose some assumptions have to be made
which materially simplify the calculations, and at the
same time give a close approximation to the true result.

Under ordinary circumstances a turbine is required to
run at a constant speed, and this speed should be so
chosen that, under normal conditions, it is that at which
the motor works with the maximum efficiency; in comparing, therefore, the merits of various systems of turbines, or the effects of varying proportions on the
efficiency of any particular class of turbine, this comparison will always be made on the assumption that the
wheels are running at the best speed.

It is evident that the maximum efficiency must be
attained when the sum of all the losses of energy is
reduced to a minimum.

All the losses, except two, depend *directly* only on
the velocity of flow of the water in the particular part of
the turbine in which they occur. When, therefore, the
velocity of flow increases, these losses also increase, but
the work done increases in the same ratio, and consequently the efficiency is not directly affected. The
velocity of flow itself depends to a certain extent on the
speed of the wheel, but the variations in the former due
to changes in the latter do not influence materially the
ratio of such losses to the total available energy.

Two of the losses, as above stated, are however *directly*
influenced by the speed of the turbine. These are, (1) the
loss due to impact or shock of the water on entering the
wheel-buckets, and (2) the loss from unutilized energy

due to the residual velocity of the water leaving the turbine (L_7).

If therefore the speed is so determined that these two losses are reduced to a minimum, it will be a sufficiently close approximation to the best speed.

As previously indicated, the loss (1) will disappear if the water enters the wheel-buckets with a relative direction parallel with that of the end of the wheel-vanes at the points of entry, while loss (2) will be smaller the less the residual velocity of off-flow u from the wheel.

Fig. 48.

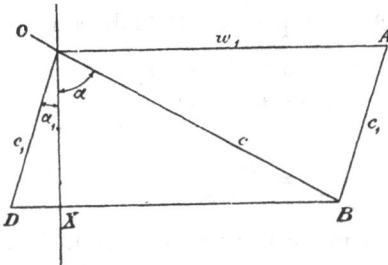

The absolute velocity of flow at the moment the water enters the wheel-buckets is the resultant of the velocity of rotation of the wheel and of the velocity of flow relative to the latter at the point of entry; hence this absolute velocity of flow must coincide with that of the water leaving the guide passages, and the angles of the guide-vanes and wheel-vanes at the point of inflow of the water be made to fulfil these conditions, if the fluid is to enter without impact.

Let $O B$, Fig. 48, represent as regards value and direction the absolute velocity of flow c of the water leaving the guide passages, and $O A$ the speed of rotation w_1 of the wheel at the point of outflow O. Join $A B$; then the

latter represents by its length and direction the relative
velocity of flow c_1, and the tangent $O D$ to the curve of the
wheel-vane at the point O should be parallel with $A B$,
while $O B$ gives the inclination of the guide-vanes where
the water leaves them. $B O X$ is the angle a formed by
the guide-vanes at the point O, with a line $O X$ at right
angles to the direction of rotation $O A$, and in the plane
defined by $O A$ and the tangent $O D$ to the wheel-vane at
the point O. In radial turbines $O X$ is radial, in axial
turbines parallel with the axis. $D O X$ is the angle a_1
formed by the wheel-vanes at the point O with $O X$.

Expressed algebraically the conditions for entrance of
the water without impact are the following :—

$$c_1 \cos a_1 = c \cos a. \qquad\qquad (A)$$
$$w_1 - c_1 \sin a_1 = c \sin a. \qquad\qquad (B)$$
$$c^2 = c_1{}^2 + w_1{}^2 - 2\, c_1\, w_1 \sin a_1. \quad (C)$$
$$\frac{w_1}{c} = \frac{\sin\,(a + a_1).}{\cos a_1} \qquad\qquad (C_1)$$

Their derivation will be clear from an inspection of the
diagram, Fig. 49.

Fig. 49.

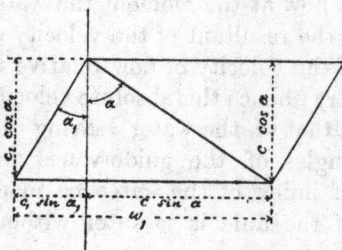

With respect to the
loss from unutilized
energy, it has been
shown that the absolute
residual velocity u with
which the water leaves
the wheel-buckets, is
the resultant of the re-
lative velocity c_2 with
which the water issues from the latter and the velocity
of rotation w_2 at the corresponding point (Fig. 50).
For any given direction of c_2 a little consideration will
show that u is a minimum when the water leaves the wheel

parallel with $O X$, that is, radially in a radial turbine and axially in an axial turbine, *vide* Fig. 51, so that c_2 is the hypothenuse of a right-angled triangle of which u and w_2

Fig. 50.

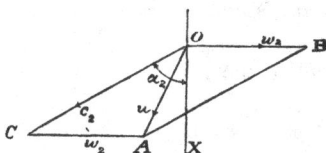

Fig. 51.

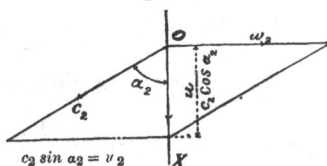

are the sides. If $a_2 = 90°$, then $u = 0$, but this is for practical reasons impossible, as in that case the water could not get away from the wheel. In fact a_2 ranges from 80° to 65°.

Algebraically the condition for radial or axial off-flow is expressed by

$$w_2 = c_2 \sin a_2. \qquad (D)$$

The two fundamental assumptions, therefore, from which the best speed is in practice determined, are : (1) Entrance of water without impact, and (2) Radial or axial exit from wheel.

These assumptions tend to simplify very considerably all the various expressions previously developed for the quantities relating to reaction turbines.

Work.

As the water leaves the wheel at right angles to the direction of rotation, it has no component of velocity in that direction, and the momentum with which the fluid enters the wheel is reduced at the time of leaving to *nil*, and therefore the force exerted on the vanes is represented solely by this momentum $m\, c\, sin\, a$, and the work done per unit of weight is

$$W_u = \frac{1}{g}\, w_1\, c\, sin\, a.$$

Now $w_1 = \frac{r_1}{r_2}\, w_2 = \frac{r_1}{r_2}\, c_2\, sin\, a_2$, since $w = c_2\, sin\, a_2$.

$c_2 = c\, \frac{A}{A}$ by the conditions of continuity;

hence $\qquad W_u = \frac{1}{g}\, \frac{r_1}{r_2}\, \frac{A}{A_2}\, c^2\, sin\, a\, sin\, a_2 \qquad$ (ii)

or $W_1 = \frac{G}{g}\, \frac{r_1}{r_2}\, \frac{A}{A_2}\, c^2\, sin\, a\, sin\, a_2$ for the total work. (iiA)

Pressure Difference.

The pressure of the water leaving the guide passages expressed in " head " is p_1, and by the time it issues from the wheel-buckets this is changed to p_2; the difference between these two pressures is determined (1) by the change in the relative velocity of flow from c_1 at the inflow to c_2 at the outflow from the wheel; (2) by losses in the wheel, including leakage through the clearance space; and (3) by the centrifugal action resulting from the change in the velocity of rotation as the water flows outwards or inwards in a radial turbine; hence

$$p_1 - p_2 = \frac{1}{2g}\, (c_2{}^2 - c_1{}^2 + \zeta_2\, c_2{}^2 + w_1{}^2 - w_2{}^2) \qquad \text{(iii)}$$

where $\zeta_2\, \frac{c_2^2}{2g}$ is the loss in the wheel as previously defined.

In an axial (or parallel flow) turbine the term $w_1{}^2 - w_2{}^4$ representing the centrifugal action disappears, since $w_1 = w_2$, but as the inflow is higher than the outflow, the difference of level (= height of the buckets) h_o appears in formula (iii), which becomes

$$p_1 - p_2 = \frac{1}{2g}\,(c_2{}^2 - c_1{}^2 + \zeta_2\,c_2{}^2) - h_o \qquad \text{(iiiA)}$$

It will be seen from (iii) that for an inward-flow turbine the pressure difference is increased by the change in the velocity of rotation (as $w_1 > w_2$), while in an outward-flow turbine the reverse is the case.

The available energy of the water on leaving the guide passages is that due to the difference of pressures $p_1 - p_2$, plus that corresponding to the velocity of flow c, or $\dfrac{c^2}{2g} + p_1 - p_2$; hence available energy for wheel, by substitution,

$$E_w = \frac{1}{2g}\,(c^2 + c_2{}^2 - c_1{}^2 + \zeta_2\,c_2{}^2 + w_1{}^2 - w_2{}^2).$$

This must equal W_u the work done plus the unutilized energy due to the velocity u and the loss in the wheel,

$$E_w = W_u + \frac{u^2}{2g} + \zeta_2\,\frac{c_2{}^2}{2g}.$$

From this relationship the velocity of flow can be calculated, but it is more convenient for practical purposes to effect the calculation in a different way.

As already shown (p. 54, formula (6)), when c is known the pressure-difference can be determined by the formula

$$p_1 - p_2 = h - (1 + \zeta^1)\,\frac{c^2}{2g},$$

and this is generally the more convenient method. As a rule the determination of the pressure-difference is not required in practice.

Velocity of Flow.

The velocity of flow c is generally determined from the equation

$$h = W_u + L_u + \frac{u^2}{2g}.$$

By (ii) $W_u = \frac{1}{g}\frac{r_1}{r_2}\frac{A}{A_2}\, c^2 \sin a \sin a_2.$

For practical purposes the loss L_u may be divided into two parts: one made up of the losses in the wheel itself, and expressed by $\zeta_2\, \frac{c_2^2}{2g}$; the other of the loss in the guide passages and other parts of the motor, expressed by $\zeta_1\, \frac{c^2}{2g}$.

Where there is a long supply-pipe, suction-tube, and sluice, the losses caused by these must also be included in L_u.

The velocity of off-flow u, as can be easily seen by reference to Fig. 51, is

$$u = c_2 \cos a_2.$$

By substituting these values in the general equation, there follows :—

$$h = \frac{1}{g}\left[\left(\frac{r_1}{r_2}\frac{A}{A_2}\sin a \sin a_2\right)c^2 + \zeta_1\frac{c^2}{2} + \zeta_2\frac{c_2^2}{2} + \frac{c_2^2 \cos^2 a_2}{2}\right]$$

and since $c_2 = \frac{A}{A_2}c,$

$$h = \frac{c^2}{g}\left[\frac{r_1}{r_2}\frac{A}{A_2}\sin a \sin a_2 + \frac{1}{2}\zeta_1 + \frac{1}{2}\zeta_2\left(\frac{A}{A_2}\right)^2 + \frac{1}{2}\left(\frac{A}{A_2}\right)^2\cos^2 a_2\right];$$

from this by suitable transposition

$$c = \sqrt{\frac{2gh}{2\frac{r_1}{r_2}\frac{A}{A_2}\sin a \sin a_2 + \zeta_1 + \left(\frac{A}{A_2}\right)^2(\zeta_2 + \cos^2 a_2)}} \qquad \text{(iv)}$$

$$\text{Let}\sqrt{\dfrac{1}{2\,\dfrac{r_1}{r_2}\dfrac{A}{A_2}\,sin\ a\ sin\ a_2 + \zeta_1 + \left(\dfrac{A}{A_2}\right)^2(\zeta_2 + cos^2 a_2)}} = K_1,$$

then
$$c = K_1\sqrt{2gh}.$$

For ordinary practice the value of K_1 is confined within comparatively narrow limits, but by altering the proportions of certain parts of a turbine, it may be made to vary through a wide range.

The values of K_1 for a variety of cases will be found in Table I. In this table for a given value of a_2 the co-efficient K_1 has been determined, which corresponds to different values of a, $\dfrac{A}{A_2}$, and $\dfrac{r_1}{r_2}$; the co-efficients ζ_1 and ζ_2 have been taken as constant; although in all probability ζ_2 varies with the dimensions of the turbine, there are not sufficient data to accurately determine this point, and for practical purposes a constant value may be used.

From the table the effect of various proportions between the parts of the turbine on the velocity c can be studied. It will be seen that for given ratios between r_1 and r_2 and A and A_2 the co-efficient K_1 *increases* as a, the angle at which the water leaves the guide passages, *decreases;* for example, for $a_2 = 75°$, $\dfrac{A}{A_2} = 2$ and $\dfrac{r_1}{r_2} = \frac{5}{4}$, with $a = 75°$, $K_1 = 0\cdot424$, while with $a = 60°$, $K_1 = 0\cdot444$; similarly for $\dfrac{A}{A_2} = 0\cdot5$ and the same ratio $\dfrac{r_1}{r_2}$, with $a = 75°$, $K_1 = 0\cdot864$, and with $a = 60°$, $K_1 = 0\cdot906$.

The effect of varying a alone is evidently not very great within practical limits.

With respect to the influence of $\dfrac{r_1}{r_2}$, reference to the table shows that with decreasing values of this ratio K_1 increases; with $a = 75°$ (a_2 being also $75°$) and $\dfrac{A}{A_2} = 1$; for $\dfrac{r_1}{r_2} = \frac{3}{2}$, $\frac{5}{4}$, and 1, $K_1 = 0.566$, 0.614, and 0.685 respectively.

The most important factor as regards its effect on the velocity of flow is the ratio $\dfrac{A}{A_2}$; for $a_2 = 75°$, $a = 75°$, and $\dfrac{r_1}{r_2} = \frac{5}{4}$; K_1 varies from 0.424 with $\dfrac{A}{A_2} = 2$ to 0.864 with $\dfrac{A}{A_2} = 0.5$.

The smaller the velocity of flow from the guide passages, the greater is the proportion of available energy existing in the form of pressure in the fluid at the moment it enters the wheel-buckets. Some writers call this proportion the *degree of reaction* with which the turbine works, so that the less the velocity of flow with a given head the greater the degree of reaction. It will be seen from the figures above given that this degree of reaction depends chiefly on the ratio of $\dfrac{A}{A_2}$, but also to a less extent on the other dimensions of the turbine.

Another Method of calculating the Value of K_1.

For ordinary practical purposes the best method of determining K_1 is based upon the following considerations :—

The useful work done can be expressed as $W_u = \epsilon\ E_u$, and substituting for W_u and E_u their values from formula (ii) p. 104, and p. 50, respectively there results

$$\frac{c^2}{g} \frac{A}{A_2} \frac{r_1}{r_2} \sin a \sin a_2 = \epsilon h$$

or

$$c^2 = \frac{\epsilon \, 2gh}{2 \frac{A}{A_2} \frac{r_1}{r_2} \sin a \sin a_2}$$

whence

$$c = \sqrt{\frac{\epsilon \, 2gh}{2 \frac{A}{A_2} \frac{r_1}{r_2} \sin a \sin a_2}} = K_1 \sqrt{2gh}.$$

Now for the efficiency ϵ in the above expression some value may be assumed which experience has proved to be probable, with a sufficient margin to allow for slight variations, so that the velocity of flow may not be too high, and consequently the area for the guide passages insufficient.

For the purpose in question ϵ may be taken as 0·81 in ordinary cases, and then

$$K_1 = 0.9 \sqrt{\frac{1}{2 \frac{A}{A_2} \frac{r_1}{r_2} \sin a \sin a_2}}.$$

It is almost superfluous to remark that where special circumstances exist tending to increase the losses, such, for instance, as a long supply-pipe or channel, these must be taken into account, and a correspondingly lower efficiency assumed; it is better, in such cases, to use the first method of calculating K_1 and add the necessary co-efficients, for the particular kind of additional resistance occurring, in the denominator.

A third expression for K_1 very usually adopted for Jonval turbines is—

$$K_1 = 0.9 \sqrt{\frac{\cos a_1}{2 \sin (a + a_1) \sin a}}.$$

This is in fact identical with the preceding formula, and
is arrived at in the following way :—

As before, $W_u = \epsilon E_u$

and from the formula on p. 104,

$$W_u = \frac{1}{g} c \, w_1 \sin a;$$

now by C_1 (p. 102) $\dfrac{w_1}{c} = \dfrac{\sin (a + a_1)}{\cos a_1};$

whence $w_1 = c \, \dfrac{\sin (a + a_1)}{\cos a_1},$

and by substitution

$$W_u = \frac{c^2}{g} \frac{\sin (a + a_1) \sin a}{\cos a_1},$$

$$\frac{c^2}{g} \frac{\sin (a + a_1) \sin a}{\cos a_1} = \epsilon \, h;$$

and therefore

$$c^2 = \frac{\cos a_1}{\sin (a + a_1) \sin a} \epsilon \, gh,$$

or $c = \sqrt{\dfrac{\epsilon \, 2gh \cos a_1}{2 \sin (a + a_1) \sin a}},$

and $K_1 = \sqrt{\dfrac{\epsilon \cos a_1}{2 \sin (a + a_1) \sin a}}.$

For the ordinary form of Jonval turbine, with stereo-
typed values of a, a_1 and a_2, the above formula is very
convenient; but it is, in the author's opinion, unsuited
for general use. In the first place, the angle of outflow
a_1 depends on all the other leading proportions of the

turbine, $\frac{A}{A_2}$, a_2 and $\frac{r_1}{r_2}$, as well as a, and cannot therefore be assumed without involving these. The ratio $\frac{A}{A_2}$ is the proportion which has more influence, for a given class of motor, than any other on the performance of a reaction turbine, and it ought therefore to form the starting-point in designing one. In the second place, a slight alteration in a_1 very materially affects the value of K_1 when calculated by the above formula, while in reality, *provided $\frac{A}{A_2}$ is not altered*, the influence of such a modification on the velocity of flow is unimportant. Of course the theoretical assumption is that if a_1 is varied, all the other dimensions are changed to correspond, so that the conditions of the inflow without shock and vertical or radial outflow are always fulfilled; in practice, however, it often occurs that a turbine has not been designed strictly in accordance with these principles, and then, if the formula in question is employed, the calculated result will be further from the truth than if the expression previously given be used, in which a_1 does not appear, but in which the correct values of $\frac{A}{A_2}$ and a_2 are sub-stituted. In such cases a deviation of the value of a_1 from that corresponding to inflow without shock implies, when used to determine K_1, a simultaneous modification of the other dimensions which does not really take place.

For these reasons, in comparing experimental with theoretical results, for K_1 the formula

$$K_1 = \sqrt{\dfrac{\epsilon}{2 \dfrac{A}{A_2} \dfrac{r_1}{r_2} \sin a \sin a_2}}$$

is always to be preferred to any other, since it gives results which approximately agree with experiment, even when the assumed conditions on which it is based are not accurately fulfilled.

Velocity of Rotation (Speed).

For vertical or radial discharge from the wheel (*vide* (*D*)) the velocity of rotation at the outflow must be

$$w_2 = c_2 \sin a_2$$

and the velocity at the inflow

$$w_1 = \dfrac{r_1}{r_2} w_2 = \dfrac{r_1}{r_2} c_2 \sin a_2$$

or since

$$c_2 = \dfrac{A}{A_2} c$$

$$w_1 = c \dfrac{A}{A_2} \dfrac{r_1}{r_2} \sin a_2 \qquad (v);$$

but

$$c = K_1 \sqrt{2gh}$$

Therefore

$$w_1 = \left(\dfrac{A}{A_2} \dfrac{r_1}{r_2} \sin a_2 \right) K_1 \sqrt{2gh}$$

If the quantity $\left(\dfrac{A}{A_2} \dfrac{r_1}{r_2} \sin a_2 \right) K_1$ be denoted by K_2, then the velocity of rotation can be written

$$w_1 = K_2 \sqrt{2gh} \qquad (vA)$$

K_2 may be termed the *co-efficient of speed*; like K_1, for a given type of turbine of normal proportions, its value lies within narrow limits.

Efficiency.

This, as has been previously shown, may be generally expressed by the ratio of the useful work done to the useful work *plus* losses from all causes, or

$$\epsilon = \frac{W}{W + L_1 + L_2, \text{ \&c.}}$$

Under the special assumed conditions as regards the inflow and outflow, this takes the form—

$$\epsilon = \frac{2\, c^2\, \dfrac{r_1}{r_2}\, \dfrac{A}{A_2}\, sin\ a\ sin\ a_2}{c^2 \left\{ 2\, \dfrac{r_1}{r_2}\, \dfrac{A}{A_2}\, sin\ a\ sin\ a_2 + \zeta_1 + \left(\dfrac{A}{A_2}\right)^2 (\zeta_2 + cos\ {}^2 a_2) \right\}}$$

from which c^2, being a factor both in the numerator and denominator, disappears, and hence

$$\epsilon = \frac{2\, \dfrac{r_1}{r_2}\, \dfrac{A}{A_2}\, sin\ a\ sin\ a_2}{2\, \dfrac{r_1}{r_2}\, \dfrac{A}{A_2}\, sin\ a\ sin\ a_2 + \zeta_1 + \left(\dfrac{A}{A_2}\right)^2 (\zeta_2 + cos\ {}^2 a_2)} \qquad \text{(vi)}$$

This expresses the fact that the efficiency of reaction turbines, designed in such a manner that the water enters without impact and leaves at right angles to the direction of rotation, is independent of the velocity of flow, and therefore the efficiency may be as great or greater in a turbine with a low velocity of flow from the guidewheels as in one where this velocity is high. When c is small, in comparison with $\sqrt{2\ gh}$, the subsequent relative acceleration due to decrease of pressure is greater than when c has a relatively high value, this acceleration being due directly to the water having to pass through passages of diminishing area.

I

Where a suction-tube is used, the last quantity in the denominator of ϵ, representing the absolute velocity u with which the water leaves the wheel, is replaced by $c_4{}^2 = \left(\dfrac{A}{A_4}\right)^2 c^2$, so that including resistance of sluice

$$\epsilon = \frac{2 \dfrac{r_1}{r_2} \dfrac{A}{A_2} \sin a \sin a_2}{2 \dfrac{r_1}{r_2} \dfrac{A}{A_2} \sin a \sin a_2 + \zeta_1 + \zeta_2 \left(\dfrac{A}{A_2}\right)^2 + \zeta_6 \left(\dfrac{A}{A_4}\right)^2} \qquad \text{(viA)}$$

where $\zeta_6 = 1 + \zeta_4$, vide p. 87.

These somewhat elaborate formulas for the efficiency are given for the reason that they show how it is affected by the leading proportions.

The simplest expression for ϵ is

$$\epsilon = \frac{W^u}{h} = \frac{w_1 c \sin a}{gh} = \frac{w_1 V_1}{gh}; \qquad \text{(viB)}$$

but to make use of this, c and w_1 must first have been determined.

If e_1 and e_2 denote respectively the clear width of the buckets at inflow and outflow, then approximately

$$\frac{A}{A_2} = \frac{r_1}{r_2} \frac{e_1}{e_2} \frac{\cos a}{\cos a_2}$$

By substituting this value in equation (vi) for the efficiency, an expression for the latter is obtained from which the influence of the proportion between the radii r_1 and r_2 can be studied.

$$\epsilon = \frac{2 \left(\dfrac{r_1}{r_2}\right)^2 \left(\dfrac{e_1}{c_2}\right) \sin a \cos a \; tang \; a_2}{2 \left(\dfrac{r_1}{r_2}\right)^2 \dfrac{e_1}{c_2} \sin a \cos a \; tang \; a_2 + \zeta_1 + (\zeta_2 + \cos{}^2 a_2)}$$

$$\left(\dfrac{r_1}{r_2}\right)^2 \left(\dfrac{e_1}{e_2}\right)^2 \left(\dfrac{\cos a}{\cos a}\right)^2$$

In a similar way the velocity of flow may be written

$$c = \sqrt{\frac{2gh}{2\left(\frac{r_1}{r_2}\right)^2 \frac{e_1}{e_2} \sin a \cos a \ tang \ a_2 + \zeta_1 + \left(\frac{r_1}{r_2}\right)^2 \left(\frac{e_1}{e_2}\right)^2}}$$

$$\left(\frac{\cos a}{\cos a_2}\right)^2 (\zeta_2 + \cos {}^2a_2)$$

Practical Application of Theory.

In order to apply the theory as hitherto developed to the determination of the leading dimensions, it will first be necessary to examine in detail the connection between the various quantities introduced into the calculations. So far, these have consisted of the three angles a, a_1 and a_2; the sectional areas of the stream at various points, A, A_1 and A_2; the radii at the points of entrance and exit, respectively r_1 and r_2.

Let $A B$, Fig. 52, represent the course followed by a thin stream during its passage along the last portion of a guide-vane, and $B_1 C$ the course of the same in a wheel-bucket; the width of the stream—*vide* Fig. 52—at B, B_1 and C is e, e_1 and e_2 respectively, e being the width of the guide passages at outflow, and e_1 and e_2 the widths of the wheel-buckets where the water enters and leaves respectively. If x be the depth of the stream at the point B, then —Fig. 53—it is easy to see that $x = \overline{Bb} \cos a$, provided $\overline{Bb}$ the portion of the circumference intercepted between $A B$ and $a b$ be so small that it can be considered as a part of a straight line. The sectional area of the thin stream at B is evidently $e x$; and as e is constant for the whole

circumference, if $B\,D$ be that portion of the latter contained between two vanes, then the *total* sectional area of
the water flowing through one guide passage is $\overline{B\,D}\,e\,\cos a$;

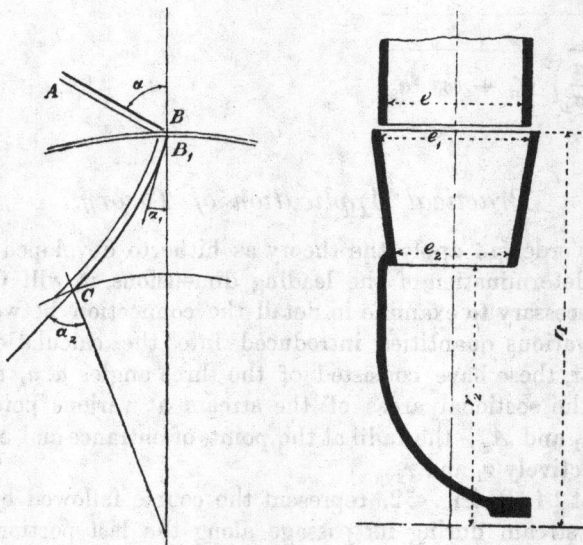

Fig. 52.

for the whole circumference the sectional area A_1 is the
preceding value multiplied by the number of guide
passages. If the vanes were extremely thin, the value of
A_1 would be (whole circumference) $\times e \times \cos a$; as
however each vane has a certain thickness, t, the aggregate
thickness of all the vanes must be deducted from the
above, whence *the measured area*

$$A^1 = 2\,r_1\,\Pi\,e\,\cos a - z\,t\,e = z\,\overline{B\,D}\,\cos a,$$

where z is the number of guide-vanes. When (r_1) the radius
of the wheel where the water enters differs considerably

from the radius at which the water leaves the guides, the latter value must be used in the above formula; this occurs when the sluice-gate passes between the guide apparatus and wheel, so that a large clearance is necessary. For purposes of approximate calculation, e_1 the width of the wheel-buckets where the water enters may be taken as equal to e the width of the guide orifices, although in practice e_1 is slightly greater than e to allow for play and inaccuracy in setting the wheel. A_1^1 and A_2^1, the *measured* inflow and outflow areas of the buckets, are determined in a manner similar to that applied to A_1, hence

Fig. 53.

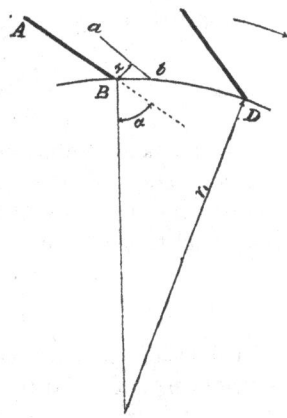

$$A^1 = 2\ r_1\ \Pi\ e\ cos\ a\ -\ z\ t\ e \qquad (b)$$
$$A_1^1 = 2\ r_1\ \Pi\ e_1\ cos\ a_1\ -\ z_1\ t_1\ e_1 \qquad (c)$$
$$A_2^1 = 2\ r_2\ \Pi\ e_2\ cos\ a_2\ -\ z_1\ t_2\ e_2 \qquad (d)$$

where t_1 and t_2 are the thicknesses of the wheel-vanes at the points of entrance and outflow respectively, and z_1 the number of vanes in the wheel.

If the depth of the guide passages measured on the circumference—$B\ D$, Fig. 53—be denoted by s, and the corresponding dimensions for the wheel passages at inflow and outflow by s_1 and s_2, then the sectional areas may also be written—

$$A^1 = z\ s\ cos\ a; \quad A_1^1 = z_1\ s_1\ cos\ a_1; \quad A_2^1 = z_1\ s_2\ cos\ a_2,$$

and hence $\dfrac{A^1}{A_1^1} = \dfrac{e\ z\ s\ cos\ a}{e_1\ z_1\ s_1\ cos\ a_1}; \quad \dfrac{A^1}{A_2^1} = \dfrac{e\ z\ s\ cos\ a}{e_2\ z_1\ s_2\ cos\ o_2}; \quad$ and for very thin vanes, or when $z\,s = z_1\,s_1$,

$$\frac{A}{A_1} = \frac{\cos a}{\cos a_1}; \qquad \text{(vii)}$$

also approximately $\qquad \dfrac{z}{z_1}\dfrac{s}{s_2} = \dfrac{r_1}{r_2};$

whence $\qquad \dfrac{A}{A_2} = \dfrac{r_1\,e_1\,\cos a}{r_2\,e_2\,\cos a_2}. \qquad \text{(viii)}$

These values are sufficiently close for first approximations, and more convenient to use than the accurate formulas.

For axial turbines it is merely necessary to substitute everywhere for r_1 and r_2 the *mean* radius r, in which case

$$\frac{A}{A_2} = \frac{e_1\,\cos a}{e_2\,\cos a_2}.$$

In practice the simplest plan is to measure off the clear depths (perpendicular to the direction of flow) $s\,\cos a$, $s_1\,\cos a_1$, and $s_2\,\cos a_2$ from a drawing.

The quantity of water Q passing through the wheel per second is generally given, and when the velocity of flow c has been found, the area A is determined by it,

since $Q = A\,c$, whence $A = \dfrac{Q}{c}$.

It is very evident from the method in which the relative velocity c_1 and direction of the water entering the wheel are determined—so that no shock shall occur—that when w_1, c and a are fixed, a_1 the angle at which the water enters relative to the wheel must also be fixed. To calculate w_1 and c, the angles a and a_2 must be known or assumed, and hence, if two of the angles are given, the third necessarily follows. It is easiest to find a_1 graphically (*vide* Fig 61). The algebraical expression is found from the condition that the speed of the wheel

stands in a certain definite relation both to the angle at
which the water enters and that at which it leaves :—

$$w_1 = c \sin a + c_1 \sin a_1,$$

and also
$$w_1 = \frac{r_1}{r_2} w_2 = \frac{r_1}{r_2} c_2 \sin a_2.$$

By substituting the values above found, so as to give c_1
and c_2 in terms of c, A, A_1 and A_2, it is easy to prove that

$$tang\ a_1 = \left(\frac{r_1}{r_2}\right)^2 \frac{e_1}{e_2} tang\ a_2 - tang\ a, \qquad \text{(ix)}$$

approximately, or accurately—

$$tang\ a_1 = \frac{A}{A_2} \frac{r_1}{r_2} \frac{\sin a_2}{\cos a} - tang\ a. \qquad \text{(ixA)}$$

The values of A_1 and $A_2{}^1$ given in formulas (*b*) and (*d*)
are the sectional areas of the guide and wheel passages,
but not the actual effective sections of the stream itself.
These will in fact be less, for the following reasons :—The
edges of the wheel-vanes passing before the outlets of
the guide passages obstruct the latter and diminish the
area available for the flow of the water; there is a
momentary diminution of this obstruction during the
time that a guide-vane covers wholly or partially a
wheel-vane, but it is *safe* to assume that the interference
is equally great at all times, and in that case it is easy to
prove that the area of the guide passages is diminished

by $z_1\ t_1\ \dfrac{\cos a}{\cos a_1}\ e_1$. In the case of a Jonval turbine, where

the guide-vanes are parallel for a portion of their length
near the outflow orifices, there is no contraction of the
stream, and therefore the effective area is represented

by $A^1 - z_1\ t_1\ \dfrac{\cos a}{\cos a_1}\ e = A$. In practice this can be most simply measured from a drawing; in Fig. 54, $A\,B$ represents the depth or thickness of the obstructed portion of the stream, and this multiplied by the width e and the number of vanes z_1 gives the total area to be deducted from A^1.

Fig. 54.

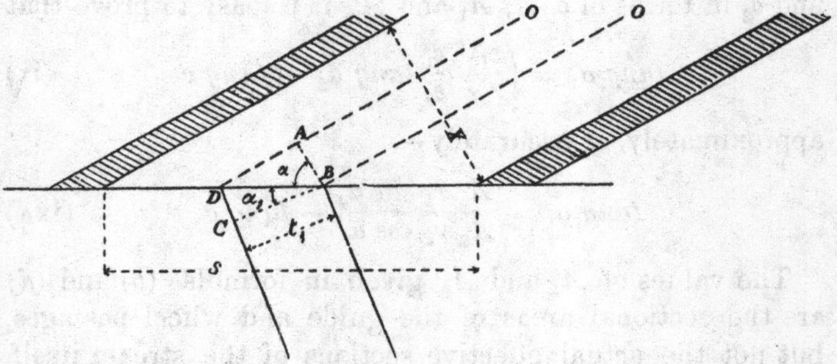

In radial inward- and outward-flow turbines, in addition to the reduction of the constructive area from the cause just explained, there is a further reduction owing to the contraction of the stream from the convergent or divergent direction of the vanes, and the value A^1 must be further multiplied with a co-efficient of contraction; for the latter the value 0·9 may be taken.

As regards the effective value A_2 of the outflow orifices of the wheel-buckets, there need be in a Jonval turbine or axial turbine no contraction, but owing to the leakage through the clearance space, less water passes through the wheel than comes out of the guide passages. Meissner, in his work on turbines, takes this loss into account in determining A_2, but there is so much uncertainty as to its amount that it appears to the author a waste of labour to do this. Hence. when there is no contraction, $A_2 = A_2{}^1$, and has the value given in formula (d). In a radial outward- or inward-flow turbine, however, contraction of the stream occurs in leaving the wheel-buckets, and, as in the previous instance, the co-efficient C may be taken as 0·9.

The true value of the ratio $\dfrac{A}{A_2}$ resulting from the above considerations would therefore be :—

$$\frac{A}{A_2} = \frac{A^1 - z_1\, t_1 \dfrac{\cos a}{\cos a_1}\, e}{A_2{}^1}.$$

Where there is contraction the same co-efficient occurs in both numerator and denominator, except in some mixed-flow wheels, so that it does not affect the proportion.

In determining A_1, $A_1{}^1$ or $A_2{}^1$ there is no necessity for introducing any trigonometrical quantities at all into the calculation ; referring to Fig. 54, the distance x ($= s \cos a$) between two guide-vanes can be measured directly from a drawing, and then the area of each passage is expressed by $x\, e$, and of the whole of the passages by

$$A_1 = z\, x\, e.$$

An exactly similar method may be followed for $A_1{}^1$ and $A_2{}^1$, the distance between the vanes, or *depth* of the passages, being respectively x_1 and x_2.

In the case of the guide orifices, if the thickness $A\,B$, Fig. 54, be denoted by y, then the effective area A, apart from contraction, is

$$A - z\, x\, e - z_1\, y\, e = e\,(z\, x - z_1\, y).$$

For radial turbines, the distance between two vanes s measured *on the circumference* may be set out on a straight

---

[1] That is, the effective area of the stream equals that of the wheel-bucket.

line, and the vane-angles measured relatively to a perpendicular to this line, so that both vane surfaces become
parallel. The mode of proceeding then becomes just the
same as for a parallel-flow turbine, except that contraction
has to be allowed for.

Another Method of Computing A, A_1, A_2.

The preceding method of determining the outflow areas
from the guide passages and buckets is based upon the
assumption that all the consecutive ideal filaments of
water follow, relatively to the wheel, *similar* paths. In
radial-flow turbines this condition is not necessarily
always fulfilled. If the vanes were extremely thin, so that
on leaving the orifices the streams from two adjacent passages or buckets were practically in contact, the assumption in question would probably be correct, but in practice
the thickness of the vanes divides the water issuing from
two neighbouring orifices into two distinct streams, which,
for an appreciable distance after leaving the orifices, are
free to follow their own course without mutual interference. Each stream will naturally take the path of least
resistance, and escape by the largest area available, provided this does not involve a sudden change of direction.
Such a path is not always that corresponding to the hypothesis of a similar course for each filament, as will be
clear from a consideration of Fig. 55, which represents an
inward-flow turbine in which the last portion of each vane
at the outflow is straight. In this case, if a section along
$A\,B$ be considered approximately at right angles to the
general direction of flow, it is clear that the portion of the
stream which has passed this section can deviate freely in
the direction indicated by the arrow, *away* from the convex
vane surface, but is prevented by the latter from deviating

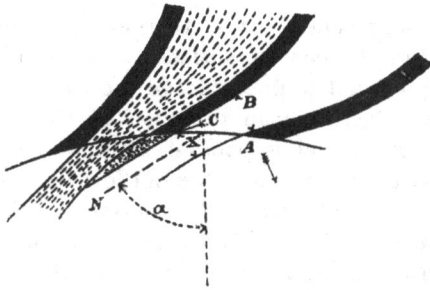

Fig. 55.

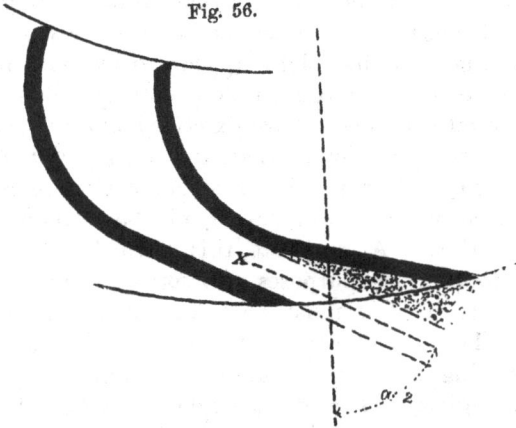

Fig. 56.

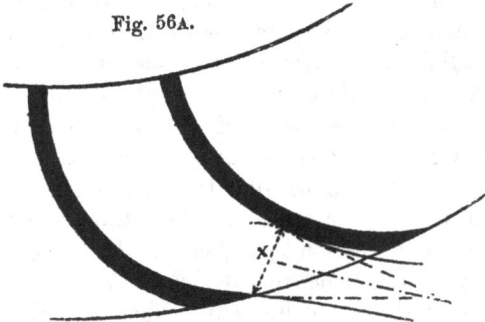

Fig. 56A.

in the opposite sense. It is therefore compelled on one
side to follow the convex surface as far as C, where the
stream may be said to finally leave the wheel. The result
is, that the general mean direction of the stream as it
issues from the wheel is parallel with the convex side of
the bucket (or passage), and the outflow angle a or a_2 is as
indicated. After passing $A B$, a contraction of the stream,
somewhat as shown in the illustration, occurs at X, where the
effective area would have to be measured if this were possible.

The measured outflow area is that corresponding to the
shortest distance $A B$ between two adjacent vanes.

There might, at first sight, appear to be some question
as to the effective outflow radius corresponding to the
method of measurement above described, but on consideration there is little doubt that the *constructive* outflow
radius—measured to the inner or outer circumference of
the vanes, as the case may be—should be introduced into
the calculations. At any rate, this method is sufficiently
accurate for practical purposes. Of course the angle of flow
and velocity of flow must be taken at the same mean
radius. The area of outflow thus obtained is *greater* than
that resulting from the assumption of similar paths as
previously given. The *mean* angle of outflow for each
stream is under these circumstances *not* the vane angle,
but the angle formed by the centre line $X N$ of the stream
with a radial line drawn to the point of intersection with
the outflow circumference ; this angle is smaller than the
vane angle. Where the outflow end of each vane is *not*
straight, a similar state of things may occur, each stream
issuing in a direction parallel to the tangent to the vane-
curve on the *convex* side, or when the vane-curve is a
volute, with inflection, parallel to the tangent at the
extremity of the concave side (Fig. 57).

For outward-flow turbines in which the vanes are constructed as shown in Fig. 56, it is clear that the effective outflow area is that corresponding to X, while the effective angle of outflow is as shown, the mean direction of flow being parallel with the concave sides of the vanes. In such cases no allowance is necessary for contraction if the various parts of the stream become parallel by the time they reach X.

When on the other hand the vanes—instead of being straight at the ends—are curved up to the outer circumference, as in Fig. 56A, the outflow takes place probably somewhat in the manner illustrated, and it is necessary to make allowance for contraction.

Fig. 57.

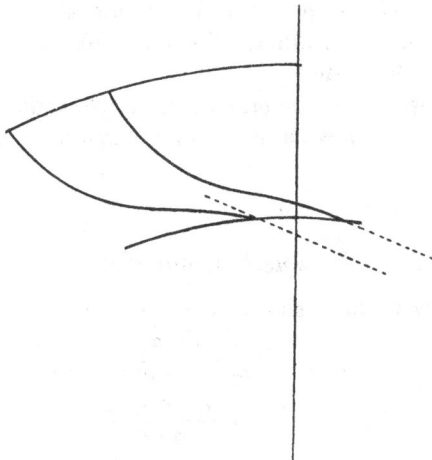

The dimensions of the radii of a turbine are purely arbitrary, fixed solely with reference to practical convenience, except, of course, in so far that the wheel must be sufficiently large to contain passages of the area necessary for the flow of the water at the required velocity.

The widths of the buckets (e_1 and e_2) have no fixed relation to one another, but must be determined with reference to the areas A, A_1 and A_2, as is evident from (*b*), (*c*) and (*d*).

The depth of the buckets in radial turbines is obviously $r_2 - r_1$ or $r_1 - r_2$, according to whether r_2 is greater than r_1, or *vice versâ*; it must be sufficient to allow of a gradual curvature of the vanes, but is in itself of little importance; with greater length there must be, it is true, increased friction in the buckets, but under all ordinary conditions this is so trifling that it may be left out of account.

The numbers of buckets and vanes z and z_1, and their thickness, are empirical, practical experience being the sole guide. The same remark applies to the depth of buckets in axial turbines, which is not limited by the radii as in radial wheels.

The following is a summary of the formulas applying to reaction turbines as used in determining their construction :—

Assumed Conditions.

(1) Inflow without shock :
$$c_1 \cos a_1 = c \cos a \qquad (A)$$
$$w_1 = c \sin a + c_1 \sin a_1 \qquad (B)$$
$$c^2 = c_1{}^2 + w_1{}^2 - 2\,c_1\,w_1 \sin a_1 \qquad (C)$$
$$\frac{w_1}{c} = \frac{\sin (a + a_1)}{\cos a_1} \qquad (C_1)$$

(2) Radial or axial outflow :
$$w_2 = c_2 \sin a_2 \qquad (D)$$
$$u = c_2 \cos a_2 \qquad (E)$$

(3) Continuity of flow :
$$Q = A\,c = A_1\,c_1 = A_2\,c_2 = A_3\,c_3, \ \&\text{c.} \qquad (F)$$
$$G = Q\,\lambda \ (\lambda = \text{weight of cubic foot of water})$$

Fundamental Relations between Dimensions.

$$w_1 = \frac{r_1}{r_2}\, w_2 \qquad\qquad (a)$$

$$A^1 = 2\, r_1\, \Pi\, e\, \cos a \; - \; z\, t\, e \; = \; e\, z\, s\, \cos a \; = \; z\, x\, e \quad (b)$$

$$A_1^1 = 2\, r_1\, \Pi\, e_1 \cos a_1 \; - \; z_1\, t_1\, e_1 \; = \; e_1\, z_1\, s_1 \cos a_1 \; = \; z_1\, x_1\, e_1 \quad (c)$$

$$A_2^1 = 2\, r_2\, \Pi\, e_2 \cos a_2 \; - \; z_1\, t_2\, e_2 \; = \; e_2\, z_1\, s_2 \cos a_2 \; = \; z_1\, x_2\, e_2 \quad (d)$$

$$A \;=\; C\left(A^1 - z_1\, t_1\, \frac{\cos a}{\cos a_1}\, e\right) \; vide \; \text{p. 121.} \qquad (e)$$

$A_2 = C\, A_2^1$ where $C = 1$ for axial and 0.9 for radial
$\qquad$ inward-flow turbines $\qquad\qquad\qquad\qquad (f)$

$$E_1 = Gh = \left(W_u + L_u + \frac{u^2}{2g}\right) G \qquad\qquad (i)$$

$$W_u = \frac{1}{g}\, w_1\, c\, \sin a = \frac{1}{g}\, \frac{r_1}{r_2}\, \frac{A}{A_2}\, c^2 \sin a \sin a_2 \qquad (ii)$$

$$W_1 = \frac{G}{g}\, \frac{r_1}{r_2}\, \frac{A}{A_2}\, c^2 \sin a \sin a_2 \qquad\qquad (iiA)$$

$$p_1 - p_2 = \frac{1}{2g}\, (c_2{}^2 - c_1{}^2 + \zeta_2\, c_2{}^2 + w_1{}^2 - w_2{}^2) \qquad (iii)$$

$$p_1 - p_2 = \frac{1}{2g}\, (c_2{}^2 - c_1{}^2 + \zeta_2\, c_2{}^2) - h_o \; \text{for axial turbines (iiiA)}$$

$$c = \sqrt{\dfrac{2gh}{2\, \dfrac{r_1}{r_2}\, \dfrac{A}{A_2}\, \sin a \sin a_2 + \zeta_1 + \left(\dfrac{A}{A_2}\right)^2 (\zeta_2 + \cos{}^2 a_2)}} \qquad (iv)$$

$$c = K_1 \sqrt{2gh} \qquad\qquad (ivA)$$

$$c = \sqrt{\dfrac{\epsilon\, 2gh}{2\, \dfrac{A}{A_2}\, \dfrac{r_1}{r_2}\, \sin a \sin a_2}} \qquad\qquad (ivB)$$

$$\left. c = \sqrt{\dfrac{\epsilon \cos a_1}{2 \sin (a + a_1) \sin a}} \quad\right\} \qquad (ivC)$$
$\qquad\qquad$ where ϵ may be taken as 0.81

$$w_1 = c \frac{A}{A_2} \frac{r_1}{r_2} \sin a_2 \qquad \text{(v)}$$

$$w_1 = K_2 \sqrt{2gh} = \left(\frac{A}{A_2} \frac{r_1}{r_2} \sin a_2 \right) K_1 \sqrt{2gh} \qquad \text{(vA)}$$

$$\epsilon = \frac{2 \dfrac{r_1}{r_2} \dfrac{A}{A_2} \sin a \sin a_2}{2 \dfrac{r_1}{r_2} \dfrac{A}{A_2} \sin a \sin a_2 + \zeta_1 + \left(\dfrac{A}{A_2} \right)^2 (\zeta_2 + \cos^2 a_2)} \qquad \text{(vi)}$$

$$\left\{ \epsilon = \frac{2 \dfrac{r_1}{r_2} \dfrac{A}{A_2} \sin a \sin a_2}{2 \dfrac{r_1}{r_2} \dfrac{A}{A_2} \sin a \sin a_2 + \zeta_1 + \zeta_2 \left(\dfrac{A}{A_2} \right)^2 + \zeta_6 \left(\dfrac{A}{A_2} \right)^2} \right.$$

with suction-tube $\qquad \text{(viA)}$

$$\epsilon = \frac{W_u}{h} = \frac{w_1 c \sin a}{gh} \qquad \text{(viB)}$$

$$\frac{A}{A_1} = \frac{z \, s \, \cos a}{z_1 \, s_1 \, \cos a_1} = \text{approximately } \frac{\cos a}{\cos a_1} \qquad \text{(vii)}$$

$$\frac{A}{A_2} = \frac{e \, z \, s \, \cos a}{e_2 \, z_1 \, s_2 \, \cos a_2} = \text{approximately } \frac{r_1 \, e_1 \, \cos a}{r_2 \, e_2 \, \cos a_2} \qquad \text{(viii)}$$

$$\tan a_1 = \frac{A}{A_2} \frac{r_1 \sin a_2}{r_2 \cos a} - \tan a \qquad \text{(ix)}$$

Application of Formulas to Axial Turbines.

To apply the preceding formulas to axial turbines, it is merely necessary to make $\frac{r_1}{r_2} = 1$, except where otherwise stated. *Strictly* speaking, the formulas under these conditions are correct only for a *very narrow* turbine, but they are usually employed with reference to the mean radius for wheels of the proportions actually occurring in practice. In this sense w $(= w_1 = w_2)$ is the velocity of rotation of the circumference taken at the middle of the width of the buckets.

It is evident that if the conditions of entrance without impact and axial exit, for given angles, are fulfilled for the mean radius, this cannot be the case for any other radius, because the velocity of rotation at any other distance from the centre must be different from that calculated, which alone complies with the assumptions made. At the outer radius the velocity will be *greater*, at the inner *less* than the calculated value. Referring to the diagram, Fig. 58, let $O\,a$ represent the value and direction of the absolute velocity c, which is assumed to be the same at every point in the width of the buckets; $a\,b$ the mean velocity of rotation as determined by calculation; then for the mean

Fig. 58.

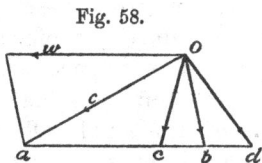

radius, $O\,b$ gives the proper inclination of the wheel-vanes at the inflow, which coincides with the relative direction there taken by the fluid.

At the inner radius the velocity of rotation is only $a\,c$, *less* than $a\,b$; the relative direction of the water entering the wheel is therefore $O\,c$, which it is clear is not the same as that of the vane. Similarly if $a\,d$ be the velocity at the outer radius, *greater* than $a\,b$, then $O\,d$ represents the relative direction taken by the entering fluid. Hence, for every other than the mean radius, the water must enter with *shock* under the assumed conditions. As the radius is either less or greater than the mean, so the impact is in the same sense as or opposed to the motion of the wheel.

What happens at the outflow is similar.

Let $O\,a$, Fig. 59, be the relative velocity c_2 and direction

K

with which the water issues from the buckets, it being

Fig. 59.

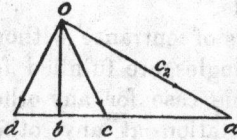

assumed that this is the same for every radius; *a b* the mean velocity of rotation, for which the water leaves the wheel in the absolute direction *O b* parallel with the axis. The velocities of rotation at the inner and outer circumferences are respectively *a c* and *a d*, and the corresponding absolute velocities of outflow *O c* and *O d*, both of which it is evident deviate from the assumed condition of vertical outflow.

The effect of these deviations from theoretical accuracy is to reduce the efficiency by causing a loss in the first place from impact or shock, and secondly by increasing the absolute residual velocity, which is least when its direction is axial. To what extent this evil may be avoided by a suitable construction of the vanes will be shown in a subsequent chapter.

In ordinary practice the angles *a* and a_2 are not constant for the whole width of the wheel, as the vanes have helical surfaces with constant pitch, so that those angles increase with the distance from the axis (except for vertical surfaces). It is easy to modify the construction shown in Fig. 59 to suit a helical vane surface. Instead of the constant angle *a O b*, the angles at the outer, mean, and inner circumferences are all different; when one of these (generally that for the mean circumference) is given, the others are easily found by construction, and for each the corresponding residual velocity can be ascertained in the usual way.

General Data and Method for Design.

In a large proportion of the cases which occur in actual practice, the head and the quantity of water available under average conditions are given, that is, h and Q.

A turbine has to be constructed which will utilize to the best advantage the power available $E_a = (Weight\ of\ water\ per\ second) \times h$, in foot-pounds per second, or

$$\frac{(Weight\ of\ water) \times h \times 60}{33,000}$$

expressed in horse-power, and if λ be the weight of a cubic foot of water, then

$$E_{hp} = \frac{Q \times \lambda \times h \times 60}{33,000} = \text{available Horse-power.}$$

Now certain dimensions or dimension-ratios can, within reasonable limits, be arbitrarily chosen, being independent of each other. Different writers have followed various methods in the choice of these quantities, and upon this choice depends the course of the subsequent calculations.

The author has for various reasons taken the following independently selected quantities, (1) the angle a at which the water leaves the guide passages, (2) the angle a_2 at which the water leaves the wheel-buckets, and (3) the ratio $\dfrac{A}{A_2}$ of the outflow-area of the guide passages to the outflow-area of the wheel-buckets, determined as previously explained.

It is usual to fix arbitrarily the ratio between the inner and outer radius, but this may also be varied to suit convenience.

When a, a_2, $\dfrac{A}{A_2}$ and $\dfrac{r_1}{r_2}$ are fixed, it is clear from formula (viii) that the ratio $\dfrac{e_1}{e_2}$ of the width of the buckets at the entrance to that at the outflow is also determined. The difference between the number and thickness of the vanes in the guide apparatus and wheel gives a slight margin for variation, as may be seen from formulas (*b*) and (*d*), but this may in practice be neglected, as the numbers of vanes and their thickness are determined purely by practical considerations—the thickness, more especially, should be as small as possible.

If, as is generally the case in Jonval turbines, the width of the buckets is the same throughout, so that $\frac{e_1}{e_2} = 1$, then the ratio $\frac{A}{A_2}$ can no longer be independently fixed, but is absolutely determined by a and a_2, since

$$\frac{A}{A_2} = \frac{\cos a}{\cos a_2} \text{ approximately,}$$

but there is no reason why the width should be constant if it is desirable to vary it, and in many cases this has been done.

The influence of the various dimensions on the efficiency of reaction turbines will be subsequently studied, but as regards the choice of a_2 it is easy to see (from the triangle of velocities, Fig. 59, for instance) that, from a theoretical point of view, the larger this is the better, as the residual velocity of the water decreases with an increase of a_2. The extreme value for a_2 in practice is about 80°, a ranges from 50° to 80° for different classes of turbines.

$\frac{A}{A_2}$ ranges from 1·8 to 0·5, but for Jonval turbines, to which class belong the majority of European reaction wheels, it is near 1 as a rule. For an inward-flow turbine (to which it may, for the sake of greater generality, be assumed that the present investigation applies) a possible value for $\frac{A}{A_2}$ would be 1·3.

Having chosen a, a_2, and $\frac{A}{A_2}$, and taking for $\frac{r_1}{r_2}$ some value between $\frac{5}{4}$ to $\frac{3}{4}$ (for inward flow), the ratio of the widths $\frac{e_1}{e_2}$ can be calculated from (viii) $\frac{e_1}{e_2} = \frac{A}{A_2} \frac{r_2}{r_1} \frac{\cos a_2}{\cos a}$. Suppose $\frac{A}{A_2} = 1·3$, $a = 75°$, $a_2 = 75°$, and $\frac{r_1}{r_2} = \frac{3}{2}$,

then $\qquad \frac{e_1}{e_2} = 1·3 \times \frac{2}{3} \times 1 = 0·866,$

that is, the width of the buckets is less at the outer than at the inner circumference. This calculation is merely necessary at the present stage in order to be satisfied that e_1 and e_2 do not assume impracticable proportions.

After these preliminary operations, the first quantity to be ascertained is the velocity of flow c; it has been shown that $c = K_1 \sqrt{2gh}$; referring to the table (I.) it will be found that for $a_2 = 75°$, $a = 75°$, $\frac{r_1}{r_2} = \frac{3}{2}$ and $\frac{A}{A_2} = 1·3$, the corresponding value of K_1 is about 0·48, hence $c = 0·48 \sqrt{2gh}$. Had the dimensions been differently chosen, K_1 and consequently c would have had a different value; by making, for instance, $\frac{A}{A_2} = 1$ instead of 1·3, K_1 would be 0·566, and therefore c proportionately greater. To utilize the same head and quantity of water per second, it is evident that turbines may be constructed which vary very considerably in their details. It has already been demonstrated that the efficiency is not directly dependent on the velocity of flow c, but as the various losses are affected by the velocities in different parts of the apparatus, and these again depend on the proportions of the latter, it is clear that to some extent the efficiency must be influenced by the same factors which determine c.

Fig. 60.

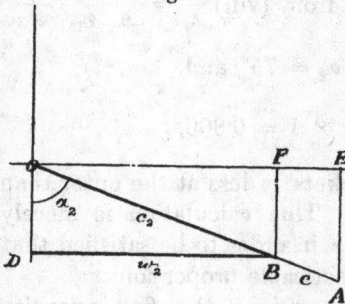

It is sufficient for the present to state that in fact the efficiency is only slightly affected by very considerable variations in the *degree of reaction* or velocity of flow from the guide passages, and consequently great latitude is admissible in the choice of dimensions without much practical modification of the results.

Having calculated c the velocity of flow, the next step is to determine w_1 the best speed of rotation at the inflow.

By (v) $$w_1 = c \, \frac{A}{A_2} \, \frac{r_1}{r_2} \, sin \, a_2 = K_2 \sqrt{2gh};$$

but this value can very easily be found graphically in the following way: Set out $O\,D$, Fig. 60, perpendicular to $O\,E$, and draw $O\,A$ forming with $O\,D$ the angle $A\,O\,1$ equal to a_2; make $O\,A = c$ to any scale, $O\,F =$ the area A, and $O\,E = A_2$ (the unit in the latter cases being the square foot, to any scale, as only the ratio is concerned); join A and E—$A\,E$ not being necessarily perpendicular to $O\,E$ as shown—and draw $F\,B$ parallel to $E\,A$ intersecting $O\,A$ in B; then $O\,B = c_2$ the relative velocity of outflow. From B draw $B\,D$ parallel with $O\,E$; then

$B\ D = w_2$ the velocity of rotation at the outflow; $w_1 : w_2 = r_1 : r_2$, from which w_1 is easily determined by construction.

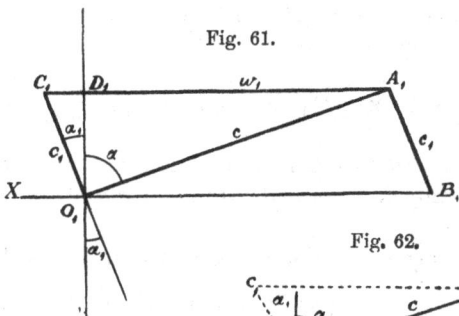

Fig. 61.

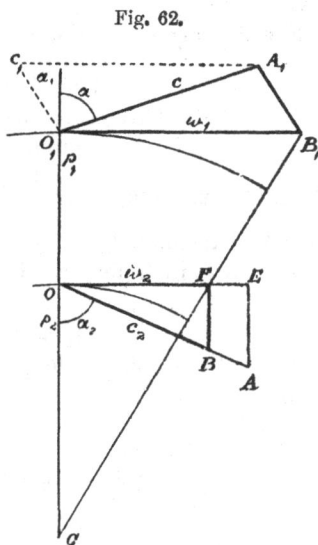

Fig. 62.

a_1 can now be graphically ascertained as shown in Fig. 61. Draw $O_1\ D_1$ perpendicular to $O_1\ X$; make angle $A_1\ O_1\ D_1 = a$, $A_1\ O_1 = c$, $A_1\ C_1$ parallel with $O_1\ X = w_1$; join O_1 and C_1, then angle $C_1\ O_1\ D_1 = a_1$ and $O_1\ C_1$ (or $A_1\ B_1$ parallel to $O_1\ C_1$) $= c_1$.

a_1 may also be calculated by formula (ix)

$$tang\ a_1 = \frac{r_1}{r_2} \frac{A}{A_2} \frac{sin\ a_2}{cos\ a} - tang\ a;$$

The two diagrams, Figs. 60 and 61, are shown combined in Fig. 62, which requires no further explanation except that $C\ O_1 = r_1$ and $C\ O = r_2$.

By (F) the quantity of water flowing through the turbine per second $Q = A\,c$, whence $A = \dfrac{Q}{c}$, and from this the effective area A can be calculated.

From (e) and (b)

$$A = C\left(2\,r_1\,\Pi\,cos\,a - z\,t - z_1\,t_1\,\frac{cos\,a}{cos\,a_1}\right)e\,;$$

of the quantities in this expression r_1, z, t, z_1 and t_1 may be empirically determined, while $cos\,a$ and $cos\,a_1$ and the rest are known.

For the former the following rules can be used :

$$r_1 = 1\cdot25 \text{ to } 1\cdot5 \sqrt{A}\,;$$

$$z = \frac{\text{Circumference corresponding to } r_1}{\text{pitch of guide-vanes}}$$

Pitch of guide-vanes $= \dfrac{\text{Diameter of wheel}}{7\cdot5 \text{ to } 9}$ (*vide* Table of rules and formulas) ;

$t = \frac{1}{2}$ to $\frac{5}{8}$ inch for cast-iron; $\frac{1}{4}$ to $\frac{3}{8}$ for wrought-iron;

$t_1 = $,,   ,,   ,,   ,,   ,,   ,,

$$z_1 = z \text{ to } 1\cdot2\,z\,;$$

c the width of the guide passages can now be calculated by (e)

$$c = \frac{A}{C\left(2\,r_1\,\Pi\,cos\,a - z\,t - z_1\,t_1\,\dfrac{cos\,a}{cos\,a_1}\right)}$$

$$= \frac{A}{C\,(z\,x - z_1\,y)}\,;\quad \text{where } C = 0\cdot9,$$

$$t_1\,\frac{cos\,a}{cos\,a_1} = y \text{ and } \frac{1}{z}\,(2\,r_1\,\Pi\,cos\,a - z\,t) = x,$$

vide Fig. 54. x and y can be measured from a drawing.

In practice the latter method must always be adopted when the outflow areas are computed as described at p. 122, *vide* Fig. 63. By rounding off the corners of the vanes at the concave side, it is certain the contraction of the stream might be much reduced, if not altogether avoided. The author is, however, not aware that this is ever done.

Fig. 63.

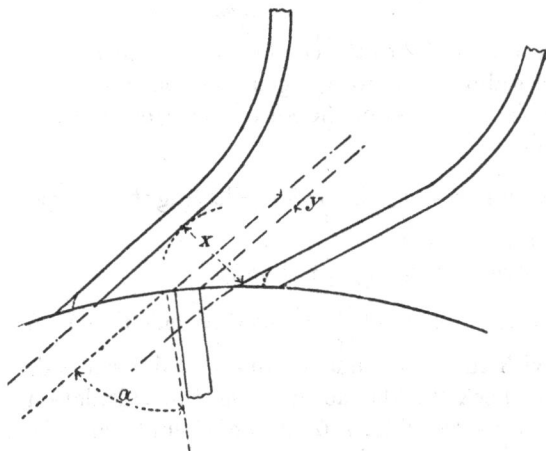

A being known, A_2 follows from the assumed ratio $\left(\dfrac{A}{A_2}\right)$, (which was taken at 1·3 for example :)

$$A_2 = \frac{A}{1·3};$$

and by (f) and (d)

$$c_2 = \frac{A_2}{(2\ r_2\ \Pi\ cos\ a_2\ -\ z_1\ t_2)\ C}$$
$$= \frac{A}{z_1\ x_2\ C}; \qquad \text{(where } C = 0·9\text{).}$$

To the calculation of e_2 exactly the same remarks apply as to the calculation of e.

When x or x_2 and y are measured as shown in Fig. 63 the formulas to be used are obviously

$$e = \frac{A}{C\,(z\,x - z_1\,y)}$$

and

$$e_2 = \frac{A_2}{C\,z_1\,x_2}$$

As a matter of fact the co-efficient of contraction C must vary according to the convergence of the vanes, but no data applicable to orifices of the form occurring in turbines are available.

r_2 follows from r_1, the ratio $\frac{r_1}{r_2}$ having been assumed at the commencement.

t_2 as a rule equals t_1, z_1 is known.

The correct value of $\frac{e_1}{e_2}$ can now be calculated and compared with the approximate value first determined.

As a check on the accuracy of the calculation, c_2 the relative velocity of flow from the buckets may be ascertained from the formula

$$c_2 = \frac{Q}{A_2}, \text{ and if } c_2 = w_1 \frac{r_1}{r_2} \frac{1}{\sin a_2},$$

the construction is correct. This can also be done graphically; it simply amounts to ascertaining whether the absolute outflow from the wheel is radial or not.

All the elements necessary for the construction of the turbine itself have now been calculated. The next step is to determine the form of the vanes ; how this is done will be subsequently described.

Where a suction-tube is employed, it is usually made of about the same diameter as the wheel where the water leaves the latter, and the residual velocity of outflow u is suddenly changed to that corresponding to the area of the suction-tube; the latter is greater, as generally constructed, than that of the wheel, and consequently there is an abrupt retardation of flow and corresponding loss of energy which might be utilized if the transition were gradual, as then the loss from unutilized energy is only that due to the velocity with which the fluid leaves the tube. There is no reason why there should be a sudden enlargement of sectional area when the water leaves the wheel; by making the suction-tube annular at the top, and of a section exactly corresponding to that of the wheel, and gradually enlarging this section towards the lower end, this evil might be avoided. The orifice through which the water leaves the suction-tube should be as large as practicable, but not sufficiently so to reduce the velocity of the water below 3 feet per second, as this is necessary to carry it off. There is, of course, no absolute rule on this point, and what the lowest admissible velocity may be must depend on circumstances, among others the sectional area and fall of the tail-race, for instance.

Instead of the available quantity of water being fixed, it may occur that the supply of water for a certain fall is practically unlimited, and that a turbine has to be constructed which with that fall will develop a certain effective power. The data are then Effective Horse-power and Head h. If—as is generally the case—the effective power is measured at the point where it is given off direct from the wheel-shaft without intermediate gearing, then to the effective power need only be added an allowance for shaft friction to arrive at the work to be done by the turbine. From the work expressed in horse-power per minute, W_1, that to be done per second is easily determined. The simplest and safest plan is then to assume a certain efficiency well within the limits of probability, say 70 to 75 per cent., and substitute this value for ϵ in formula (viii)

$\epsilon = \dfrac{W_1}{Gh}$; from this the weight of water required per second can be determined, $G = \dfrac{W_1}{\epsilon \, h}$, and then $Q = \dfrac{G}{\lambda}$ the quantity in cubic feet per second. After this the method of calculation is the same as before, both Q and h being now given.

CONSTRUCTION OF THE VANES OF TURBINES.

With given angles of entrance and exit at the ends of the vane, it has been shown that the intermediate form of the vane-curve is a matter of indifference, provided the curvature is not too abrupt, and that there are no sudden changes of sectional area in the passages and buckets.

Guide-Vanes. In an axial or parallel-flow turbine the water generally speaking enters the guide apparatus in a direction parallel with the axis of the wheel, and the tangent to the vane-curve at the point of entry should be parallel with this direction. In Fig. 64, $A\,B$ and $C\,D$ represent the curves of the vanes at their intersection with a cylindrical surface of the mean diameter in the middle of the width. Assuming the axis of the turbine to be vertical, then the first portions at A and C of the vane-curves should also be vertical. At the lower end, where the water leaves the guide passages, the vanes should form the angle a with a vertical line, or strictly speaking with a direction parallel to the axis, and the lower end of the curve should be continued as a straight line for a certain length $B\,5$ and $D\,4$, so that the surfaces of the vanes run parallel with each other for some distance in order to guard against contraction of the stream.

In an inward-flow radial turbine this is not possible, as the vane-curves terminate at the points B and D, &c., on a circle, and if they form equal angles with the radii at those points must of course converge; in an outward-flow turbine they will diverge.

Supposing, as is often the case, that the width of the passages is uniform, and that $M\,N$ represents the mean course of the stream, then the depth of the latter, measured

Fig. 64.

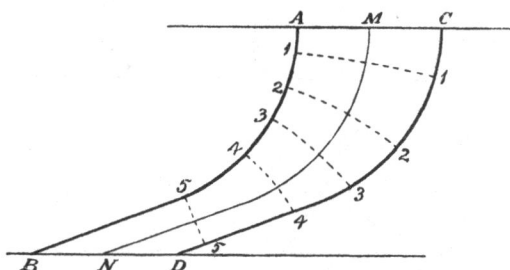

normally to the direction of flow at all points as shown by 1 1, 2 2, 3 3, &c., at any point is proportional to the sectional area. The curves $A\,B$ and $C\,D$ should be so chosen that the lengths 1 1, 2 2, 3 3, &c., *gradually and uniformly* diminish down to the shallowest part.

A simple method of construction, Fig. 65, is to describe a portion of a circle with some radius $O\,A$ from the centre O on the continuation of the line $A\,C$, and draw a tangent to this circle at E, forming the angle a with a direction at right angles to $A\,C$, parallel with the axis. It is usual to make the construction such that E is the point of intersection of a line joining O and D, and the method of pro-

ceeding is as follows:—Draw BE at the proper inclination
a; set out BD the pitch of the vanes on the mean circumference; from D draw DO at right angles to BE,
intersecting the latter in E, and the continuation of AC—
the level of the top of the guide-vanes—in O; from O as
centre with the radius OE describe the arc of the circle
AE. Of course each vane-curve is exactly similar. It is
often preferable, although involving a little more trouble,
to draw the curves by hand, trying various forms until
that which gives the most gradual change in the area of
the stream is found.

Fig. 65.

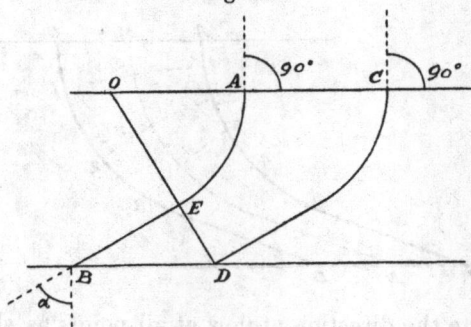

The usual construction of the surface of the vanes is
such that a series of radial lines drawn from any points of
the curve at the mean circumference—in the middle of
the passage—to the axis, touch the surface for its whole
width; in other words, the surfaces are *helical*. In this
case it is clear that the angles at the inner and outer circumferences are not the same as those at the mean circumference, and the greater the width of the passages the
greater will be the difference between the angles. There
is, however, no reason whatever why the angle should not

be the same throughout the whole width of each vane, the lower portion of the surface of the latter, instead of being helical, is then simply a plane surface, that is, the part of it which is parallel with the adjacent vane, of the length represented by *E B* in Fig. 66.

Wheel-Vanes. Similar remarks to those respecting the guide-vanes apply to the wheel-vanes. The angles a_2 and a_1 being determined, the lower end of the curve *B b*, Fig. 66 (for the mean circumference), is set out at the proper angle a_2; *D* being the end of the next vane-curve, *D E* is drawn at right angles to *B b*, and from some centre

Fig. 66.

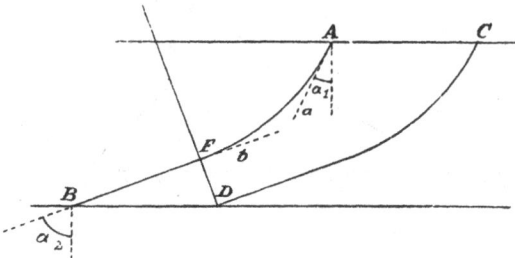

on the prolongation of *D E* an arc of a circle touching *B b* in *E* may be drawn; the line *A a* at the upper end of the curve should be inclined at the angle a_1 and form a tangent to the circle previously referred to. As in the case of the guide-vanes, the curvature is best found by trial. Many designers use a parabolic curve, but there is no special merit in this, and in some cases it is unsuitable.

In arranging the curves so as to insure a gradual change in the sectional area of the passages, any increase or diminution in the width of the buckets must be taken into account; as a rule axial turbines are of equal width

throughout, but in the variety known by the name of
Henschel, the width is increased towards the outlet.

As was before pointed out, with given values of a and
a_1, if the vanes are designed for entry of the water without
impact at the mean circumference, this condition will not
be fulfilled for the inner and outer circumferences; it is,
however, possible to overcome this difficulty by giving the
vanes varying angles at the points of entry. The mean
velocity being fixed, the velocities at the inner and outer
diameters are thereby determined, and must be in the
same ratio as the latter.

Fig. 67

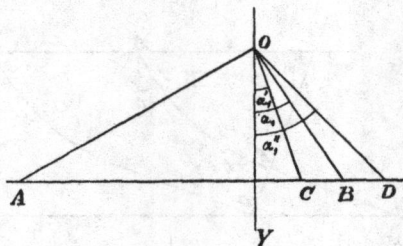

Let $O A$, Fig. 67, represent the value and direction of
the velocity c with which the water leaves the guide passages, $A B$ the velocity w of the wheel at the mean circumference; then $O B$ is the relative velocity c_1 with which
the water enters the wheel-buckets at that circumference,
and $B O Y$ the suitable angle for a_1. At the inner circumference, instead of $A B$, the velocity of rotation is
$A C$, and if the water is to enter without impact there
also, $O C$ must be the direction of inflow, and the angle
$a_1{}^1 = C O Y$; similarly $A D$ is the velocity at the outer
circumference, and $D O Y$ the proper angle $a_1{}^{11}$ for the
vane. By varying the angles a_1 according to the diameter

in the manner explained, it would be possible to insure the entry of the water without impact at all points of the width.

The differences in the speed at different diameters also affects, as previously explained, the absolute velocity of outflow u, so that with a constant angle a_2 for the whole width of the buckets, the direction of outflow is vertical only at the middle, for which the calculations were made, and the velocity of outflow is consequently greater at the inner and outer circumference than at the middle, thus causing a loss of energy in excess of that assumed.

With a helical vane surface the angle a_2 will be less at the inner and greater at the outer circumference than at the middle.

The losses arising out of the deviations from the assumed conditions, both as regards the entry and exit of the water, are inconsiderable when the width of the buckets is moderate in proportion to the diameter of the wheel, but for wide buckets may have a very appreciable influence on the efficiency.

The problem then is, how to construct the turbine vanes so that both these sources of loss may be partially or wholly avoided.

It is evident that the pressure p_1 at which the water leaves the guide passages does not vary at various diameters, as otherwise there would be a motion of the water *across* the buckets, a state of things which could not continue, and must almost instantly correct itself. The sum of the energy due to the pressure p_1 plus that represented by the velocity of flow c, is under any given circumstances a fixed quantity depending on the head of water and the proportions of the wheel; hence if p_1 is constant, so must also be c.[1] The velocity of flow has

[1] The ratio $\dfrac{c_2}{c}$ is not modified by the fact that the relative velocity

therefore to be taken as invariable for all parts of the
width of a guide passage, and may be calculated in the way
shown with the angles chosen for the mean circumference.

Fig. 68.

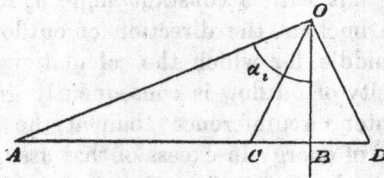

c_1 of inflow varies at different points, *provided* the inflow takes place
at all of them without shock, since the area of any narrow segmental
strip of the orifice varies inversely as the relative velocity of flow.

If, for instance, $d\,A_1$ be the area of any segmental strip of the
inflow orifices A_1, and $d\,A_2$ that of the corresponding strip of the
outflow orifice A_2, the relative velocity of inflow at any diameter

$$c_1 = \frac{Constant}{d\,A_1},$$

since $d\,A_1\,c_1 = $ Constant ; further,

$$d\,A_2\,c_2 = d\,A_1\,c_1 = \text{Constant},$$

and therefore

$$d\,A_2\,c_2 = \text{Constant}.$$

The only *workable* assumption is that the relation between c and c_2 is
governed by the *total* areas A and A_2, so that in effect

$$\frac{d\,A}{d\,A_2} = \frac{A}{A_2},$$

and $$c_2 = \frac{d\,A}{d\,A_2}\,c = \frac{A}{A_2}c.$$

In practice the ratio $\frac{d\,A}{d\,A_2}$ at different diameters varies only slightly

from $\frac{A}{A_2}$, and when, as is frequently the case, a and a_2 have the same

value, $\frac{d\,A}{d\,A_2}$ is absolutely constant. It is assumed throughout the

preceding investigation that all particles of water entering at a certain
diameter leave at the same diameter in a Jonval wheel of constant
width.

The relative angles of exit at the inner and outer circumferences have now to be so determined that the water there shall leave the wheel parallel with its axis, as is the case at the mean diameter, *if possible.*

Let $O A$, Fig. 68, represent the value and direction of the relative velocity of outflow, and $O B$ the absolute velocity of the water leaving the wheel—at the mean circumference, parallel with the axis. If $A C$ and $A D$ be respectively the speeds at the inner and outer circumferences, then $O C$ and $O D$ are the corresponding absolute velocities of outflow. It will be seen that these are divergent and both greater than $O B$, in other words, only at the mean circumference is the speed of rotation the best.

With a helical vane surface for which the angles at the different diameters vary, the divergences in the directions of outflow are not so great as when a_2 is constant.

Fig. 69.

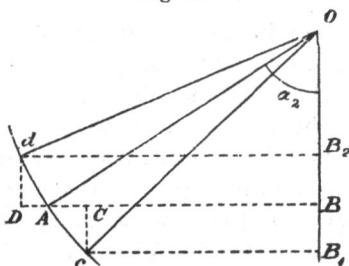

The corrections required to insure a parallel outflow at all diameters are effected as follows: Let $O A$, Fig. 69, as before denote the value and direction of the relative velocity of outflow c_2, and $A B$ the velocity of rotation w at the mean circumference. (Since c is constant, c_2 must be so as well, as the two velocities have a fixed ratio to each other.) The absolute velocity of outflow for the mean circumference is $O B = u$.

Make $B C$ and $B D$ respectively equal to the speed of the inner and outer circumference; draw $C c$ and $D d$ parallel with $O B$ until they intersect a circle described from the centre O with the radius $O A$ in c and d respectively ; join $O c$ and $O d$; then $B O c$ and $B O d$ are the appropriate angles for the end of the vane at the inner and outer circumferences, and the corresponding absolute residual velocities are $O B_1$ and $O B_2$, both vertically directed. When the correc-

Fig. 70.

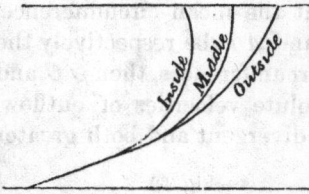

tions just described are applied to the angles of inflow and outflow, a_1 and a_2, the vane-curves at the inner, mean, and outer circumference will have the form shown in Fig. 70.

The whole of the formulas as applied to the design of turbines are based upon the assumption that the water leaves the wheel at right angles to the direction of rotation and enters without impact, and if these conditions are not fulfilled the results obtained from them are more or less vitiated ; on the other hand, varying the angles in the manner indicated also causes a deviation from the assumed conditions, as may be seen by reference to the Table I. for K^1, which varies for different angles a_2, but not to a very great extent. This would affect the value of the velocity of flow c, but in using the angle at the mean diameter for determining the latter no error of practical importance is incurred. Similar remarks apply to the ratio $\dfrac{A}{A_2}$, which is also influenced from the same cause.

Note.—Meissner in his work on turbines, in dealing with the correction of the vane angles, treats the velocity of flow at different diameters, as though the wheel were actually divided by concentric divisions into separate parts, and calculates the velocity of flow independently for each, so that it varies with the diameter according to the angles of the vane. This method the author considers incorrect, as, for the reasons given, *c* must be at any rate *approximately* constant for the whole width of the guide-passage orifices.

Vane Curves and Angles.

The effect of the angles a and a_2 on the efficiency of a turbine has been already considered, and the angle a_1 is fixed when the former have been determined on.

It was shown that for a given ratio $A : A_2$ efficiency is promoted by making a and a_2 as great as possible, but that they may vary through a very considerable range without material detriment to the performance of the motor.

Axial Turbines. For Jonval turbines with constant width of bucket, when a and a_2 are equal, the relative direction of entrance of the water is vertical or parallel to the axis, and the vane angle $a_1 = 0^\circ$. For purposes of manufacture it is very convenient to adopt this arrangement, and from a theoretical point of view there is little to be said against it. When the angles at which the water leaves the guide-vanes and wheel-vanes respectively are equal, the ratio $\dfrac{A}{A_2}$ is approximately 1. The best value for that ratio, as apparent from the Table No. II. of efficiencies, is about 0·75, but the difference is so small between the efficiency for the latter value and that for $\dfrac{A}{A_2} = 1$, as not to be worth practical consideration, so that the proportions in question may be adopted without hesitation.

The usual limits to the values of a and a_2 are stated in the summary of rules and formulas for the design of Jonval turbines; but it is worthy of notice that the greater the angles the larger must be the turbine for a given quantity of water.

In parallel-flow wheels with buckets of constant width the ratio $\dfrac{A}{A_2}$ may be varied by making a_2 greater or less than a; in the former case a_1 is positive, in the latter negative. In turbines of the Henschel type, in which the buckets are wider at the outflow than at the inlet, a_1 may be negative even when a_2 is greater than a; this is evident from the formula

$$tang \; a_1 = \frac{e_1}{e_2} \; tang \; a_2 - tang \; a,$$

where, provided $\dfrac{e_1}{e_2}$ is a sufficiently small fraction, $tang \; a_1$ may always become negative.

For radial inward-flow wheels a_1 is greater than for parallel-flow wheels with the same angles a_2 and a; the formula in this case is

$$tang \; a_1 = \frac{e_1}{e_2} \left(\frac{r_1}{r_2} \right)^2 tang \; a_2 - tang \; a,$$

where $\dfrac{r_1}{r_2} > 1$.

Under ordinary circumstances with Jonval turbines the simplest plan is to make a and a_2 equal, but there are cases in which for some reason it may be desirable to have as high a speed of wheel as possible, and to effect this a high ratio of A to A_2 is required, which renders it necessary to make a_2 greater than a. Where the buckets are not restricted to a constant width, the same end may be attained without this expedient by a suitable ratio $\dfrac{e_1}{e_2}$ between the widths at the inflow and outflow.

For axial turbines the vane-curves are frequently constructed in the manner shown in Fig. 71.

Let $A D$ be the depth of the wheel—or guide passages—parallel with the axis; $C D$ the circumferential length of a vane.

The outflow end of the vane is made straight for a certain length $B C$, and set out at the proper angle a_2 (or a). $C B$ is continued to C_1, its point of intersection with the line $A C_1$; $A C_1$ being the direction of inflow, $B C_1$ and $A C_1$ are then each equally divided into an equal number of parts by the points a, b, c, d, e, f and

Fig. 71.

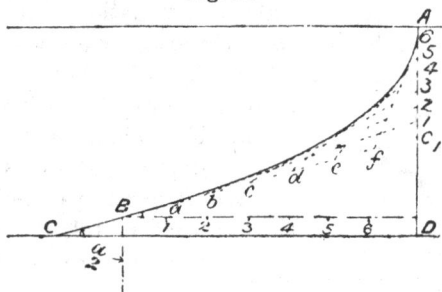

1, 2, 3, 4, 5, 6 respectively, and the lines $1a$, $2b$, $3c$, $4d$, $5e$, and $6f$ are drawn. These lines form tangents of the desired vane-curve, which can then be filled in.

In the illustration the angle a_1 is taken as $0°$, but the construction remains the same with any other value, only then $A C_1$ is not perpendicular to $C D$.

The ratio of the horizontal length $C D$ to the depth $A D$ may be from $1\frac{1}{2}$ to 2. As elsewhere explained, the vane-curves must always be drawn with due regard to as gradual a change as possible in the sectional area of the streams passing between them; and, apart from this, there is little merit in any particular form of curve.

Back Vanes. In cases where the relative angle of entrance is negative, the form of the passage between the vanes of a turbine may be such as to cause great and sudden changes in the section of the stream. This will be clear on reference to Fig. 72. Assuming the width

Fig. 72.

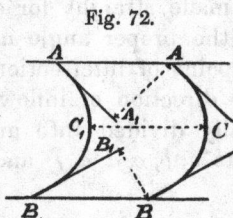

of the turbine to be uniform, the area of the passages is proportional at various points to the normal depth, and the latter increases from $A A_1$ at the entrance to $C C_1$ in the middle, and decreases again at the outflow to $B B_1$. To avoid these extreme variations, the buckets have, under such circumstances, been made with so-called *back vanes*, as illustrated in Fig. 73. This construction

Fig. 73.

amounts simply to giving the vanes a variable thickness to suit the desired depth of the passages; but to avoid excessive weight and thickness of metal, the vanes are cored out, and are, so to speak, double.

This construction is now very unusual, as the relative angle of entrance a_1 is generally positive for reaction turbines, and in impulse wheels back vanes have no object.

Where back vanes are used the two surfaces forming the sides of a passage are of different shape, and the courses taken by the particles of fluid at various points of the stream are not parallel with each other nor similar in form. This leads to varying velocities and disturbances within the buckets resulting in internal friction, whirling motion, and consequent loss; back vanes should therefore be avoided.

In turbines intended to work sometimes with free deviation and sometimes drowned, according to the tail-water level, back vanes are necessary.

Radial Turbines. With radial turbines the difficulties occurring in axial turbines owing to the varying speeds at different points of the vanes are absent; all points at

Fig. 74.

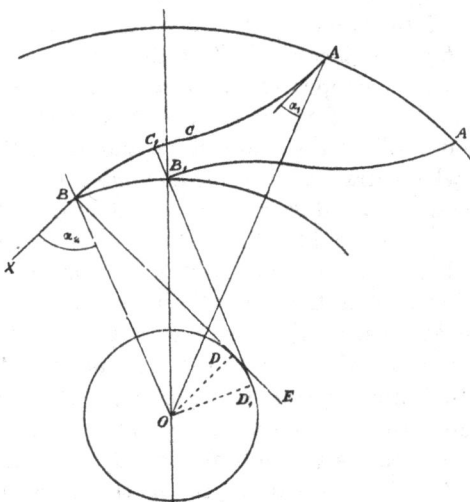

which the water enters move with the same speed, and all points at which the water leaves have an equal velocity.

A different drawback, already referred to, is incidental to the use of inward-flow wheels, and that is the contraction of the streams issuing from the buckets; to obviate this to some extent the form of wheel-vane shown in Fig. 74 has been devised with double curvature. The result of this is that the passage between the vanes remains constant in depth for some time before the inner circumference is reached; this of course does not prevent the convergence of the streams after leaving the various buckets and the resulting loss of energy due to the increased pressure at the orifices. The form of vane in question is used in the well-known "Vortex" turbine, and was originally suggested by Professor James Thomson.

The construction of the vane-curves in question is shown in Fig. 74. $A B$ represents the vane-curve, the tangent $B X$ to which at B forms the angle a_2 with the radius $O B$; the portion $B C$ of this curve is a volute, described by development from a circle of the radius $O D$; to find $O D$, from B draw $B E$ at right angles to the tangent $B X$ of the vane-curve, and from the centre O of the wheel, draw $O D$ at right angles to $B E$, which will be the desired radius of a circle touching $B E$ in D.

The true curve would be described by the end of a very fine thread as it is unwound from the circumference of the circle $O D$ and kept in tension during the process, so that in all positions it forms a tangent to the circle. It is easy to determine any number of points of the curve on paper by imitating this construction. The point C_1 of the volute-curve is such that a tangent drawn from it to the circle $O D$ passes through the end B_1 of the next vane-curve $A_1 B_1$; the normal depth of the stream here is $B_1 C_1$, and it is evident that as far as both volute-curves of the adjacent vanes are intersected by the same tangent to the circle $O D$, the normal depth of the passage between them remains constant, and there is no contraction.

The two points B_1 and C_1 of the same imaginary thread both describe similar volutes parallel with each other and separated by the distance $B_1 C_1$. The volute $B C$ should be carried to C a short way beyond C_1 to insure the parallel motion of all parts of the stream, and the vane then continued by a curve of another form terminating in

A with the inclination a_1. Care must be taken that the transition is gradual.

The guide-vanes of the "Vortex" turbine are constructed in such a manner that the passages between them are long, and of a gradually convergent form. Thus the velocity of flow gradually increases from the outer towards

Fig. 74A.

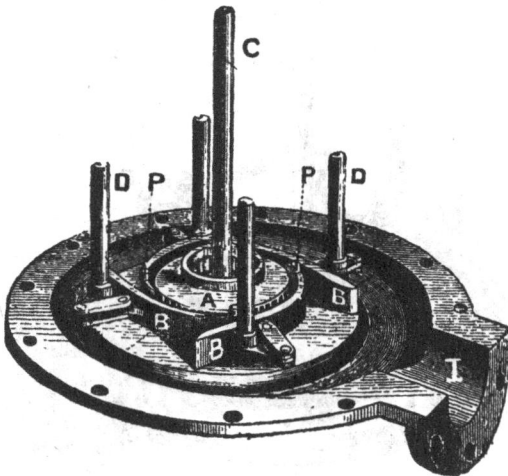

the inner circumference, causing the water to take the form of a vortex.

The construction is illustrated in Fig. 74A—in which the turbine is shown with the cover of its outer casing removed—and in 74B, showing the exterior of the casing complete.

A is the wheel keyed to the shaft *C*; *B, B,* &c. are the guide-vanes, pivoted at *P, P,* &c., nearer to their inner than to their outer ends; *D, D,* &c., shafts and bell-cranks connecting the guide-vanes with outside bell-cranks and coupling-rods *E*, *E*, &c. Suitable gear acting on one of the shafts, *D*, actuates simultaneously all the guide-vanes, increasing or diminishing the outflow area of the guide passages in a manner sufficiently clear from the illustration. *I* is the supply-pipe; *H* the wheel-cover; *G* a bracket and screw for adjusting the pivot (not shown in the illustration).

The absolute angle of outflow *a* from the guide passages

Fig. 74B.

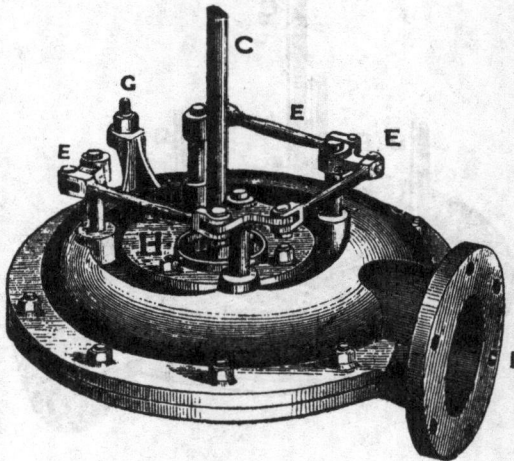

is very large, so that the water enters the wheel almost tangentially, and very few guide-vanes are necessary. It is, however, a drawback to this form of turbine, that owing to the great length of the guide-vanes the dimensions of the casing for large wheels become excessively great.

In order to obtain, with a given diameter of wheel, a greater outflow area than would otherwise be possible, every alternate wheel-vane is made shorter than the rest, so that it occupies only the outer half of a bucket where

the passages are widest, as shown in Fig. 75. This construction also tends to reduce the loss by friction in the
passages.

In other systems of radial inward-flow turbines the
curves of guide and wheel vanes both terminate in
straight lines at the outflow (*vide* Fig. 22). The outer
extremities of the guide-vanes are very often radial. The
radial height of the guide passages in such turbines is then
about the same as that of the wheel-buckets, and the
guide-casing has a much smaller diameter relatively to the

Fig. 75.

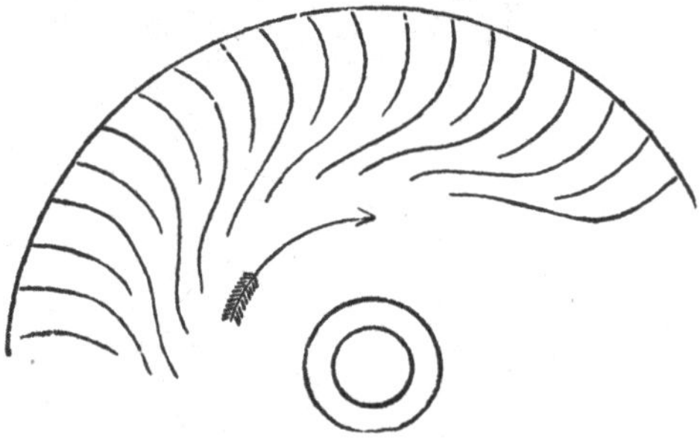

wheel than is required for the "Vortex" turbine; on the
other hand, a much greater number of vanes is necessary
in order to give the flow the desired direction. Judging
by results, as far as efficiency is concerned, there appears
to be no difference between the two systems, with both of
which admirable results may be obtained.

For the vanes of radial outward-flow turbines the construction shown in Fig. 76 is convenient, and may be
applied to either guide or wheel vanes.

Let m_1, m_2, m_3, &c., be the points at which the vanes
are pitched on the mean circumference of outflow. Taking
the point m_2—for example—set out $m_2 t_2$, forming the

outflow-angle a_2, with a radial line from the centre of the
wheel to m_2.

Fig. 76.

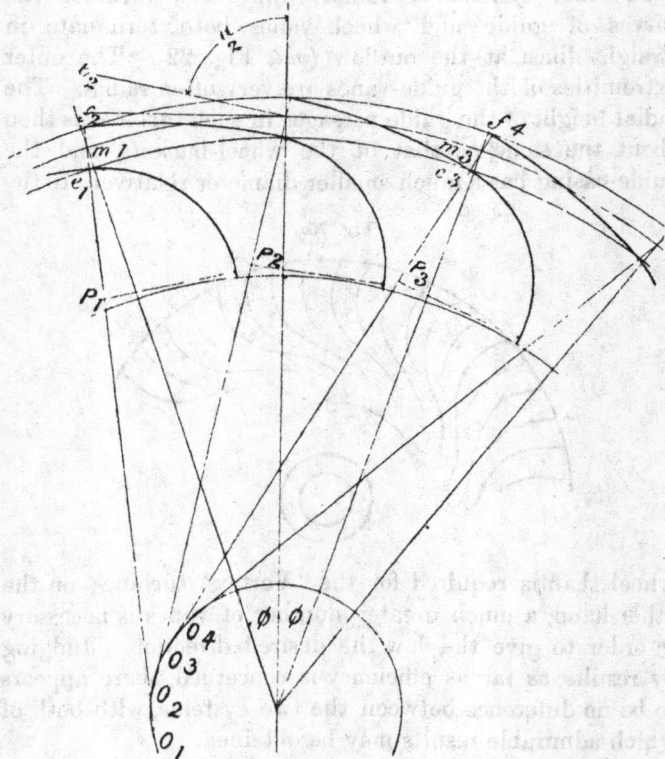

Perpendicular to $m_2 t_2$ draw $m_2 O_3$, and on either side
of m_2 set out—on $m_2 O_2$—the length

$$m_2 e_2 = m_2 f_3 = \tau_2 \cos a_2 \, tang \, \frac{\phi}{2},$$

where ϕ is the angle subtending the pitch.

From the adjacent pitch-point m_3, draw $m_3 \ O_4$ in the same way as $m_2 \ O_3$.

$m_2 \ O_3$ and $m_3 \ O_4$ intersect in O_3, and O_3 is the centre from which the arc $e_2 \ f_2$, forming the outer portion of the vane-curve, is described.

From O_4, with the radius $O_4 \ e_3 = O_3 \ e_2$, the corresponding part $e_3 \ f_3$ of the adjacent vane-curve is also drawn. The centres O_1, O_2, O_3, &c., for all the outer vane-curves, lie in a circle. If the construction is correct, the arcs drawn through the points e_1, e_2, e_3, &c., must pass through the corresponding points f_1, f_2, f_3, &c., lying on the outer circumference.

The inner portions of the vane-curves are also arcs described from points P_1, P_2, P_3, &c., as centres, on the radii $O_2 \ m_1, O_3 \ m_2, O_4 \ m_3$.

The result of this construction is, that the tangents to the adjacent vane-curves at e_1 and f_2, e_2 and f_3, &c.—where the outflow area is least—are *parallel*, and thus there is no *contraction* of the streams issuing from the buckets.

It is generally easy to find by trial the proper position for each of the centres P_1, P_2, P_3, &c., in such a manner that the relative angle of inflow is correct. The inflow end of each vane may terminate in a straight line, if necessary, forming a tangent to the inner portion of the vane-curve. When the distances $e_1 \ f_2$, $e_2 \ f_3$, &c., are determined as described, and the thickness of the vanes is settled, the clear depth of the outflow passages is of course fixed, and to obtain the desired outflow area, the *width* of the passages must be calculated accordingly.

General Remarks on the Construction of Vanes.

In designing and constructing the vanes of a turbine everything possible should be done to avoid unnecessary friction or resistance; to this end the surfaces of the vanes should be as smooth as practicable, and the ends carefully tapered and rounded off where the water enters.

As thin vanes offer less resistance to the passage of the water, wrought iron or steel vanes are preferable from the theoretical point of view, although vanes cast in one with the rest of the wheel are more substantial, and not so soon affected by rust.

A practical confirmation of the statement that, with given proportions and angles, the form of the vane-curves is of minor importance, is to be found in the fact that turbines with widely-differing construction and necessarily very variously-shaped vanes give nearly equal efficiencies.

Experiments made by Weisbach tend to show that *abrupt curvature* has a greater effect in increasing the loss in the buckets than the *length* of the passages. With a large relative angle of outflow a_2, under otherwise similar conditions and with a given depth of wheel, the curvature of the vanes is necessarily more abrupt than when a_2 has a smaller value; but, on the other hand, the loss from unutilized energy is greater, so that the gain in one direction is more or less balanced by the loss in another. Possibly differences of 2, or at most 3 per cent, in the efficiency may result from variations in the vane-curves apart from the proportions and angles, but it must be evident that an intimate connection exists between the latter and the former.

Determination of Absolute Path of Stream.

It is often desirable, as a check upon the form of the buckets and vanes, to draw the curve of the mean absolute path followed by the water. In the course of this there

should be no abrupt changes, but the curvature should
be gradual and continuous; if this is not the case, some
change must be made in the form of the vanes. To
describe the actual path of the stream it is only necessary to make use of the fact that the absolute velocity of
a particle of water in any position is the resultant of the
velocity of rotation at that point and the relative velocity
with respect to the wheel. Taking, for example, the
case of an axial turbine as being somewhat simpler than
that of a radial wheel.

At the point where it
enters the buckets the
absolute course of the
water is that with which
it leaves the guide-
vanes; at any other
point P, Fig. 77, when
the shape and width of

Fig. 77.

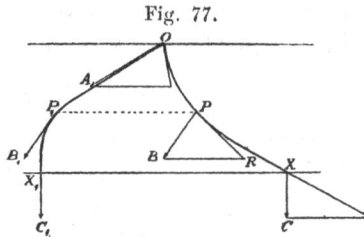

the vanes and buckets are given, the relative velocity of
flow at that point can be ascertained from the conditions of
continuity—the relative velocity of inflow c_1 is to the
relative velocity at the point in question in the inverse
ratio of the corresponding sectional areas of the stream.
Let $P R$ be the relative velocity at P, then $P B$, the resultant of this and the velocity of rotation, gives the absolute
direction and velocity of the stream. While, however, the
water has moved relatively to the vane from O to P, the
point P has advanced to the position P_1, so that O and
P_1 are *two* points of the absolute path; $P_1 B_1$ parallel to
$P B$ is the tangent to the path at P_1.

A third point is that at which the water leaves the
wheel, denoted by X on the vane; the direction of
absolute flow $X C$ here is axial if the wheel is correctly
designed, and while the water has moved relatively from

M

O to X, X has advanced to X_1, so that the latter is the last
point of the absolute path.

If the shape of the buckets is such that the velocity
with which the water traverses the wheel, parallel with the
axis, remains *constant*, it is a very simple matter to construct
the curve of the absolute path. On entering the wheel
the water has the absolute velocity $O\,A$, Fig. 78; the
vertical component is $O\,V$, and this, by hypothesis, must
remain the same at every point during the passage through
the buckets. $A\,P$ is the velocity of rotation of the wheel,
so that while the water moves vertically through a distance
$O\,V$, the wheel has advanced by $A\,P$.

Fig. 78.

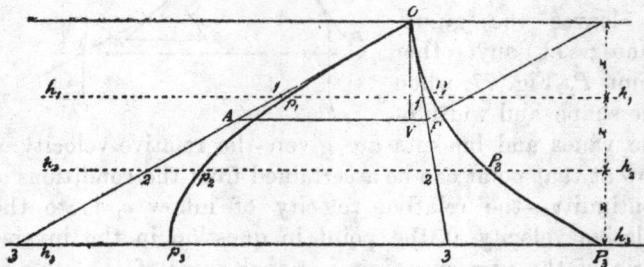

If a series of lines $h_1\,h_1$, $h_2\,h_2$, $h_3\,h_3$, parallel with $O\,X$,
be drawn at equal distances x apart, the interval 1 1 intercepted on $h_1\,h_1$ by its intersection with $O\,A$ and $O\,P$
represents the distance traversed by the wheel while the
water has passed vertically through x and arrived at the
point P_1 on the vane-curve; similarly 2 2 on $h_2\,h_2$ is the
length by which the wheel has advanced while the fluid
has traversed the vertical distance 2 x and arrived at the
point P_2; and at the point P_3 the corresponding advance
is 3 3. To determine the points of the absolute path

it is only necessary to mark off from P_1, P_2, and P_3 the distances $P_1 p_1$, $P_2 p_2$, and $P_3 p_3$ equal respectively to the intervals 1 1, 2 2, and 3 3; then p_1, p_2, and p_3 are the points required.

In constructing the absolute path for a radial turbine it is necessary to take into account the fact that the speed of the wheel varies at different points of the vane according to the distance from the axis of rotation. The result of this on the curve of the absolute path, as compared with that of an axial turbine having the same velocity of rotation at the inflow, is to produce a greater deflection for an inward-flow and a less deflection for an outward-flow wheel. Let $O A$, Fig. 79,

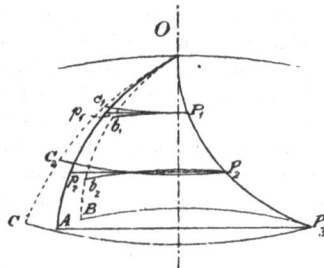

Fig. 79.

be the absolute path for an axial turbine, then for an inward-flow wheel the path will be $O B$, for an outward-flow wheel $O C$. The distance $P_1 p_1$ which is traversed by the point P_1 of the vane while the water moves relatively from O to P_1 in an axial turbine, is reduced to $P_1 b_1$ in the case of the inward-flow turbine, $P_1 b_1$ bearing to $P_1 p_1$ the ratio of the radius of P_1 to that of the starting-point O; similarly $P_2 b_2$ is to $P_2 p_2$ as the radius at P_2 to the radius at O; as the radii of P_1 and P_2 are less than that of O, $P_1 b_1$, and $P_2 b_2$ are smaller than $P_1 p_1$ and $P_2 p_2$.

In the case of an outward-flow wheel where c_1 and c_2 are points of the absolute path, $P_1 c_1$ and $P_2 c_2$ are greater than $P_1 p_1$ and $P_2 p_2$ respectively.

CHAPTER VI.

THE DESIGN OF REACTION TURBINES (*continued*).

Influence of angle of outflow from guides and area ratio on efficiency.—Methods of regulating turbines.—Correct system of regulation.—The distribution of water in turbines.—Helical vanes —Constant angles of outflow.—Loss from unutilized energy.—Loss by shock.—Comparison of axial with radial turbines.—Reasons for theoretical superiority of inward-flow wheels.—Practical drawbacks. —Mixed-flow turbines.—On the use of the suction-tube and Boyden diffusor.—Best shape for suction-tube.—Boyden diffusor.—Necessary adjuncts.—Losses in tubes and diffusors.—Experiments.

Effect of Proportions and Dimensions on the Performance of Reaction Turbines.

IT has been already shown that under otherwise similar conditions the efficiency is greater the larger the relative angle of exit a_2, as the residual velocity of off-flow decreases, the more nearly the relative direction with which the water leaves the buckets approaches to the direction of rotation. It is therefore always desirable to make the angle a_2 as great as practicable in all kinds of reaction turbines. If, however, the direction of outflow approach too closely to that of rotation, it is obvious that the depth of the passages becomes very small, and to allow a given quantity of water to pass through the wheel, either the width must be very great or the diameter large, so that the number of buckets may be increased. In practice

a_2 rarely if ever exceeds 80°, and is generally less; it is not desirable to exceed a certain proportion between the width and diameter of the wheel, and an increase in the diameter means increase of cost and of space required. If the residual velocity of the water as it leaves the wheel is not low enough, the efficiency can be improved by the use of a *diffusor* or suitably constructed *suction-tube.*

In considering the effect of the other dimensions and proportions on the efficiency of turbines, it will be convenient to investigate first the case of axial turbines and afterwards to compare these with those of the radial type.

Before this can be done, some assumptions must be made as to the co-efficients of resistance, and the value of any conclusions arrived at necessarily depends on the accuracy of these assumptions. It has so far been taken for granted that the co-efficients ζ_1 and ζ_2 for the resistances in the guide apparatus and wheel respectively are constant, but this has been done merely as a matter of convenience for practical purposes.

As regards ζ_1, the co-efficient for the guide passages, there is no reason to suppose that under average conditions its value varies appreciably for surfaces of a given character and a given class of workmanship.

The loss in the guide wheel is made up—as has been stated—of the friction of the walls of the passages, the resistance of the bend, and the obstruction offered by the edges of the vanes to the entry of the water. The first depends on the nature of the internal surface of the passages and their proportions. For wheels of about the same power, but varying as regards the vane-angles and the ratio of the areas A and A_2, the proportions of the passages will not differ sufficiently to seriously affect the co-efficient of resistance. The second, the resistance of the bend, will also vary very slightly for unit velocity for different wheels. The third component of the loss, due to the obstruction of the vanes, depends on the proportion which the obstructive area bears to the total area of the stream entering the guide passages; this proportion is not necessarily affected in any way by the angles a and a_2 nor by the ratio $\dfrac{A}{A_2}$, so that as far as that is concerned a constant value of ζ_1 is justified. On the whole, therefore, for purposes of comparison, a constant value of ζ_1 is a reasonable assumption where turbines not differing very widely in dimensions are concerned.

The losses from the moment the water leaves the guide passages to that at which it quits the wheel have been hitherto all included under the co-efficient ζ_2. In reality they are of a very complicated character, which it would be impossible to determine theoretically with any approach to accuracy. It is quite certain, however, that the value of ζ_2 in fact is not constant. It includes, among other factors, the loss by leakage through the clearance space between guide passages and wheel, and from this it might be supposed that its value would increase with an increasing ratio $\dfrac{A}{A_2}$, that is, the greater the degree of reaction and the higher the pressure at the mouth of the guide passages, since the amount of leakage must—it would be inferred—depend on that pressure.

Experiments which have been carried out and calculations made by Professor Fliegner of Zürich, tend to show that this is not the case, but that, on the contrary, with increasing reaction and greater pressure at the mouth of the guide passages, the co-efficient ζ_2 decreases. The number of data available on this subject is, however, very limited, and quite insufficient to form any basis for rules or formulas.

For the present the assumption of a constant value of ζ_2 will be adhered to, and the conclusions to which it leads compared with practical results.

The efficiency of a reaction turbine is expressed from (viii) by :

$$\epsilon = \frac{2 \, \dfrac{r_1}{r_2} \, \dfrac{A}{A_2} \, \sin a_2 \sin a}{2 \, \dfrac{r_1}{r_2} \, \dfrac{A}{A_2} \, \sin a_2 \, \sin \, a \; + \; \zeta_1 \; + \; \left(\dfrac{A}{A_2}\right)^2 (\zeta_2 + \cos^2 a_2)}.$$

In Table No. II. the efficiencies corresponding to various values of $\dfrac{r_1}{r_2}$, $\dfrac{A}{A_2}$, a_2, and a will be found. For an axial turbine $\dfrac{r_1}{r_2} = 1$, so that

$$\epsilon = \frac{2 \, \dfrac{A}{A_2} \sin a_2 \sin a}{2 \, \dfrac{A}{A_2} \, \sin a_2 \, \sin \, a \; + \; \zeta_1 \; + \; \left(\dfrac{A}{A_2}\right)^2 (\zeta_2 + \cos^2 a_2)}.$$

Influence of a on Efficiency. The numerator of the fraction expressing the value of ϵ is proportional to the effective work done *per unit of velocity of flow* by the wheel; the same quantity also appears in the denominator, but with other quantities added; for any given values of $\dfrac{A}{A_2}$ and a_2 these additional quantities are constant, and only the first term of the denominator changes with the angle a, to the sine of which it is proportional. The greater, therefore, the value of $2 \, \dfrac{A}{A_2} \, \sin \, a_2 \, \sin \, a$, the more closely will the fraction ϵ approach unity, and hence the efficiency is increased by making a larger.

This will be most clearly seen by reference to Table No. II.; for instance, with $\dfrac{A}{A_2} = 2$ and $a_2 = 70°$, the efficiency increases from 0·674 for $a = 50°$ to 0·722 for $a = 75°$; with $\dfrac{A}{A_2} = 1$ the corresponding figures are respectively 0·764 and 0·804. It will be noted that the effect is not very marked even for a considerable variation in the angle, but still it is evident that to obtain the highest efficiency a should be as large as possible.

This conclusion is quite independent of the value of the co-efficients of resistance ζ_1 and ζ_2.

In practice it is not usual to make a greater than 75°, occasionally 80°.

Influence of Ratio $\dfrac{A}{A_2}$ *on Efficiency.* An inspection of the Table No. I. for K_1—which is proportional to the velocity of flow from the guide passages—shows that for a decreasing ratio $\dfrac{A}{A_2}$, with any given value of a and a_2, the velocity of flow c becomes greater; at the same time the relative velocity c_2 with which the water leaves the wheel-buckets decreases, but more rapidly than c increases. The loss in the wheel, to which the co-efficient ζ_2 corresponds, is assumed to be proportional to the square of the relative velocity c_2, and this is true also of the residual velocity of off-flow $u = c_2 \cos a_2$. On the other hand the work done depends only on the product of c and c_2, and is therefore proportional only with the first power of c_2; consequently with a diminishing ratio $\dfrac{A}{A_2}$ the work done decreases less rapidly than the losses in the wheel and from unutilized energy. The result is that the efficiency varies with the ratio $\dfrac{A}{A_2}$, and attains a maximum, as is evident from an examination of Table No. II. for a value of $\dfrac{A}{A_2}$ about 0·75. Between the efficiency corresponding to this value and that for $\dfrac{A}{A_2} = 1$, there is a practically unappreciable difference, and for axial turbines the most usual ratio is in the neighbourhood of these, ranging from somewhat above to somewhat below unity. The greatest occurring difference in the efficiency for values of $\dfrac{A}{A_2}$ between 2 and 0·25 is, however, only 0·096 or 9·6 per cent., the efficiency diminishing for an increasing ratio $\dfrac{A}{A_2}$ above 0·75. If it is taken into account that ζ_2 tends to become less for increasing reaction, the actual efficiencies for high ratios of A to A_2 would be greater than those contained in the table. The value 1·75 for $\dfrac{A}{A_2}$ has seldom been exceeded in practice, and for this the theoretical efficiency is 0·80, only 4·9 per cent. less than the maximum. In view of these facts it may safely be stated that for practical purposes the efficiency is only slightly affected by varying degrees of reaction.

This conclusion is fully borne out by experimental results, in so far as they are available for comparison ; the latter is rarely the case, for various reasons. In order to

compare two or more turbines as regards the influence of any
particular proportion or dimension on their performance,
they should in other respects be as nearly alike as possible, and the workmanship of all should be of the same
quality; this condition can only be insured when the
motors tested are by the same maker; otherwise the
results may be affected by subordinate details to such an
extent as to vitiate any conclusions drawn. In spite of
this, the best performances recorded of a large number
of turbines of different kinds varying very widely in
dimensions and proportions, agree sufficiently well to
confirm the theoretical deductions.

Three experimental turbines of Rittinger's—previously
mentioned—afford an exceptionally good opportunity of
observing the influence of various proportions of A to A_2
on the performance of the motors. All three wheels were
of the same outside diameter and approximately the same
power, and were tested with as nearly as possible an
equal head of water; the angles a and a_2 differed somewhat, and the ratio $\dfrac{A}{A_2}$ very greatly in the three motors.

The following are the necessary data and results :—

	a	a_2	$\dfrac{A}{A_2}$	Efficiency.	No. of revolutions per minute.
(1)	78°	64°	0·522	0·673	99
(2)	73°	74°	1·274	0·697	135
(3)	68°	75°	1·661	0·632	151

In each instance the speed given is the *best* speed.
There is, it will be noted, a maximum variation of only
6·5 per cent. in the efficiencies, while between the
efficiencies of motors (1) and (3), which differ most in
construction, the variation is only 4·1 per cent.

It should be observed that in Rittinger's turbines the

differences in the ratio $\dfrac{A}{A_2}$ were obtained almost entirely by varying the angles, and the effect of the latter on the efficiency cannot be separated from that of the ratio in question; in general, however, this is the only method in which the proportion between the outflow areas is varied in axial turbines, for which the width of buckets remains constant.

Methods of Regulating Reaction Turbines.

A reaction turbine only works to the best advantage when water is admitted to all the buckets; all systems of regulation, except one, reduce the efficiency of the turbine when not working full. This one correct method of regulation is attended with practical difficulties and is very seldom applied. It will be subsequently explained ; but before this is done it is desirable to describe the other systems adopted, and show in what consist their disadvantages. These systems are the following :—

(1) Regulation by means of the head-race sluice.
(2)   ,,   ,,   ,,   tail-race sluice.
(3)   ,,   ,,   ,,   a throttle valve.
(4)   ,,   ,, partially closing the guide passages.
(5)   ,,   ,, means of concentric subdivision.

(I.) *Regulation by Head-race Sluice.* The orifice through which the water is admitted to the reservoir or space immediately above the turbine is altered by means of an ordinary sluice, thus allowing more or less water to pass through, means being provided for the passage of the surplus fluid by another channel when the whole available quantity is not required.

If by partially closing the sluice the quantity of water admitted is reduced to half that with which the wheel works at its full power, the result will be that the head will sink until it corresponds approximately to half the velocity of flow at full power, which—as the head is proportional to the square of the velocity—would be ¼ of the original head. At this reduced velocity of flow the speed should be also reduced in proportion if the motor is to work to the best advantage, that is, to about half the speed at full power; as, however, it is generally necessary to run at a constant speed, this latter will under the assumed conditions be double as great as the best speed, and as this is approximately the velocity with which the wheel revolves without a load, doing no work, the efficiency is reduced to zero.

From this extreme case it will be seen that the method of regulation by the head-race sluice is very defective.

(II.) *Regulation by Tail-race Sluice.* This system is generally applied to turbines with suction-tubes. The sluice is usually a circular one, surrounding the lower end of the suction-tube, as shown in Fig. 24, p. 35. Suppose, again, in this case, that the sluice is so far closed that only half the quantity of water required at full power is allowed to pass. The velocity of flow will be reduced to one-half, but, as the work done per unit of weight is proportional to the *square* of the velocity, the efficiency will be only ¼ of that at full power with the same speed, and the *total* work done only ⅛.

By the regulating action of the sluice, therefore, the efficiency is very much reduced, and the system in question is consequently bad.

Apart from this, it has another drawback, which consists in the fact that a very considerable reduction in the outflow orifice produces a comparatively slight effect on the velocity of flow. The effect of regulating by a sluice to the suction-tube is illustrated by the following figures, taken from the results of experiments made many years ago by Messrs. A. Koechlin and Co. of Mulhouse.

Open height of sluice.	Quantity of water per second.	Efficiency.
0·426	296·10	0·718
0·176	271·71	0·631
0·095	253·90	0·349

It is clear that as the opening of the sluice is changed the speed of greatest efficiency is altered in proportion as the velocity of flow is reduced, and it is assumed that for practical reasons the speed must be maintained constant under all conditions.

For raising and lowering the sluice of a suction-tube an arrangement of rack and pinion, acting on rods by which the sluice is suspended, is generally employed.

(III.) *Regulation by Throttle Valve.* The effect of a throttle valve applied either before the water enters the guide passages or in the suction-tube is similar to that resulting from the methods of regulation (I.) and (II.) previously described, and is therefore objectionable ; owing to its shape it offers a greater resistance to the passage of the water than the sluices ; the lower edge of a sluice may be rounded off with a curve of large radius, but with a throttle valve this is impracticable.

(IV.) *Regulation by partially closing Guide Passages.* This system is now most generally adopted for regulating reaction turbines. The mechanisms employed are various, but the result produced is in all cases much the same ; the following is a description of some of the more important arrangements used :—

(1) **Vertical slides to each guide passage.** This construction is shown in Figs. 80, 81, and 82; every slide is

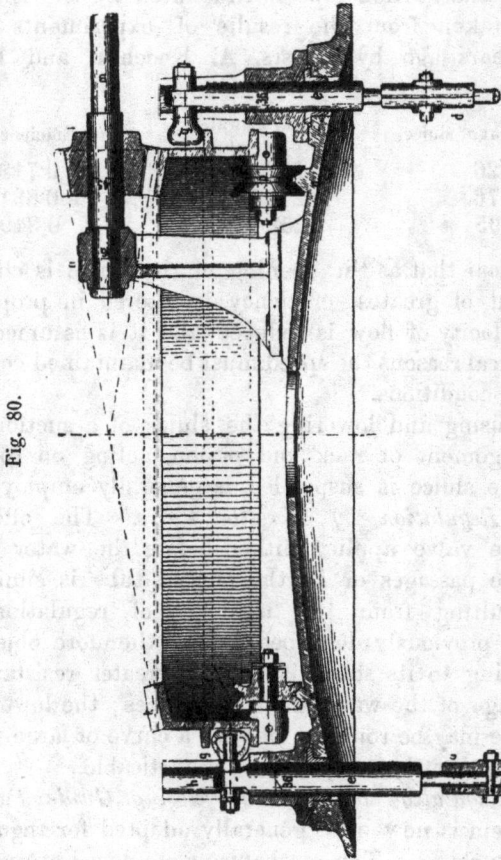

Fig. 80.

attached to a rod, which in some cases is worked independently, while in others several are coupled together and connected by a cross head to a single rod. Each rod

may be actuated by a hand wheel and nut acting on the
screwed upper end, or all the rods can be worked from
a sort of horizontal cam (*vide* also Fig. 30), which is furnished with a groove on the outer circumference into
which project rollers carried on pins attached to the upper
ends of the slide rods ; as the cam is made to revolve,
each of the rods, with the slides connected to it, is raised
or lowered in succession.   Referring to the illustrations,
Figs. 80 and 81 show the cam with the grooves formed
by projecting flanges i_2 and i_3; at i_4 the groove is inclined,

Fig. 81.

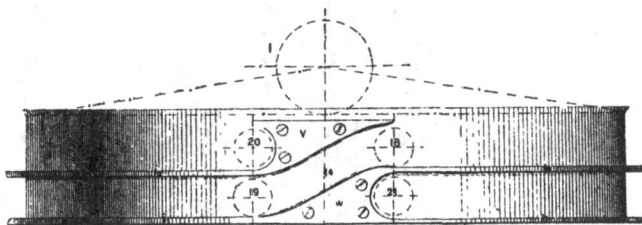

and as the cam revolves from left to right, a roller which
has arrived in the position indicated by 18, is forced down
or drops by its own weight into the relative position 19,
thus lowering the slides connected with it and closing the
corresponding guide passages.   18 and 20 show the positions of the extreme rollers when all slides are open, 19
and 21 the same when all are shut.

Fig. 82 shows the construction of the slides *a, a, a,*
attached to the rods *b, b, b,* which are connected by the
cross head *c*; the central rod is continued upwards, *vide*
Fig. 80, and at its upper end *e* is enlarged in diameter,
and carries the roller *g* revolving on the pin *f*. At the
top of the cam *i* is a toothed bevel ring *k*, which gears

with the pinion *l* keyed to the shaft *m*, by means of which
motion is imparted to the cam; *n* is the bearing for the
shaft *m*. The cam is supported by rollers *p*, which run on
a circular rail *q* cast on the top of the guide-wheel cover *o*.
The dimensions given in the illustration are in millimètres.

In Fig. 83 another form of slide is shown, in which the
faces *c* are of cast-iron with curved wooden blocks *f* on the
back. The slides are worked by the rods 1, 2, and 3;
b, b, b are the guide, and *a, a, a* the wheel vanes.

Fig. 82.

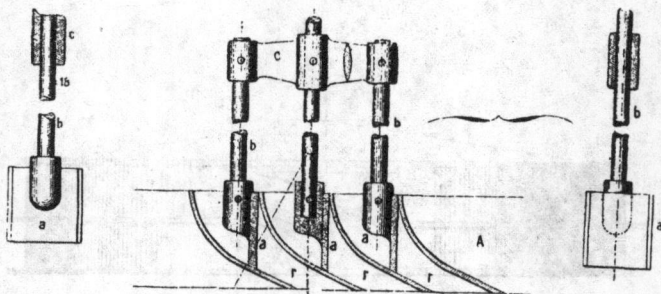

The backs of the slides are curved, in order that they may
offer less resistance to the passage of the water beneath them.

This type of slide is unnecessary where the regulation
is effected, as it should be, by *entirely* closing some of the
guide passages.

(2) **Horizontal slides on the top of guide passages.** The
slides, each of which generally covers two or three guide
passages, are moved radially inwards or outwards by a cam,
in a manner similar to that previously described; as the
mechanism is all under water, it is liable to become choked
and otherwise interfered with by mud, sand, and other
material carried by the water, and is on this account not to
be recommended for practical purposes.

(3) **Pivoted flaps in the guide passages.** Fig. 84 illustrates this arrangement : two guide passages *B* are covered by one flap *R*, which is pivoted over the central vane between the two passages. Each flap is worked by a rod *a* attached by a pin at

Fig. 83.

u. The rods can be actuated by a cam in the manner described under (1) ; another method of effecting this is by prolonging the spindle on which each flap turns through the

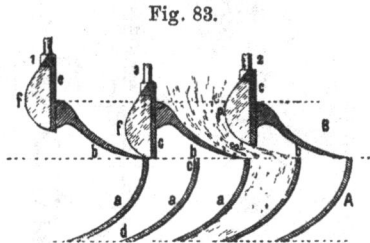

inner wall of the guide-wheel, and on the end of this prolongation fixing a crank, the pin of which projects into a groove in the surface of a cylinder or cam concentric with the turbine shaft. When this cam is made to revolve, the pins are forced to travel in succession up or down an inclined portion of the groove, thus closing or opening the flaps.

Fig. 84.

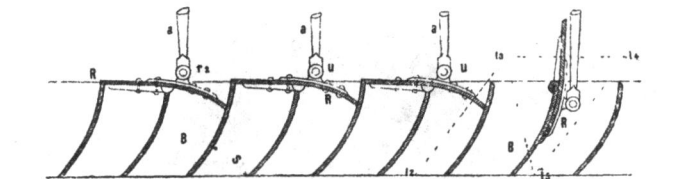

This latter arrangement has the drawback of being inaccessible. The use of flaps as described is open to the objection that they do not close tight, and are rather weak in case, as often happens, they should stick and require the application of considerable force to move them.

(4) **Scroll regulator.** This arrangement is shown in

N

Fig. 85.

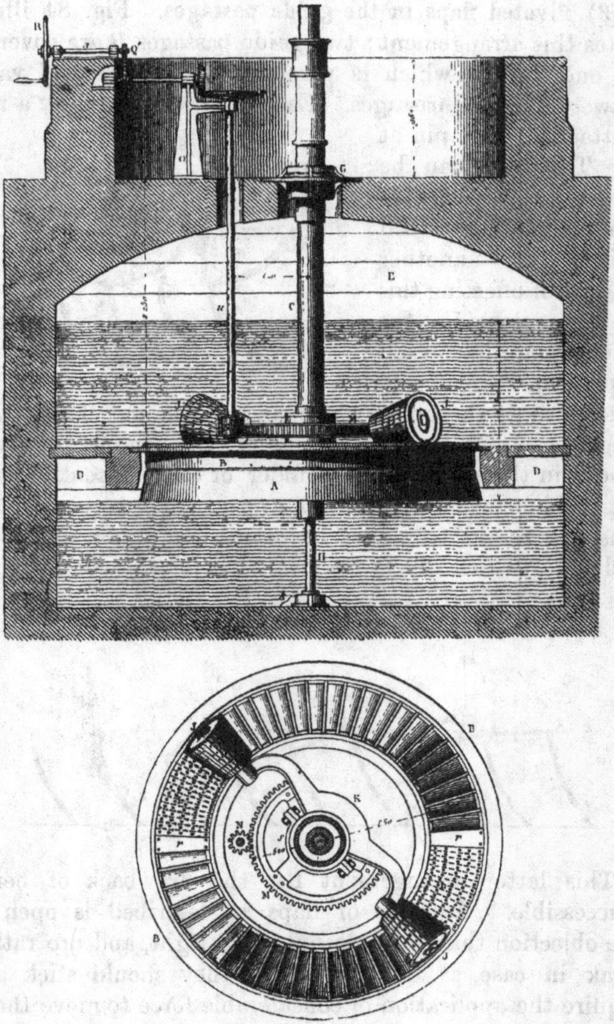

JONVAL TURBINE WITH SCROLL REGULATOR.

Fig. 85; it consists essentially of two scrolls or bands of leather or india-rubber *L L*, on the face of which a series of metal strips are riveted, wound upon conical rollers *J J*. The rollers are connected by a casting *K*, which turns loose about the turbine shaft *C*. Each band or scroll is fastened down to the guide apparatus at one end *r*, and when the casting *K* with the rollers is made to revolve in one direction, the bands are unrolled and laid down over the guide passages; when the motion is in the opposite direction they are wound up again and the passages un-

Fig. 86.

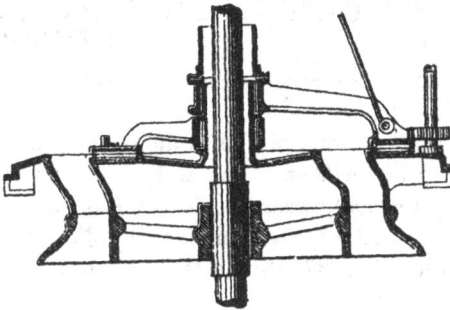

covered. The mechanism is actuated by the wheel *M* and pinion *N*, from the hand-wheel *R* through the spindle *u*. This method of regulation is over twenty-five years old, but is still used by some engineers, and has given very satisfactory results; leather has been found to be the best material for the scrolls.

(5) **Ring slide or sluice.** With this system two concentric half-rings of different diameters are employed, each of which is sufficient to cover one-half of the guide passages; the entrances to the latter are situated on two concentric annular surfaces, half the orifices being in one

of these and half in the other diametrically opposite.
When all the guide passages are open, the two slides

Fig. 87.

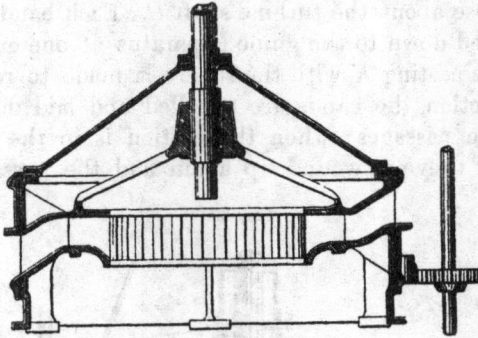

rest over those portions of the annular surfaces not occupied by the guide orifices. The slides may be either flat,
conical, or cylindrical, according to circumstances. Fig. 86
shows this method of regulation as applied to an axial
turbine, while Fig. 87 illustrates its adaptation to an
inward-flow wheel.

(6) **Movable guide-vanes.** The late Professor James
Thomson introduced a system of regulation for his
inward-flow turbines which consists in attaching the guide-
vanes to pivots, so that by changing their inclination with
respect to the wheel the area of the guide passages can be
varied. This construction has already been described and
illustrated in connection with the "Vortex" turbine, Figs.
74A and 74B.

Various means of actuating the vanes are employed, the
simplest being by means of levers or cranks fixed on the
ends of the spindles about which the vanes turn.

The drawback to this method is, that in all positions of

the vanes except one the water enters the wheel with impact, and there is a corresponding loss of energy.

(7) **Circular slide between guide passages and wheel.** This system, shown in connection with "the Humphrey" turbine (*vide* Chapter XI.), is applied to radial inward-flow turbines, and consists of a ring or cylinder concentric with the wheel, fitted between the latter and the guide apparatus, and containing a number of openings corresponding to the orifices of the guide passages; the spaces between the openings are at least equal to the latter, so that by turning the regulator cylinder the guide passages can be covered. This arrangement necessitates a great thickness of the guide-vanes measured on the inner circumference of the guide-wheel, and in a reaction turbine interferes very much with the continuity of flow essential to its correct action.

(8) **Sluice inside wheel.** For Fourneyron (outward-flow) turbines a cylindrical sluice inside the wheel shutting off the guide passages vertically—that is, reducing the height of the passages all round—has been employed; this arrangement, of course, when the wheel is not working at its full power, allows the wheel-buckets to be only partially filled, and therefore interferes with the correct action of the turbine when working drowned; if, as sometimes occurs, the off-flow takes place above water, then the influence of the sluice would not be so detrimental, as in that case the motor becomes practically an impulse turbine with free deviation. A similar method of regulation is employed with some inward-flow wheels.

There are besides the mechanisms enumerated a large number of other arrangements and modifications, which it would occupy too much space to describe.

The system of regulation by partially closing guide passages reduces the efficiency, but not to the same extent as when the exit of the water is controlled by a throttle-valve or sluice at the lower end of the suction-tube.

Theoretically it is essential that a reaction turbine should work *full* of water in every part, and by closing some of the guide passages this condition is of course interfered with, as water is only admitted to a portion of the wheel-buckets at a time. When, however, a good arrangement is adopted, the result on the efficiency is not as bad as might be anticipated. It is best to close a certain number of the guide passages entirely in such a manner that all the closed passages are contiguous, as this interferes less with the efficiency than when *every* passage in the turbine is *partially* closed. It has been found that to obtain good results air should be admitted to the closed passages, so that the water in the corresponding buckets of the wheel may be able to escape freely. This is done in some cases through tubes connecting the interior of the guide-buckets with the atmosphere. When the passages are closed the tubes are open.

The effects of regulation by the systems in question will be best illustrated by the examples following, selected from the results of actual experiment.

Jonval Turbine at Goeggingen.

This turbine[1] is regulated by flaps hinged on the outside of the guide-wheel, each flap covering three passages, or $\frac{1}{12}$ of the whole area.

Proportion of passages open.	Efficiency.	No. of revolutions.
All.	83·2 per cent.	45·52
$\frac{3}{4}$	79·5 ,,	45·29
$\frac{7}{12}$	74·9 ,,	44·35
$\frac{1}{2}$	75·1 ,,	45·63

[1] *Vide* Chap. XI.

Humphrey Turbine at the Tremont and Suffolk Mills.

The "Humphrey"[1] turbine is of the mixed-flow class, the water entering radially, flowing inwards, and leaving axially. In the wheel in question regulation took place by means of a circular slide between guide passages and wheel as described under (7).

Proportion of passages open.	Efficiency.	No. of revolutions.
100 per cent.	81·9 per cent.	47·42
82·5 ,,	76·69 ,,	51·60
77·24 ,,	76·54 ,,	48·00
52·90 ,,	61·02 ,,	50·64
40·66 ,,	56·14 ,,	45·59

A comparison of these results with those given in the preceding example demonstrates the superiority of the regulating arrangement adopted in the Jonval turbine ; with 50 per cent. of the passages open the efficiency was reduced by 8·1 per cent. as compared with that obtained with all open, while in the case of the Humphrey turbine, with the whole area of the passages and with 52·9 per cent., the efficiencies were respectively 81·9 per cent. and 61·02 per cent., showing a reduction of 20·88 per cent.

Fourneyron Turbine.

The following figures are given by Professor Unwin as the result of experiments with a Fourneyron turbine regulated by a cylindrical sluice working out of water :—

[1] *Vide* Chap. XI.

Opening of sluice	Efficiency.
Full	62 per cent.
$\frac{7}{8}$	60 ,,
$\frac{5}{8}$	43 ,,
$\frac{3}{8}$	30 ,,

With the turbine drowned the efficiencies were lower.

(V.) *Regulation by concentric subdivision.* This method, which is applicable to various kinds of turbines, is generally adopted only in those of large dimensions; it consists, in the case of an axial turbine, in dividing the whole guide apparatus and wheel by one or more concentric partitions into two or more parts, each part forming, so to speak, a separate turbine. As less power is required these parts can be shut off in succession, and those remaining act each as a complete turbine, the efficiency of which is not affected by its neighbours being put out of action.

In radial turbines the subdivision for purposes of regulation is effected by partitions at right angles to the axis.

For large powers this system of regulation is one of the best in use; the efficiency is only slightly reduced by shutting off one of the divisions. In the case of a double Jonval turbine, of which particulars are given in Meissner's work on Turbines and Water-wheels, the efficiency was about 6 per cent. less with one part only in use than when the whole wheel was in action.

(VI.) *Correct system of regulation.* The only mode of regulating a reaction turbine which is theoretically correct, is by reducing the areas of both guide and wheel passages simultaneously, so that the proportions remain the same. This is not practicable in any way for axial turbines, but has been applied in exceptional cases to radial-flow turbines.

The turbine of Messrs. Nagel and Kämp is constructed on this principle. The wheel (Fig. 88) is of the outward-

Fig. 88.

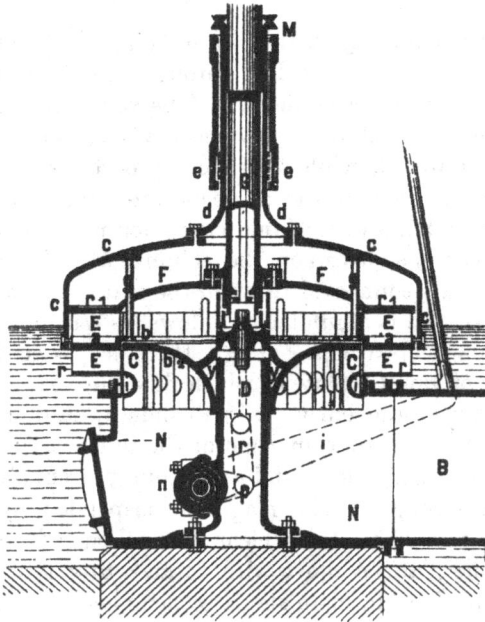

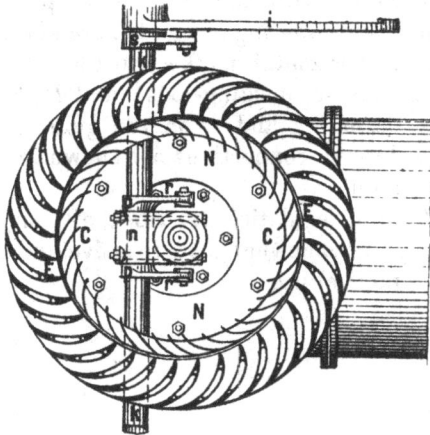

flow (Fourneyron) type, the water being supplied to the guide apparatus from below through the tube B, from which it enters the chamber N. The vanes of the guide-wheel are attached to a movable cover b_2, which can be raised or lowered, while their lower portions are guided vertically by slots in the projection or flange i i which extends round the whole inner circumference. The boss y of the cover b_2 slides upon the column D. The wheel E itself is rigidly connected through the plate or arms F with the shaft G, and is regulated as follows:—Within the upper and lower wheel-casings r and r_1, there is placed between each pair of vanes a plate a, which fills the space between them, but fits sufficiently easily to slide up or down freely. All these plates a are connected at the inner circumference to a ring b of angle-iron, and at the outer circumference to a circular casing c. The ring b is suspended by rods g from the top of c. It is clear that when c is raised or lowered the plates a move with it and alter the depth of the wheel. The mechanism is so arranged that c and b_2 rise or fall together, and the width of the wheel-buckets always corresponds with that of the guide passages.

The raising and lowering of c and b_2 are effected through the lever l and horizontal shaft n; to the latter are keyed levers p and s and connecting-rods r and D; of these, two levers p act through r and D on the guide-casing b_2, and two levers s through connections not shown in the illustration, on the casing c.

The method of regulation described, although perfect in principle, appears somewhat too expensive and complicated for general practical use.

The Distribution of Water in Turbines.

Not merely the effect of a given weight of water acting on a turbine is influenced by the form of the vanes, but also the distribution of the water flowing through the guide passages and buckets, and this again may modify to some extent the efficiency, so that to ascertain the connection between the latter and the construction of the vanes, the distribution as well as the direction and velocity of flow of the water must be taken into account.

In radial inward- or outward-flow wheels, in which all the water enters or leaves at the same diameter, and the edges of the vanes are parallel with each other, while the angles of inflow or outflow are constant for the whole width of the vanes, the distribution is necessarily uniform, on the assumption that the velocity of flow does not vary with the width, but in a parallel-flow turbine the case is different. It may be supposed that a Jonval wheel is divided—by imaginary cylindrical surfaces—into a large number of very narrow concentric rings of equal width, and the quantity of water flowing through each of these with a given construction of vanes can then be ascertained ; if through each ring the same quantity of water passes in a given time, then the distribution is *equal* or *uniform*, but if more (or less) flows through the outer than through the inner rings, then the distribution is *unequal*. Obviously, to avoid internal friction, the distribution should be similar in all parts of the buckets and guide passages.

The most usual form of vane surface both for guides and buckets is the *helical,* and the distribution of water resulting from this construction will therefore be first considered, beginning with the guide passages. With helical vane surfaces the angle *a decreases* from the outer to the inner diameter, since the pitch of the helices defining the surface at all diameters is constant. If s be the clear distance, Fig. 89, between two vanes measured parallel with the direction of rotation, then the normal depth X, or clear distance measured at right angles to the direction of flow, is $s \cos a$, and to this product the outflow area of any of the narrow concentric rings of equal width is proportional. The value of s, however, is proportional to the diameter or mean radius r of the given ring, while $\cos a$ increases as the radius decreases; the effect of this

Fig. 89.

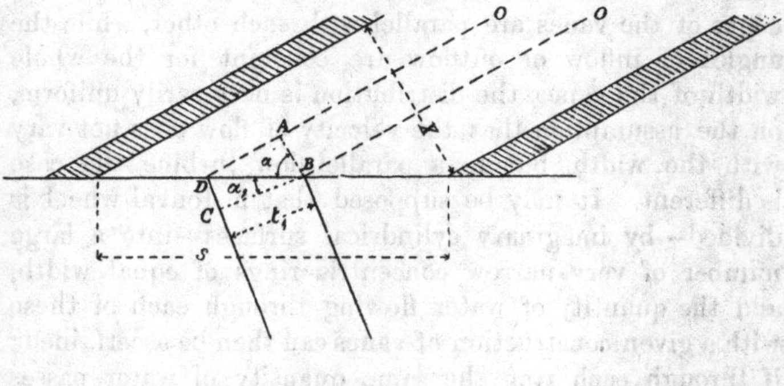

is, that the product $s \cos a$, for moderate widths of the passages, is nearly the same at all diameters, and therefore the distribution of the water almost equal.

As an example a turbine may be taken in which the width of the buckets is $\frac{1}{4}$ of the mean diameter—a rather large proportion; the inner, mean, and outer radii are then in the ratios $3 : 4 : 5$.

Assuming the angle a at the mean diameter to be $70°$, those at the inner and outer diameters respectively will be $64° \ 15'$ and $74°$, and

$$cos \ 64° \ 15' = 0·4345$$
$$cos \ 70° \ \ \ \ = 0·3420$$
$$cos \ 74° \ \ \ \ = 0·2760$$

The distribution is then represented at the three diameters by

$$3 \times 0·4345 = 1·3035$$
$$4 \times 0·342 \ = 1·368$$
$$5 \times 0·276 \ = 1·380$$

the quantity of water passing through the wheel at the outside being very slightly greater than at the inside.

If, as is usual, the inflow into the buckets only takes place without shock at the mean diameter, then owing to the shock at other points the distribution will probably be somewhat disturbed, in a manner which it is impossible to accurately determine; but it is reasonable to assume that the relative velocity of inflow c_1 will be inversely proportional to the normal depth of the inflow orifices at various points, and with a constant velocity of flow c, a little consideration shows that the relative velocity of outflow c_2 must remain constant also, since *in any one of the narrow concentric rings* before referred to, c_1 is inversely proportional to A_1, and hence $A_1 c_1 = $ constant.

If this is not the case, then either a part of the water entering at one diameter is forced radially towards another or escapes through the clearance space between the guides and wheel.

When the angle a_1 is constructed at all points for entry without shock, the distribution of the water remains unchanged on leaving the guide passages.

It is of course evident that with a constant value of c_2, and with the angle a_2 very nearly the same as a, the distribution of the water on leaving the buckets is practically the same as on leaving the guides.

If, instead of helical vane surfaces, the angles a and a_2 are constant for the whole width of the wheel, then the product $s \cos a$ (or $s_2 \cos a_2$) varies proportionately to the radius, and the quantity of water flowing through a given small width at the outer circumference is greater than at the inner; the distribution increases from the inside to the outside in the ratio of the diameters (or radii).

With the turbine of the dimensions before assumed, but with a constant angle a, the distribution of the water at the inner, mean, and outer diameters respectively would be as $3 : 4 : 5$.

Fig. 90.

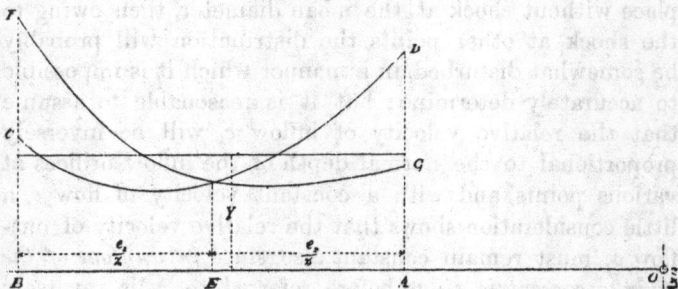

In determining the loss from unutilized energy both the residual velocity and the distribution of the water must be taken into account. In any of the narrow rings into which the wheel is supposed to be divided, the total loss from unutilized energy is proportional to the square of the residual velocity u and the quantity of water q passing through that ring, that is, to the product $q \times u^2$.

If for every point of the width this product be determined, or a value proportional to it, and the results plotted as ordinates with the corresponding radii as abscissæ, a curve is obtained, and the area $A B C D$ (Fig. 90), of which it forms one boundary, represents the total loss Σ $(q\,u^2)$, and the *mean* ordinate Y the average value of $u_m{}^2$, or average loss from unutilized energy per unit of weight.

When the distribution is uniform the ordinates of the curve are simply proportional to u^2 at each point, since q is constant.

Under any given conditions as regards the vane angles, it is easy to find graphically the residual velocity at any point, and by multiplying this with a value proportional to the distribution at the same point, to determine the ordinates for the curve in question. It will be found convenient to take the quantity of water passing through the wheel at the mean diameter as *unity*.

For a helical vane surface the curve representing the unutilized energy has the form shown in Fig. 90 by the line $C\,D$, $O\,E$ being the mean radius, $O\,A$ and $O\,B$ respectively the inner and outer radius.

With a constant angle a_2, and consequent distribution proportional to the radius at any point, the curve of unutilized energy is shown by $F\,G$. The distribution of loss, it will be seen, differs very much in the two cases considered, but in each the total loss is very nearly the same.

The distribution of loss by shock over the width of a turbine might be determined by a method analogous to that followed in the case of the unutilized energy. For *uniform* distribution the loss *per unit of weight* of the fluid, on either side of the mean circumference, would be theoretically equal, so that the curve representing that loss would be symmetrical with respect to a vertical line through the middle of the width of a bucket, assuming that, as before, the abscissæ are measured parallel with the edge of a vane and the ordinates perpendicular to it.

In the absence of experimental data showing in what

way and to what extent theory and facts regarding loss by shock differ, it would be unsafe to draw any definite conclusions from an investigation of the kind indicated.

Comparison of Axial with Radial Turbines.

An investigation of the relative merits of Axial as compared with Radial and Mixed-flow turbines when based on theoretical grounds is comparatively easy, but any attempt to form a judgment on this point from existing experimental results labours under disadvantages similar to those mentioned in discussing the effect of different proportions on the efficiency of turbines of the same class. The matter is if possible rendered more difficult by the fact that the manufacture of radial-flow turbines—more especially with inward flow—is confined almost entirely to America, where these motors are generally of the mixed-flow type, the water entering radially and leaving axially.

In dealing with the subject from a theoretical point of view, it will be assumed in the first instance that the co-efficients of resistance ζ_1 and ζ_2 are the same for radial as for axial turbines. In Table No. II. will be found the efficiencies of radial wheels for different ratios between the radii of inflow and outflow.

For an outward-flow turbine this ratio $\dfrac{r_1}{r_2}$ is *less*, for inward-flow *greater* than unity.

The expression $\dfrac{1}{g}\dfrac{A}{A_2}\dfrac{r_1}{r_2}\sin a \sin a_2$, which represents the work done on the wheel per unit of weight, is directly proportional to the ratio $\dfrac{r_1}{r_2}$, and consequently is greater for an inward-flow than for an outward-flow turbine. This is explained by the fact that as a particle of water flows outwards in a radial wheel the successive parts of the vane surface with which it comes in contact have an increasing speed, and the deflection of the stream produced by any given curvature of the vane becomes less the greater the distance from the axis. With an inward-flow turbine the *reverse* of this is the case. *The same vane curve*—referred to a line normal to the direction of rotation—will cause a greater absolute deflection of the stream in an inward-flow wheel than in an axial wheel, and in the latter again a greater deflection than in an outward-flow turbine.

At different distances from the centre the speed with which, so to speak, the vane is *getting out of the way* of the water, varies, so that when the latter approaches the circumference that speed increases, and the obstruction to a straight course becomes less; when the water approaches the centre, the opposite occurs.

Provided the losses remain constant, the greater the work done per unit of velocity of flow the higher the efficiency of a turbine must be, hence *for the same values of* $\dfrac{A}{A_2}$, *a and* a_2, the efficiency of a radial inward-flow wheel is greater than that of an axial turbine, and—apart from certain considerations to be afterwards discussed—increases with the ratio of the outer to the inner radius, or, for a given outside diameter, as the inner diameter diminishes. This is shown clearly in Table No. II.: with $\dfrac{A}{A_2} = 2$, $a_2 = 70°$, and $a = 75°$;

for $\dfrac{r_1}{r_2} = \frac{2}{3}$, the efficiency is $0\cdot796$; for $\dfrac{r_1}{r_2} = \frac{5}{4}$, $\epsilon = 0\cdot765$;

for $\dfrac{r_1}{r_2} = 1$, $\epsilon = 0\cdot722$, and for $\dfrac{r_1}{r_2} = \frac{2}{3}$, $\epsilon = 0\cdot634$.

A given angle of outflow a_2 in an inward-flow turbine is equivalent to a greater angle in a wheel with parallel

o

or outward-flow, because owing to the decreasing speed of the vane towards the centre a greater deflection of the stream is obtained.

Mathematically the highest efficiency would be attained by an inward-flow turbine with the inner radius *nil*. This of course is a practical impossibility, and it is easy to perceive that the reduction of the inner diameter must soon reach a limit. In the first place, after leaving the buckets the water has to be deflected through an angle of 90° degrees in order that it can flow off into the suction-tube or tail-race, and if this is too abruptly effected a loss of energy results which makes itself felt by retarding the flow. In the second place—apart from the above consideration—if the inner radius be too small, the converging streams leaving the buckets will interfere with each other, causing a disturbance of the water and choking the exit. In the third place, for a *constant width* of the buckets, the ratio $\dfrac{A}{A_2}$ varies as that of the *outer* to the *inner* radius, and if the latter is reduced, it becomes necessary to increase the width of the buckets at the out-flow to maintain the same ratio $\dfrac{A}{A_2}$. This widening of the buckets cannot however be carried to any great length without loss, as it involves a change in the direction of flow of parts of the stream at right angles to its general course. On the other hand, if this be not done, and the ratio $\dfrac{A}{A_2}$ increased in proportion to $\dfrac{r_1}{r_2}$, then, as a reference to Table II. will show, the efficiency is diminished.

Difference between Performance of Inward- and Outward-flow Turbines.

A radial outward-flow turbine, under given conditions, consumes more water than an inward-flow wheel of the same proportions and with the same angles of inflow and outflow. This is often ascribed to the action of centrifugal force, which in the one case is supposed to *accelerate* while in the other it *retards* the flow. That is an inaccurate way of explaining the phenomenon, and likely to lead to misunderstanding. The true explanation is, that owing to causes already stated, *less work is done on the wheel* by a given weight of water *per unit of velocity* in the outward-than in the inward-flow turbine, and consequently the fluid leaves the buckets of the former with a larger proportion of unutilized energy and correspondingly higher velocity. Centrifugal force, so called, cannot possibly produce any acceleration of flow, since the rotary motion of the fluid to which it is due is itself the result of the pressure caused by the head, and to credit the centrifugal force with increasing or reducing the velocity of the water is to confound cause with effect.

Mixed-flow Turbines compared with Radial-flow Turbines.

In a mixed-flow turbine the two objections to a small inner diameter, previously noticed, are to some extent removed by continuing the buckets into that part of the wheel-casing where the water assumes a vertical course; the change in the flow from a horizontal to a vertical direction is thus made *during the passage through the buckets,* and the exit of the water near the centre does not necessitate such an abrupt alteration of its course as when the wheel is of the purely radial type. Unfortunately these apparent advantages of the mixed-flow turbine are to a great extent neutralized by accompanying drawbacks; the chief of these is, that for a small mean outflow-radius r_2, except with extremely narrow buckets, the variation in the speed from the inner to the outer circumference of the points at which the water leaves the wheel, is so great as cause losses which cannot be materially mitigated by any modifications in the form of the vanes. The cause of these losses has already been explained in connection with axial turbines, but in the case of mixed-flow wheels they are more serious, owing to the greater ratio of the width of the buckets to the diameter of the wheel. Another point to be noticed as bearing upon the construction of inward-flow or mixed-flow turbines is, that where the difference between the inner and outer diameters is great, the mean angle a_2 cannot be made as large as when the diameters are more nearly equal, and this again reduces the efficiency.

Taking all these facts into account, it may be safely stated that there can be no advantage in mixed-flow turbines as generally constructed over a well-designed simple inward radial-flow wheel, with a moderate ratio of outer to inner radius. There are very few reliable experimental results available from which a comparison can be made between the various classes of turbines, and these show no marked difference between the efficiency of radial inward-flow, and axial- or parallel-flow wheels; radial outward-flow turbines are somewhat inferior to the latter. A large number of tests have been carried out in America, chiefly with mixed-flow turbines, but although useful for comparing these among themselves, there is reason to believe that in some instances the co-efficients used for the calculation of the quantities of water consumed were too small, thus giving too high an efficiency. Many of the wheels in question of obviously irrational design are reported to have shown an efficiency only attained by very good European turbines.

In spite of this, however, a test carried out by Mr. Francis, a well-known American engineer and careful experimenter, with a "Humphrey" mixed-flow turbine, shows that an efficiency equal to that of the best Jonval turbines may be obtained by a motor of this class, although in the case in question the inner diameter, where the water leaves the wheel, is relatively to the outer diameter very small. The loss due to the deviation from the best speed towards the centre is up to a certain point compensated by the advantage arising from the difference in the diameters.

It is obvious that, as in the case of radial inward-flow wheels, to maintain a given ratio of A to A_2, the less the mean outflow-radius r_2 the greater must be the width of the buckets, and therefore also the greater the divergence from the best speed, and a limit must soon be reached at which the values of $\dfrac{A}{A_2}$ and a_2 can no longer be chosen, but are restricted within a very narrow range.

As compared with axial or parallel-flow turbines, both inward-flow radial and mixed-flow wheels have a disadvantage in the fact that the water must change its direction of flow twice—once before entering the wheel from a vertical to a horizontal course, and again on leaving from the horizontal to the vertical.

Against this may be set the greater friction on the bearings of an axial turbine due to the vertical reaction and pressure on the vanes, which tends to force the shaft downwards. In a mixed-flow turbine the downward reaction also exists, but is approximately balanced by the upward force exerted on the casing, arising from the deflection of the water in passing through the buckets, from a horizontal to a vertical direction.

As far as experience goes, it shows that advantages and disadvantages practically compensate each other in the different systems.

It seems not improbable that a mixed-flow turbine with comparatively narrow buckets, so that there should be no great variation from the best speed, and a moderately large ratio of outer to mean inner diameter, might give results considerably superior to those obtained with a parallel-flow turbine of otherwise similar proportions (that is, with the same values of $\frac{A}{A_2}$, $sin\ a$ and $sin\ a_2$).[1]

The question naturally presents itself to the practical man: Why have American engineers so generally adopted the mixed-flow type of turbine if there is no great advantage as regards efficiency to be derived from so doing?

The explanation would appear to be partly given in the conditions under which water-power in the United States is generally available, namely, in the form of large rivers with an immense volume of water. Under such circumstances the power at the disposal of the engineer or manufacturer in any given case is vastly in excess of his requirements, and economy in consumption of water is not so much an object as low first cost of the motor. The aim of many turbine manufacturers has therefore been, not to produce an economical turbine, but one which, with a given outside diameter, will do the largest amount of

[1] Vide *American Turbines.*

work regardless of the quantity of water consumed per horse-power.

It is easy to ascertain what are the conditions necessary to obtain this result. The work done by a turbine can be expressed as—

$$W = Q \lambda \epsilon h = A c \lambda \epsilon h,$$

or substituting for c its value from (vi$a$)

$$W = A \lambda \epsilon h \left/ \sqrt{2 \frac{r_1}{r_2} \frac{A}{A_2} \sin a \sin a_2 + \zeta_1 + \left(\frac{A}{A_2}\right)^2}\right.$$

$$\overline{(\zeta_2 + \cos {}^2 a_2)}.$$

Under given conditions the total work done is directly proportional to the velocity of flow c. Taking r_1 the outer radius (of an inward-flow wheel) and A the area of the guide passages as fixed, it is clear that the *larger A_2* and the *smaller sin a* and *sin a_2*, the greater will be the value of c. On the other hand, as may be seen by reference to Table II., the efficiency decreases as $\sin a$ and $\sin a_2$ are reduced, and beyond a certain point this applies also to A_2. The *diminution of the efficiency*, however, is *less rapid* than the *increase* of the velocity of flow, so that, within certain limits, by sacrificing efficiency the total work obtained from a wheel of a given diameter and with a given area of the guide passages may be increased. The mixed-flow type of turbine appears to lend itself most readily to this purpose, more particularly to enlarging the area of exit A_2 without increasing the outer diameter at any part.

The preceding remarks are not intended to apply to the *best* American wheels, which have undoubtedly been shown by experiment to have a distinct advantage as regards efficiency over the best Jonval wheels. This advantage, however, is chiefly due, not to the *mixed* flow, but to the fact that the flow is radially *inward ;* with similar proportions a purely radial-flow wheel would probably have an equally high efficiency.

On the Use of the Suction-Tube and Boyden Diffusor.

It has been previously demonstrated that, in addition to its primary function of enabling the turbine to be placed at any convenient height above the tail-water level, a suction-tube may also improve the efficiency of the motor by relatively reducing the velocity with which the water finally leaves the apparatus, as the loss from unutilized energy depends on this.

Suppose, in the first instance, a given wheel is used without a suction-tube; the water leaves the buckets with a certain absolute residual velocity u, which is entirely lost. If now a suction-tube be added, so constructed that from the moment the water leaves the wheel to that at which it finally issues from the tube there is a gradual enlargement of the area of the stream, the velocity after leaving the buckets will be gradually reduced from u to c_4 in the inverse ratio of the sectional areas. The velocity u, however, after the suction-tube has been added, will not be the same as before, but *greater ;* the gradual enlargement *increases the velocity of flow c,* and improves the efficiency by making the residual velocity less in proportion to c than it was previously. Of course the increase in the velocity of flow requires a correspondingly greater *best speed.*

When *no* suction-tube is used

$$c^2 = \cfrac{2gh}{2\dfrac{A}{A_2}\dfrac{r_1}{r_2}\sin a \sin a_2 + \zeta_1 + (\zeta_2 + \cos^2 a_2)\left(\dfrac{A}{A_2}\right)^2}$$

when a suction-tube is used

$$c^2 = \cfrac{2gh}{2\dfrac{A}{A_2}\dfrac{r_1}{A_2}\sin a \sin a_2 + \zeta_1 + \zeta_2\left(\dfrac{A}{A_2}\right)^2 + \left(\dfrac{A}{A_4}\right)^2},$$

where A_4 is the area of the outlet from the tube. In the latter expression $\left(\dfrac{A}{A_4}\right)^2$ takes the place of $\left(\dfrac{A}{A_2}\right)^2 \cos^2 a_2$, all other quantities remaining the same; if, therefore, $\left(\dfrac{A}{A_4}\right)^2$ be less than $\left(\dfrac{A}{A_2}\right)^2 \cos^2 a_2$, then the velocity of flow will be increased by the use of a suction-tube.

The effect on the efficiency can be calculated from the formula

$$\epsilon = \cfrac{2\dfrac{A}{A_2}\dfrac{r_1}{r_2}\sin a \sin a_2}{2\dfrac{A}{A_2}\dfrac{r_1}{r_2}\sin a \sin a_2 + \zeta_1 + \zeta_2\left(\dfrac{A}{A_2}\right)^2 + \left(\dfrac{A}{A_4}\right)^2}$$

and compared with that obtained without a suction-tube as given by

$$\epsilon = \cfrac{2\dfrac{A}{A_2}\dfrac{r_1}{r_2}\sin a \sin a_2}{2\dfrac{A}{A_2}\dfrac{r_1}{r_2}\sin a \sin a_2 + \zeta_1 + \zeta_2\left(\dfrac{A}{A_2}\right)^2 + \left(\dfrac{A}{A_2}\right)^2 \cos^2 a_2}$$

Expressed in terms of the velocity of flow

$$\epsilon = \cfrac{c^2 \, 2\dfrac{A}{A_2}\dfrac{r_1}{r_2}\sin a \sin a_2}{2gh}$$

from which it is clear that the improvement in efficiency

will be exactly proportional to the square of the increased velocity c. In the preceding formula for the efficiency with suction-tube, it has been assumed that the additional friction and resistance caused by the tube is negligible.

As an example, assume $\dfrac{A}{A_2} = 1$, $\dfrac{r_1}{r_2} = 1$, $a = 65°$ and $a_2 = 65°$, then without a suction-tube

$$\epsilon = 0.765 \, ;$$

with a suction-tube having an outflow area four times as great as the area A of the guide passages, so that $\dfrac{A}{A_4} = \frac{1}{4}$, the efficiency would be

$$\epsilon = 0.809,$$

an improvement of 4·4 per cent.

Where the relative out-flow angle a_2 is large, say 75° to 80°, there is not much to be gained by a gradually enlarged suction-tube area, as the absolute velocity of the water on leaving the wheel is already low. The action of the tube is, however, better when thus formed than when of equal sectional area throughout, as the column of water in it is less liable to pulsations arising from the presence of air, which causes irregularities in the running of the wheel. If, as is generally the case, the sectional area of the suction-tube immediately below the wheel is much in excess of that of the stream leaving the buckets and is the same as the outflow area A_4, there is theoretically no gain of efficiency secured by its use. For a parallel-flow turbine a form of suction-tube similar to that indicated in Fig. 90A would probably be best.

It has been shown that the less the angle a_2 the smaller the width of the buckets for a given quantity of water, but at the same time the lower the efficiency. It may in some cases be convenient, when a suction-tube is used, to keep down the size of the wheel in the manner indicated, and compensate for the loss of efficiency by giving appropriate dimensions to the tube.

The so-called Boyden diffusor is an arrangement designed for use with radial outward-flow wheels, solely to improve their efficiency in a manner similar to that in

Fig. 90A.

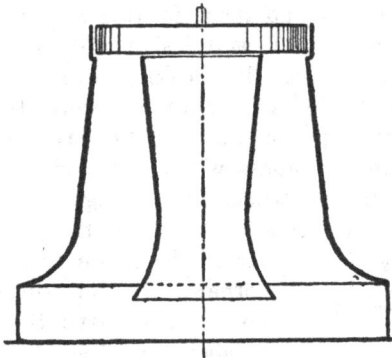

which this is effected by an enlarged suction-tube. It has the form shown in Fig. 91, and consists of a circular casing, concentric with the wheel, having the cross-section (through the axis), represented at *A B*, of a bell-mouthed shape. It forms, so to speak, a continuation of the wheel-casing, but is stationary; at *A* the area of the inner

Fig. 91.

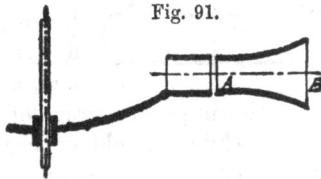

orifice is the same as that of the wheel measured over the circumference, while at *B* it is greater. The effect is exactly similar to that of a gradually enlarged suction-tube, and can be calculated in the same way; for A_4, the area at B through which the water leaves the diffusor must be substituted in the formulas. It has been used to a considerable extent in America to improve the efficiency of Fourneyron turbines; a corresponding apparatus might, where necessary, be applied to Jonval wheels working without suction-tube in the tail-water, consisting of a short bell-mouthed tube below the wheel, but in many cases there would be no room for this without alterations to the tail-race. For a well-designed turbine appliances of this kind are, generally speaking, unnecessary, or the improvement effected by them is not worth the increased first cost, but experimental results have proved their efficacy in many instances where the residual velocity of the water leaving the wheel was too high.

In designing the suction-tube of a turbine care should

Fig. 92.

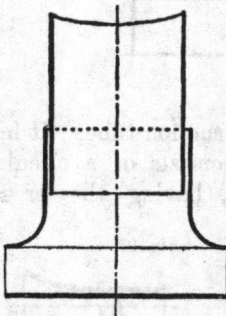

always be taken to round off the corners of the sluice-gate, or, better still, curve the lower end as shown in Fig. 92; sharp corners and sudden turns much increase the hydraulic resistance, and reduce the efficiency.

Another point of great practical importance when a suction-tube is employed, is the provision of an air escape-cock at the upper end, as the presence of air interferes with the proper action of the tube; it is also essential that the sluice should entirely and effectually close the outflow orifice when required, so that the tube before starting may be *quite filled* with water. Owing to a neglect of these precautions, in some cases suction-tubes have given unsatisfactory results, and fallen into disrepute with many makers.

Loss in enlarged Suction-Tubes and Diffusors.

It has been thus far assumed that in flowing through a gradually enlarged tube or diffusor the water suffers no loss of energy other than that due to the friction of the walls; this is, however, not strictly correct. When water flows out of a bell-mouthed or inverted conical tube, under a certain head measured to the centre of the orifice, the velocity of flow from the latter is *theoretically* the same whatever may be the form of the tube or mouthpiece. If the area of the outflow orifice is, for instance, four times as great as that of the smallest part of the tube, the velocity of flow through this latter must—if continuity is preserved —also be four times as high as in the orifice.

It is clear, however, that there must be an absolute limit to the velocity of flow, since this depends primarily on differences of pressure, and with a given head attains its maximum when the pressure at the corresponding point is *nil*, that is, with a vacuum in the smallest section of the tube.

If, therefore, the ratio between the area of the orifice and that of the smallest section is so great that the maximum possible velocity of flow would be exceeded if continuity were maintained, the assumed conditions no longer hold good; the water, instead of filling out the mouthpiece, flows through it as an independent stream surrounded by *dead* water, and the case is practically similar to that of a sudden enlargement of the area of the tube, and is accompanied by a corresponding loss of energy.

This is an extreme instance, but it has been found by experiment, that unless the enlargement of the tube is very gradual and the velocity of flow small, a very considerable loss results from the decrease of the latter, and this loss may become as great as, or greater than that which would be theoretically due to a *sudden* change from the least to the greatest sectional area of the tube.

The effect of losses of energy of this kind on the action of a Boyden diffusor or enlarged suction-tube is to reduce the advantage obtained by their use.

Francis found by experiment that the efficiency of a Fourneyron turbine was increased 3 per cent. by the addition of a Boyden diffusor, theoretically the advantage should have been 4 per cent.

Venturi, who was among the first to make accurate observations on the discharge of water through a diverging tube, found that with an angle of divergence of 4° 27′ the velocity through the smallest section was increased in the ratio of 2 to 1 as compared with the discharge through the same orifice when no tube was employed. According to theory (Bernoulli's), the increase should have been in the proportion of 3·03 to 1, showing a very considerable loss of energy, proportional to $(3·03)^2 - (2)^2$ or $9·18 - 4$, over 50 per cent.

Experiments carried out by Francis on the same subject demonstrated that the greater the ratio of the final discharge orifice to the smallest section, the greater the loss. Fliegner's investigations, subsequently to be described, confirm this result, and also prove that beyond a certain point a greater head (and consequently increased velocity) is accompanied by increased loss.

Francis noted that the losses occurred in spite of the tubes being quite full and the theoretical conditions apparently complied with. The fact of a tube being full, however, does not necessarily indicate true continuity of flow, since, as will be explained in describing Fliegner's experiments, a part of the water round the sides of the tube may be in a relatively nearly stationary eddying condition, the true form of the discharging jet not being defined by the walls of the tube.

A very interesting series of experiments, made by Professor Fliegner of Zürich, on the outflow of water through submerged inverted conical mouthpieces, shows that in every case investigated by him a loss of energy took place over and above that due to friction. The experiments were carried out with mouthpieces of various proportions and under different pressures. The results proved that with increasing pressure (or velocity) the loss of energy

Fig. 93.

decreased up to a certain point and then became greater, so that there appeared to be in every instance a minimum resistance at some particular velocity of flow.

Professor Fliegner is of opinion, that when the natural angle of the conical jet of water, issuing from the smallest section, is less than that of the mouthpiece, the conditions of continuity of flow are not completely fulfilled; the jet follows, so to speak, an independent course, its surface not being in contact with the walls of the mouthpiece, *vide* Fig. 93. The effect is then the same as if the tube were abruptly enlarged in sectional area, and the loss is as great or even greater.

True continuity of flow is only possible when the natural angle of the fluid jet is greater than that of the mouth-

piece, so that the surface of the former comes into contact
with the walls of the latter.

Between the assumed theoretical conditions with no loss
and abrupt change of flow, various proportions of loss
occur, but in no case does the proportion appear to be less
than ⅛th of the theoretical loss due to a sudden change
from the maximum to the minimum velocity of flow.

If A_3 and A_4 denote respectively the least and greatest
sectional areas of a conically divergent or bell-mouthed
tube, p_3 and p_4 the corresponding pressures, and c_3 and c_4
the velocities of flow, then when there is no loss—

$$\frac{c_3{}^2}{2g} + p_3 = \frac{c_4{}^2}{2g} + p_4;$$

or since
$$c_3 = \frac{A_4\, c_4}{A_3}$$

$$[\left(\frac{A_4}{A_3}\right)^2 - 1]\, \frac{c_4{}^2}{2g} = p_4 - p_3.$$

p_4 the pressure in the orifice is generally given, c_4 is
determined by the effective head h—

$$c_4 = \sqrt{2gh}.$$

The smallest possible value of p_3 is O, and the maximum
possible velocity of flow c_3 in the smallest section of the
tube $max\ c_3 = \sqrt{2g\,(h + a)}$

(where $a = 33\cdot91$ feet, the column of water corresponding
to atmospheric pressure).

Assume for example $h = 30\cdot09$ feet, then
$$max\ c_3 = 8\cdot025 \sqrt{64} = 64\ 200\ \text{feet per second},$$
and $c_4 = 8\cdot025 \sqrt{30\cdot09} = 43\cdot973.$

In this case the greatest ratio of A_4 to A_3 with which
continuity of flow is theoretically possible is

$$max\ \frac{A_4}{A_3} = \frac{max\ c_3}{c_4} = \frac{64\cdot20}{43\cdot97} = 1\cdot46;$$

with a smaller value of h it would be greater; for instance, with $h = 1$,

$$c_3 = 47 \cdot 476 \text{ and } c_4 = 8 \cdot 025, \text{ whence } max \frac{A_4}{A_3} = 5 \cdot 916.$$

When, as is actually the case, a loss of energy occurs in the conically divergent tube or mouthpiece, the hydro-dynamic equation takes the form

$$\frac{c_3^2}{2g} + p_3 = \frac{c_4^2}{2g}(1 + \zeta_4) + p_4,$$

or

$$[\left(\frac{A_4}{A_3}\right)^2 - (1 + \zeta_4)]\frac{c_4^2}{2g} = p_4 - p_3,$$

where ζ_4 represents the proportion of the energy lost to that due to the velocity c_4, the total loss of head in question being $\zeta_4 \frac{c_4^2}{2g}$.

In calculating the efficiency of a turbine using an enlarged suction-tube or Boyden diffusor, in the formula previously employed, it would be necessary to introduce, instead of merely the ratio $\left(\frac{A_2}{A_4}\right)^2$, the expression $(1 + \zeta_4)$ $\left(\frac{A_2}{A_4}\right)^2$.

Practical Limits for Position of Wheel in Suction-Tube.

Theoretically the limit of height h_2 at which a turbine may be placed above the tail-water level in the suction-tube is

$$h_2 = \overset{\text{Feet.}}{33 \cdot 91} - \frac{c_4^2}{2g},$$

where c_4 is the velocity with which the water leaves the tube.

As c_4 bears some proportion, varying generally from $\frac{1}{4}$ to $\frac{1}{6}$, to the theoretical velocity of flow $\sqrt{2gh}$, the theoretically admissible height of suspension or suction h_2 will vary with the head h.

In practice, however, with the usual cylindrical suction-tube, the height of suspension for various diameters of tube should not exceed the values given in the following table, which is taken from Meissner's work.

Diameter of suction-tube. Feet.	Admissible height of suspension. Feet.	Diameter of suction-tube. Feet.	Admissible height of suspension. Feet.
13	9·84	3·3	26·24
11·5	11·15	2·3	27·88
9·8	12·46	1·6	27·88
8·2	13·77	0·98	29·52
6·5	14·76	0·49	31·16
4·9	19·68		

From the above heights the value of $\dfrac{u^2}{2g}$ must be deducted when the head is greater than 33 feet. If the suction-tube, instead of being cylindrical, is gradually enlarged towards the outflow orifice, as previously explained, the suction height h_2 may be greater than indicated in the table, chiefly because with that form the column of water is less liable to pulsations than when the tube is cylindrical.

CHAPTER VII.

IMPULSE TURBINES.

Essential principle of impulse turbines.—Ventilation of buckets.
—Cases in which impulse turbines are preferable.—Advantages —
Limits within which hitherto employed.—Theory of impulse turbines.
—General equation.—Velocity of flow.—Useful work.—Available
energy.—Relative velocities of flow.—Efficiency.—Axial impulse
turbines.—Full admission.—Area of guide passages.—Width.—
Radius of wheel.—Determination of vane angles.—Course of water
through buckets.—Increased width at outflow.— Factors determining
width at outflow.—Partial admission —Effect of increasing width of
buckets.—On the form of vanes for impulse turbines.—Relative
angle of inflow negative.—Correction of vane angles for different
diameters.— Correction of inflow angle.—Correction of outflow angle.
—Extra source of loss with partial admission.—Absolute path of
water.—Turbines to act both as impulse and reaction wheels.

THE distinction between reaction and impulse turbines has
already been more than once referred to. The essential
point in connection with impulse turbines is the principle
of *free deviation,* that is, freedom for the water after leaving
the guide passages to follow in one direction its own course,
and during its transit through the wheel to be at all times
under atmospheric pressure only. To insure this result,
the buckets of an impulse turbine should never be *filled*
by the water, but must be so constructed as to allow free
access to the atmosphere. To attain this end with greater
certainty, Girard introduced the ventilated buckets shown

in Fig. 94; with these, apertures *c c* are made in the sides
of the wheel communicating with each bucket *b* to admit
the air, the apertures being so placed that they can never
be obstructed by the water.

It has been shown that to act to the best advantage
a reaction turbine must work with every portion full of
water, and consequently the dimensions of the wheel are
to a great extent dependent on the quantity of water the
motor is intended to utilize, and when the quantity is very
small the area of the guide and wheel passages assume
inconveniently or even impossibly minute dimensions, and
to obtain the necessary speed at the circumference a very

Fig. 94.

high number of revolutions
per minute is required. An
impulse turbine is subject
to no such restrictions;
the water may be admitted
to all the buckets round
the circumference of the
wheel or to only one or
two, and the efficiency is
not materially affected thereby. The dimensions of an
impulse turbine may therefore be chosen to suit practical
convenience within very wide limits, more especially as
regards the diameter, so that for very high falls and small
quantities of water comparatively large wheels running at
a moderate number of revolutions can be employed. As
for a given fall the quantity of water does not influence
the efficiency, impulse turbines can be regulated to any
desired extent by closing the guide passages.

Owing to these advantages, *Girard* turbines, as impulse
wheels are now very generally called, have to a great
extent superseded reaction turbines on the continent of
Europe, where, in a large proportion of cases in which

water power is available, the *quantity* is variable, and a motor is required which will better adapt itself to changeable conditions than the reaction turbine.

Girard turbines are now to be found of almost all possible varieties, axial, inward-flow, outward-flow, with vertical, horizontal, and inclined axes, and utilizing falls ranging from 4 to 1,800 feet, and quantities of water from 140 to 0·04 cubic feet per second.

The Theory of Impulse Turbines.

The fundamental difference between the theory of reaction and impulse turbines lies in the fact that the condition of continuity of flow in all parts does not apply to the latter, and therefore there is not a constant ratio between the velocities of flow in any two parts of a given impulse turbine.

The general equation (1) applies—as previously stated—equally to reaction and impulse turbines:

Available energy = useful work + work lost on friction, &c., + energy remaining in water due to residual velocity, but owing to the fundamental difference above-mentioned, it leads when applied in detail to different results in the two cases.

Velocity of Flow.

The velocity of flow of the water from the guide passages is for impulse turbines always that due to the head less the equivalent of losses arising from friction and other hydraulic resistances; these latter in any given case are a fixed quantity, and hence the velocity of flow

$$c = K \sqrt{2gh_1},$$

where K is a constant co-efficient for any particular conditions. Hence for a certain fall the velocity of flow remains the same whatever the speed of the wheel may be, and is independent of the angles of the wheel vanes or the area of the passages between them.

The head h_1 to be used in calculating c is of course the effective head measured from the orifices of the guide passages, which is always somewhat less than the total head, as the impulse wheel is placed above the water. As the water issues from the guides and leaves the wheel at atmospheric pressure, the available energy on entering the wheel is represented by the velocity of flow c, and is all in the *kinetic form*, while in a reaction turbine, in most instances, a considerable part of the energy available before the water enters the wheel buckets exists in the *potential* form as pressure; hence for an impulse wheel

$$(I) \qquad\qquad E_w = \frac{c^2}{2g}$$

and

$$(II) \quad 2gh_1 = (1 + \zeta_1)\, c^2, \text{ whence } E_w = h_1 - \zeta_1 \frac{c^2}{2g} \quad (Ia)$$

where ζ_1 is a co-efficient giving the proportion of energy lost by friction, &c., before the water leaves the guide passages.

Between K and ζ_1 the following relation exists—

$$K^2 = \frac{1}{1 + \zeta_1}.$$

The independence of c of the proportions of the motors very much simplifies the theory of impulse as compared with reaction wheels.

In calculating c of course all the special circumstances of the case must be taken into account in estimating the losses which reduce the effective head for producing velocity. If, for instance—as very frequently happens with high falls and small quantities of water—the water has to be conveyed through a long pipe from the collecting tank or reservoir to the wheel, *vide* Fig. 95, the loss from friction in this pipe must be estimated by the well-known formula. If, again, the fluid passes through a bend on its way to the wheel, the resulting loss should be calculated. Where the fall is low or moderate, so that the water flows directly from a large tank or reservoir into the guide buckets, the only appreciable loss is that occurring in the guide passages, to which the co-efficient ζ_1 is applicable. As a result of long experience for properly formed guide vanes, the value of ζ_1 is found to range from 0·234 to 0·018, or

Fig. 95.

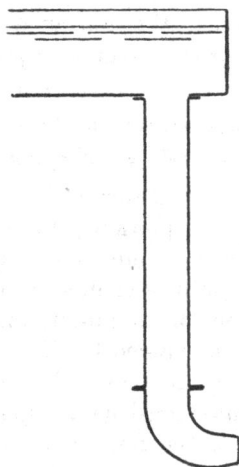

$$K_1 = \sqrt{\frac{1}{1 + \zeta_1}} = 0·9 \text{ to } 0·95; \text{ whence}$$
$$c = 0·9 \text{ to } 0·95\sqrt{2gh_1}.$$

Useful Work.

The work done on the vanes of an impulse turbine is expressed in precisely the same manner, and is due to the same causes, as in the case of a reaction turbine; the formula applicable is—

$$(8) \qquad W = M \left[w_2 \left(c_2 \sin a_2 - w_2 \right) + w_1 c \sin a \right]$$

for radial turbines, which for axial turbines becomes

$$W = M \left[w \left(c_2 \sin a_2 - w + c \sin a \right) \right].$$

When the entrance of the water takes place without impact, it has been proved that

$$W = w \left(c_2 \sin a_2 - c_1 \sin a_1 \right) M$$

for axial (or parallel-flow) turbines, and

(8a) $W = \left(w_2 c_2 \sin a_2 - w_1 c_1 \sin a_1 + w_1^2 - w_2^2 \right) M$

for radial turbines.

In equation (8) the quantity c can be calculated at once when the effective head is given, but c_2 the relative velocity of outflow cannot be directly determined from the proportions of the motor, and so long as the entry of the water is accompanied by shock it is not possible to ascertain it with any accuracy. For constructive purposes, however, the water must be assumed to enter *without* shock, and in that case the formula (8a) can be applied for the useful work done.

The available energy of the water on entering the wheel is

$$E_w = \frac{c^2}{2g} \text{ (per unit of weight)},$$

during its passage through a parallel-flow wheel with vertical axis or a radial wheel with horizontal axis, as shown in Figs. 96 and 97, a further accession of energy takes place, due to the fall h_o corresponding to the depth of the buckets, which results in an acceleration of the relative velocity of the water during its passage through the wheel; the total available energy for the latter is therefore $\frac{c^2}{2g} + h_o$.

This is expended in doing useful work, overcoming frictional resistance in the buckets, and producing the residual absolute velocity of outflow u, or—otherwise expressed—the useful work done is equal to the available energy E_w less the loss from friction (and other hydraulic resistances) and the energy due to the residual velocity or unutilized energy.

The loss in the buckets may be expressed by

$$\zeta_2 \frac{c_2^2}{2g};$$

and hence

(III) $$E_w - \zeta_2 \frac{c_2^2}{2g} - \frac{u^2}{2g} = W_u.$$

From this equation it is possible, by substituting the detail values of the various quantities, to ascertain the relation

Fig. 96.

Fig. 97.

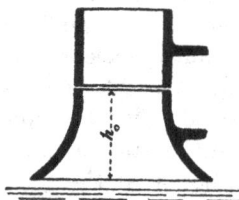
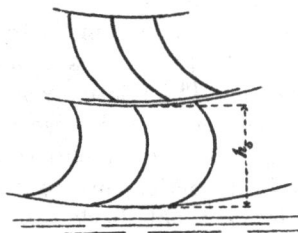

between c_2 the relative velocity of outflow and other known values. It may be noted that the method is exactly similar in principle to that employed to calculate the velocity of flow in the case of reaction turbines, only the quantity now required is not the absolute velocity of flow c, but the relative velocity c_2. As will be subsequently explained, the determination of the latter is of importance for the design of impulse wheels.

When the water enters the buckets without shock

(by C) $$c^2 = c_1^2 + w_1^2 - 2\,c_1\,w_1\,\sin\,a_1,$$
and $$u^2 = c_2^2 + w_2^2 - 2\,c_2\,w_2\,\sin\,a_2.$$

If, for the present, it be assumed that $h_o = O$ (as, for instance, with a radial wheel with vertical axis), then

$$E_w = \frac{c^2}{2g} = \frac{1}{2g} (c_1{}^2 + w_1{}^2 - 2\, c_1\, w_1\, sin\, a_1).$$

Relative Velocities of Flow.

By substituting in (III) the values above given, and simplifying and transposing the equation, the following relation between the relative velocities c_1 and c_2 and the velocities of rotation w_1 and w_2 is arrived at—

(IV) $(1 + \zeta_2)\, c_2{}^2 - c_1{}^2 = w_2{}^2 - w_1{}^2.$

The same result can be arrived at by making $p_1 = p_2$ in formula (iii), p. 127, for reaction turbines, that is, when the assumption is made that the pressure in the buckets is constant. In this equation every quantity except c_2 is known. The relative velocity with which the water enters the buckets can be ascertained graphically in the now familiar way, or algebraically, precisely in the same manner as for a reaction turbine, on the assumption that the fluid enters without shock.

w_2 and w_1 the speeds at the circumferences are given in any particular case.

Neglecting, for the moment, the friction in the buckets, equation (IV) becomes

$$c_2{}^2 - c_1{}^2 = w_2{}^2 - w_1{}^2.$$

For an *axial turbine* $w_2 = w_1$, and consequently $c_2 = c_1$, that is, the relative velocity of flow remains constant during the passage of the water through the buckets.

This result it was easy to foresee without the use of mathematics. In an axial wheel every part of a vane curve moves with the same speed, and as, after entering the bucket, no accelerating or retarding force (apart from friction and the depth of the wheel h_o) acts on the water, the relative velocity must necessarily remain the same throughout.

The effect of a varying speed of the vane on the relative velocity is not so easy to follow, but it will be clear on consideration that in this case the relative velocity must change as the fluid traverses a vane and approaches or recedes from the centre, as the case may be. This change is not the result of an accelerating or retarding force acting on the water, but merely of the fact that the latter passes over portions of the wheel moving with increasing or decreasing velocity in which the fluid does not participate.

The following extreme case will best illustrate this. Assume that the relative motion of the water in a radial outward-flow wheel is such that the absolute path is a radial straight line $A B$, Fig. 98, and the absolute velocity—under these conditions constant—represented by $A C$. In this case no work is done, and the velocity with which the water leaves the wheel is the same as that with which it enters. The *relative* velocity of entrance c_1 is found by the well-known parallelogram $A E C C_1$, in which $A E = w_1$ the velocity of the circumference where the water enters, and $A C_1 = c_1$.

Similarly the relative velocity of outflow c_2 is given by $B D_1$ in the parallelogram $B F D D_1$, in which $B F = w_2$ the velocity of the circumference where the water leaves

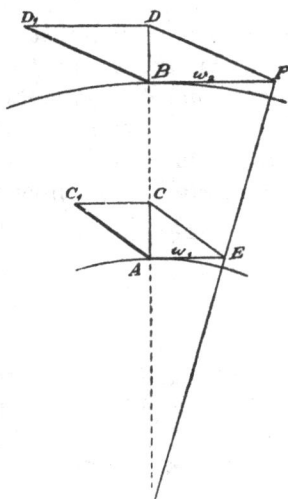

Fig. 98.

the wheel. It is obvious that in this instance $B D_1$ is greater than $A C_1$, since $B F$ is greater than $A E$, that is, an increase in the relative velocity has taken place.

From equation (IV) it may be seen that when w_2 is greater than w_1, that is, for an outward-flow wheel, the relative velocity is increased during the passage of the water through the buckets; on the other hand, for an inward-flow wheel, where w_2 is less than w_1, the relative velocity is reduced.

When in traversing the vanes the water has an additional fall h_o (as before explained), the relative velocity is increased by the amount due to this height reduced to some extent by friction. The complete formula for calculating c_2 under these conditions is—

$$(\text{IV}a) \qquad c_2{}^2 = (c_1{}^2 + w_2{}^2 - w_1{}^2 + 2gh_o)\,\frac{1}{1 + \zeta_2}.$$

It must be noted, that in the case of a radial wheel with horizontal axis, h_o is not necessarily equal to the full depth of the buckets, but is the vertical distance through which the water falls from the moment it enters to that at which it leaves the wheel.

This will be clear from Fig. 99, in which $A B$ represents

Fig. 99.

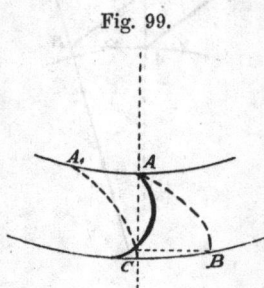

the absolute path of the water in traversing an outward-flow radial wheel; the point B at which the water leaves the vanes is considerably above the lowest point C. It is easy, however, to avoid this by placing the inlet so that the outflow takes place at the lowest point, for instance at A_1, $A_1 C$ being the absolute path.

This construction is generally only adopted where the quantity of water is so small that it can all be admitted through two or three guide passages at the lowest part of the turbine.

For a parallel-flow turbine, where $w_2 = w_1$, equation (IV) becomes simply

(IVb) $$c_2{}^2 = (c_1{}^2 + 2gh_o) \frac{1}{1 + \zeta_2}$$

Efficiency.

The hydraulic efficiency of an impulse turbine is expressed in the first instance in precisely the same way as that of a reaction wheel. The useful work done is by (8), p. 58:

$$W = M\left[w_2\left(c_2 \sin a_2 - w_2\right) + w_1 c \sin a\right];$$

the total available energy

$$E = Mgh,$$

whence the efficiency

$$\epsilon = \frac{W}{E} = \frac{w_2\left(c_2 \sin a_2 - w_2\right) + w_1 c \sin a}{gh},$$

or since

$$w_1 = \frac{r_1}{r_2} w_2,$$

(V) $$\epsilon = \frac{w_2\left(c_2 \sin a_2 - w_2 + c \dfrac{r_1}{r_2} \sin a\right)}{gh}.$$

On the assumption that the velocities c_2 and c are constant—independent of the velocity of rotation—the best speed is the same as for a reaction wheel—

$$w_m = \tfrac{1}{2}\left(c_2 \sin a_2 + \frac{r_1}{r_2} c \sin a\right).$$

Here again, however, the formula is incorrect, although somewhat nearer the truth than in the former case, as at any rate c is constant.

The relative velocity of outflow c_2 varies with the speed as a result of the impact or shock which occurs at all speeds except the best, and it is impossible in general to calculate it with any certainty.

Assuming, as previously, entry without shock and radial or axial outflow, there follows for the efficiency

$$\text{(VI)} \quad \epsilon = \frac{w_1\, c\, sin\, a}{gh} = \frac{w_2\, c\, \dfrac{r_1}{r_2}\, sin\, a}{gh} = -\frac{c\, c_2\, \dfrac{r_1}{r_2}\, sin\, a\, sin\, a_2}{gh}.$$

If, instead of comparing the work done with the *whole available energy*, it is compared only with the *energy available for the wheel*, represented by the velocity of flow c, and the efficiency thus measured be denoted by ϵ_1, then

$$\text{(VII)} \quad \epsilon_1 = \frac{w_1\, c\, sin\, a}{\dfrac{c^2}{2}} = \frac{2\, w_1\, sin\, a}{c}.$$

By (A) and (B)—for entrance without shock—

$$\frac{w_1}{c} = \frac{sin\, (a - a_1)}{cos\, a_1};$$

and this substituted in (VII) gives

$$\text{(VIII)} \quad \epsilon_1 = \frac{2\, sin\, a\, sin\, (a - a_1)}{cos\, a_1}.$$

As the losses incurred *previous* to the water entering the wheel are entirely independent of what takes place *afterwards*, the expression ϵ_1 really represents the efficiency of the wheel as distinguished from that of the supply-pipe or channel and guide apparatus. From this it will be seen that the maximum efficiency can be expressed entirely in terms of the two angles a and a_1, provided always that the conditions of entry without shock and radial exit are fulfilled. The equations (VII) and (VIII) apply only when after entering the wheel the water is subject to no further fall h_o during its passage through the buckets. Denoting by h_2 Fig. 100, the mean height of suspension of

a radial wheel with vertical axis, or the height of the lower
edge of the vanes of an axial wheel above the tail-water
level, then the losses incurred which are independent of
the proportions of the wheel are (1) those due to friction
and other hydraulic resistances in the supply-pipe and
guide apparatus, already considered, and (2) the loss arising
from the unutilized portion
of the head h_2 necessitated
by the condition that the
turbine must run free of
the water. In the latter,
particular reaction turbines
have an advantage, as owing
to their working drowned
the whole head is utilized.

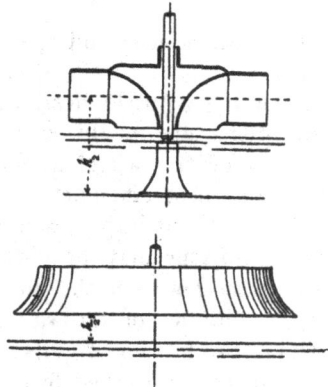

Fig. 100.

The first loss is expressed
by

(IX) $L_1 = \zeta_1 \dfrac{c^2}{2g}$;

the second by
(X) $L_2 = h_2$.

For an axial wheel with vertical axis or a radial outward-
flow wheel with horizontal axis of the kind previously
referred to on page 216—

$$h - h_2 = h_1 + h_o;$$

by (II) the available energy on entering the wheel

$$E_w = h_1 - \zeta_1 \frac{c^2}{2g},$$

and to this must be added h_o the fall in the buckets;
hence the total energy available for the wheel,

(XI) $E_w{}^1 = h_1 + h_o - \zeta_1 \dfrac{c^2}{2g} = h - h_2 - \zeta_1 \dfrac{c^2}{2g},$

and the efficiency of the apparatus *apart from the wheel*

(XII) $$\epsilon_2 = \frac{h - h_2 - \zeta_1 \dfrac{c^2}{2g}}{h},$$

or where $h_o = O$,

$$\epsilon_2 = \frac{h_1 - \zeta_1 \dfrac{c^2}{2g}}{h}.$$

The product of ϵ_1 and ϵ_2 equals the total efficiency ϵ—
(XIII) $\epsilon_1 \epsilon_2 = \epsilon.$

In dealing with the design of impulse turbines the same assumptions are made with respect to the maximum efficiency of the wheel as for reaction turbines, namely, that the highest efficiency will result from a combination of inflow without impact and radial outflow. In order that these conditions may be complied with, certain relations must exist between the speed of the wheel, velocity of flow, and angles of the vanes.

The formulas for impulse turbines have already been stated in the simplified form resulting from the assumptions in question, and can be applied directly to the calculation of the dimensions.

It will be convenient to consider first the conditions governing the design of parallel-flow or axial impulse wheels, and subsequently how these are modified in the case of a radial turbine.

Axial Impulse Turbines.

In the first instance it will be assumed, that the turbine the dimensions of which are required is to be so arranged that the water enters over the whole circumference.

The guide passages of an impulse turbine are in all respects like those of a reaction turbine, and must of course be entirely filled with water, but as soon as the streams issuing from the guide passages enter the wheel-buckets there should be an excess of sectional area in the latter, so that from the very commencement of its course in the wheel the water may be freely accessible to the atmosphere.

Since the guide passages are quite full, their sectional area A, where the water leaves them, is determined as soon as the quantity of water Q and velocity of flow c are known.

The velocity of flow can be calculated from formula (II) when the various resistances are known, and these can be determined with sufficient accuracy from the dimensions of the conducting pipes and guide apparatus with the help of the various empirical co-efficients applicable to each case.

The orifices of the guide passages are obstructed by the edges of the wheel-vanes moving in front of them, and to compensate for this the area of the passages, as obtained by dividing the quantity of water Q by the velocity c, must be increased by multiplying with a certain experimental co-efficient, or (what comes to the same thing and is the usual practice) using a value of c smaller than the true one to calculate A. If, after deducting all losses previous to entering the guide passages, H represents the effective head, the actual velocity of flow for well-made guide vanes will be

$$c = 0.95\sqrt{2gH}; \qquad (A)$$

to determine the area A, however, a value of c represented by $\qquad c^1 = 0.85\sqrt{2gH}$

must be employed, equivalent to multiplying the true value by 0·894, so that

Q

$$A = \frac{Q}{0 \cdot 894 \, c} = \frac{Q}{0 \cdot 85 \, \sqrt{2g \, H}}. \qquad (B)$$

Retaining the same notation as for reaction turbines

$$A = (2 \, \Pi \, r \, cos \, a - z \, t) \, e, \qquad (C)$$

and from this, when a, r, z and t have been fixed upon, e the width of the guide passages can be determined.

r, z, and t are, as before, empirically chosen quantities, for which the usual rules will be subsequently given; a the angle at which the water leaves the guide passages must be chosen, and it is easy to see from formula (VIII) that the efficiency increases with the angle a. In practice a ranges from 60° to 70°.

Transposing the above equation for A,

$$e = \frac{A}{2 \, \Pi \, r \, cos \, a - z \, t}, \qquad (D)$$

and the leading proportions of the guide apparatus are settled.

The dimension e should in practice not exceed a certain proportion of the radius r, and this proportion ranges from $\frac{1}{4 \cdot 5}$ to $\frac{1}{6}$; for the sake of brevity the ratio $\frac{e}{r}$ may be denoted by y. Neglecting the influence of the thickness of the vanes, the area A may be expressed by

$$A = e \, 2 \, r \, \Pi \, cos \, a,$$

or introducing the ratio y,

$$A = 2 \, y \, r^2 \, \Pi \, cos \, a,$$

from which r may be calculated by

$$r = \sqrt{\frac{A}{2 \, y \, \Pi \, cos \, a}} \qquad (F)$$

It is in some circumstances preferable to *assume* the

value of y, and determine the mean radius r by the above formula, instead of from the empirical rule elsewhere given; r being known, e follows approximately from

$$e = y\, r,$$

but should be calculated accurately as before.

Where the graphical method can be applied, it is exactly the same as for reaction turbines; the denominator in formula (D) is simply the sum of all the clear distances x between the guide-vanes at the outflow, of course measured at right angles to the direction of flow,

and $$e = \frac{A}{z\, x}. \qquad (E)$$

The next step is to determine the proportions of the wheel itself, and the first value required is that of the relative angle of inflow a_1.

Upon that angle depends the relative velocity c_1 with which the water enters the wheel; this determines the speed of rotation w and the relative velocity of outflow c_2 (by formula IVb); the latter, again, bears a definite relation to the speed, owing to the condition that the water must leave the buckets parallel with the axis. By a combination of the various formulas expressing these relations, the following approximate connection between the angles a and a_1 is arrived at

$$a_1 = 2\,a - 90^\circ; \quad (a_1 \text{ negative}). \qquad (G)$$

Geometrically interpreted, this means that the absolute direction in which the water enters bisects the angle formed by the relative direction of inflow with the direction of rotation.

In Fig. 101 let $O\,B$ be the absolute direction of inflow forming the angle $a = B\,O\,Y$ with a vertical $O\,Y$; $O\,A$ the relative direction of inflow forming the angle $a_1 = A\,O\,Y$

with $O\,Y$; then $O\,B$ should bisect the angle $A\,O\,C$, so
that $< A\,O\,B = < B\,O\,C$; then $A\,O = A\,B$ or $c_1 = w$.

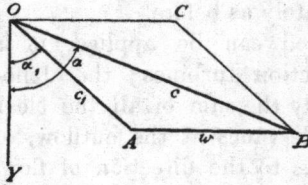

As above intimated, this rule is based only on an
approximation, and when, as
is sometimes the case, the
depth of the wheel h_o forms
a considerable portion of
the total fall, it is not sufficiently accurate as a basis
for the subsequent calculation
of c_2. The better plan under
any circumstances is to temporarily fix a_1 as given by the
approximate rule in question, then by the graphical method
determine c_1 and simultaneously w.

The relative velocity of outflow can next be calculated
by (IVb)

$$c_2{}^2 = (c_1{}^2 + 2gh_o)\,\frac{1}{1 + \zeta_2},$$

in which h_o the depth of the wheel must be empirically
chosen and subsequently corrected if necessary; it may be
taken as

$$h = \frac{r}{4} \text{ to } \frac{r}{5\cdot5}, \tag{H}$$

r having been already determined.

The co-efficient ζ_2 ranges from $0{\cdot}05$ to $0{\cdot}1$.

When c_2 is known the angle a_2 is easily constructed, or
determined by the relation

$$sin\ a_2 = \frac{w}{c_2}.$$

The construction is shown in Fig. 102. $O\,B = w$; from O,
with $O\,A = c_2$ as radius, describe an arc $A\,C$, and at B

draw BA at right angles to OB, intersecting the arc in A; join O and A, then OAB is the required angle a_2.

It is desirable that a_2 should be as great as practicable, in order to reduce the absolute velocity of outflow, and if by the above construction a_2 should turn out too small, the assumed value of a_1 can be modified—reduced—so as to increase the speed of the wheel, and this will have the desired effect on a_2.

Fig. 102.

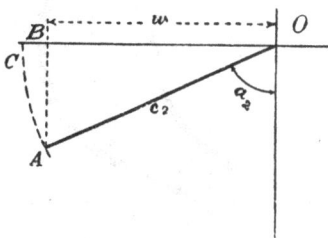

If a_2 is made greater without altering a_1 the result will be that the absolute direction of outflow is no longer parallel with the axis; provided the deviation is not very considerable, this is not of any practical importance as long as the velocity of outflow is sufficiently low.

It may also happen that the value of a_2 obtained is too great—in practice as a rule it does not exceed 72°—and then a modification in the opposite sense to that previously considered is necessary.

Before all the points involved in finally determining the angle a_2 can be thoroughly appreciated, it is essential to follow more closely the course of the water in the wheel-buckets. To insure free deviation and avoid the possibility of the water coming into contact with the convex surface of the vane next to that which it is traversing, no bucket should be filled beyond a certain proportion of its depth, about 0·7 to 1·0 of the latter at the inlet, and 0·5 to 0·75 at the outlet. Assuming, in the first instance, that the width of a bucket is constant throughout, the depth of the stream passing through it must vary inversely as the relative velocity of flow; if the latter remains unchanged, so

Fig. 103.

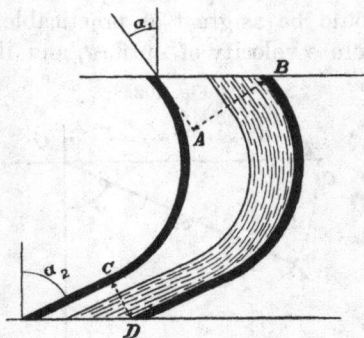

will the depth. In fact, a slight increase in the relative velocity takes place as the water traverses a parallel-flow wheel, and therefore under the assumed conditions the depth will somewhat decrease towards the outflow. As a_1 is less than a_2 it is very obvious that the depth of the passage between the vanes is considerably greater at the inlet AB, Fig. 103, than at the outlet CD; it may, therefore,

Fig. 104.

easily happen, that although there is sufficient clearance between the stream and the convex vane-surface where the water enters, the outlet area of the bucket is much too small, and the wheel becomes choked and cannot work as it is intended. This is illustrated in Fig. 104. The greater the angle a_2 the smaller will be the outflow area, and a little consideration shows that for a certain depth of stream the inclination of the vanes is restricted within certain limits. To extend these limits the depth of the stream must be reduced, and the only means for effecting this is to widen the buckets towards the lower (outflow) end. This allows the water to spread out, and the same sectional area is obtained with less depth by increased width. Girard turbines are now almost always constructed in this manner.

When a_2 has been fixed on, it is therefore essential to

make the width of the buckets, $e_2 = YX$, Fig. 105, at the
outflow end sufficiently great to allow
a certain proportion of clearance
between the water and the convex
surface of the adjacent vane. If δ_2
Fig. 106, denote the depth of the
stream in each bucket where it leaves
the wheel, then the sectional area of
the stream is $e_2 \times \delta_2$; this multiplied

Fig. 105.

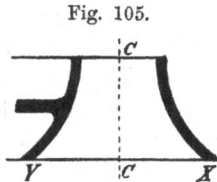

by the relative velocity of flow c_2 must equal the *quantity*
of water flowing through each bucket, and if z_1 denotes
the total number of buckets, then the *total* quantity of
water per second $Q = e_2\, \delta_2\, z_1\, c_2$, and from this

$$e_2 = \frac{Q}{\delta_2\, z_1\, c_2}. \qquad (J)$$

The depth of the stream δ_2 must bear a certain ratio
x_2 to the depth d_2 of the passage between the buckets at
the corresponding point, so that

$$\delta_2 = x_2\, d_2. \qquad (K)$$

Substituting this in (J) there follows—

$$e_2 = \frac{Q}{x_2\, d_2\, z_1\, c_2}. \qquad (L)$$

d_2 can be measured from a drawing when the pitch of
the vanes is known, but to determine this the number of
vanes z_1 must be first settled. As regards the relative
number of vanes in the guide apparatus and wheel two
different views may be adopted, both of which when put
into practice appear to give good results.

According to one of these, the number of guide-vanes
should be greater than the number of wheel-vanes, since
only a certain proportion x of the inflow area is to be filled
by the water, and, to ensure this, the outflow area of the

guides must be less than the corresponding relative inflow area of the buckets; this is illustrated in Fig. 107.

According to the other view of the subject, the pitch of the guide-vanes may be greater than that of the wheel-vanes. The buckets are then, in certain relative positions of the guide and wheel, *filled* at the inflow, but owing to the widening of the wheel casing are only partially filled at other parts. It is assumed in this case, that the actual outflow area from the guide passages is reduced by the edges of the wheel-vanes passing before the orifices, and that consequently no more water can leave the guide passages than the wheel is capable of passing. Besides

Fig. 106.   Fig. 107.

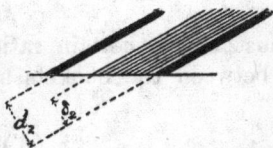
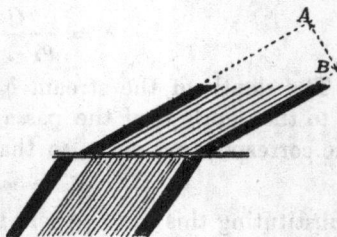

this, a certain clearance is always allowed in the width of the buckets which gives additional area.

If the view first stated is adopted, the number of guide-vanes z is purely empirical, but the number of wheel-vanes z_1 must be so determined that only the desired proportion x_1 of the buckets at the inflow is filled.

It is easy to arrange this by trial graphically, and when z_1 has been thus fixed, its value can be substituted in formula (L) and c_2 calculated.

The width e_1 of the wheel-buckets at the inlet is nearly the same as that e of the guide passages, there being a slight addition for clearance; the width at any point is inversely proportional to the velocity of flow, the cosine of the vane angle and the proportion of the sectional area of the bucket filled,[1] hence

$$\frac{e_2}{e_1} = \frac{x_1\ c_1\ cos\ a_1}{x_2\ c_2\ cos\ a_2}. \qquad (M)$$

By substituting in this expression the values of c_1 and c_2 resulting from the conditions of inflow without shock and vertical outflow respectively, and making use of the relations arising from the equation (G)

$$2\ a\ -\ a_1 = 90°,$$

the ratio of the widths of the buckets can be written as follows—

$$\frac{c_2}{e_1} = \frac{x_1}{x_2}\ sin\ 2\ a\ tang\ a_2; \qquad (N)$$

and if x_1 be made the same as x_2,

$$\frac{e_2}{e_1} = sin\ 2\ a\ tang\ a_2.$$

From this formula the effect of the angles a and a_2 on the proportions of the widths of the wheel-buckets can be easily calculated; the following are the results for a few values—

$$a\ =\ 30°, 50°, 70°, 70°.$$
$$a_2\ =\ 70°, 70°, 70°, 60°.$$
$$\frac{e_2}{e_1}\ =\ 2·38,\ 2·65,\ 1·75,\ 1·10.$$

If x_2 is less than x_1, these ratios for the same angles are greater in the inverse proportion.

The accurate formula for the value of $\frac{c_2}{e_1}$ is

$$\frac{e_2}{e_1}\ =\ \frac{x_1}{x_2}\frac{c_1}{c_2}\left[\frac{2\ r\ \Pi\ cos\ a_1\ -\ z_1\ t_1}{2\ r\ \Pi\ cos\ a_2\ -\ z_1\ t_2}\right], \qquad (O)$$

where t_1 and t_2 are the thicknesses of the wheel vanes at inflow and outflow.

[1] When the influence of the thickness of the vanes is neglected

In practice $\dfrac{c_2}{c_1}$ ranges from 2 to 3.

If the radius r as determined by the preceding method turns out impractically small, the idea of admitting the water over the whole circumference of the wheel must be abandoned and partial admission resorted to.

In this case the radius has no necessary relation whatever to the other parts of the wheel, but is purely empiric.

It may sometimes occur that the number of revolutions of the wheel is prescribed, and in that case r must be calculated from the relation

$$w = \frac{\Pi \, r \, n}{30},$$

where n is the number of revolutions to be made per minute; thence

$$r = \frac{30 \, w}{n \, \Pi}.$$

Let the proportion to the whole circumference of that portion over which the water is admitted be denoted by ϕ, Fig. 108, then the total sectional area of the stream

$$A = (\phi \, 2 \, \Pi \, r \, \cos a - z \, t) \, e, \quad (P)$$

where z denotes the number of guide *passages* and t the thickness of the vanes.

Fig. 108.

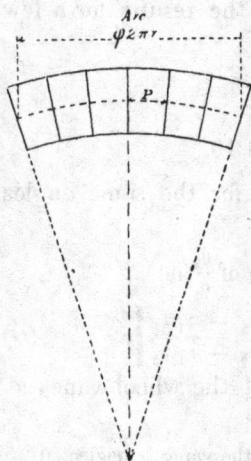

The ratio y of e to r must be assumed—it ranges from 0.1 to 0.2—and then since r is known e follows.

From the above equation for A the proportion of the circumference ϕ over which water is admitted can be ascertained: in the first instance, neglecting the quantity $z\,t$,

$$\phi = \frac{A}{2\,\Pi\,r\,e\,\cos\,a}; \qquad (Q)$$

then, choosing an appropriate pitch P, the number of passages z can be calculated from

$$z\,P = \phi\,2\,\Pi\,r,$$

$$z = \frac{\phi\,2\,\Pi\,r}{P}; \qquad (R)$$

and a corrected value of ϕ determined by again using formula (P) in its complete form, whence

$$\phi = \frac{A + z\,t\,e}{e\,2\,\Pi\,r\,\cos\,a}. \qquad (S)$$

It is not necessary to make use of trigonometrical quantities when the guide passages can be drawn out; the quantity $2\,\Pi\,r\,\cos\,a$ is simply the aggregate depth of all the guide passages, *including* thickness of vanes ($A\,B$, Fig. 107, being the depth of one passage), and this can be measured direct from the drawing.

A considerable margin should always be allowed, over and above the area A absolutely necessary, in turbines with partial admission, as they are always regulated, and the size of the wheel is not increased thereby.

Summary of Formulas for Girard Turbines.

(A) $\qquad c = 0\cdot 95\,\sqrt{2g\bar{H}}.$

(B) $\qquad A = \dfrac{Q}{0\cdot 85\,\sqrt{2gH}}.$

(C) $\qquad A = e\,(2\,\Pi\,r\,\cos\,a - z\,t).$

(D) $\qquad e = \dfrac{A}{2\,\Pi\,r\,\cos\,a - z\,t}.$

(E) $\qquad e = \dfrac{A}{z\,x}.$

(F) $\qquad r = \sqrt{\dfrac{A}{2\,y\,\Pi\,\cos\,a}}$, where $y = \dfrac{c}{r}$.

(G) $\qquad a_1 = 2\,a - 90°$ (a_1 negative).

(H) $\qquad h_o = \dfrac{r}{4}$ to $\dfrac{r}{5 \cdot 5}$.

(J) $\qquad c_2 = \dfrac{Q}{\delta_2\,z_1\,c_2}$.

(K) $\qquad \delta_2 = x_2\,d_2$.

(L) $\qquad c_2 = \dfrac{Q}{x_2\,d_2\,z_1\,c_2}$.

(M) Approximately $\dfrac{e_2}{e_1} = \dfrac{x_1}{x_2}\dfrac{c_1}{c_2}\dfrac{\cos\,a_1}{\cos\,a_2}$.

(N) Approximately $\dfrac{c_2}{e_1} = \dfrac{x_1}{x_2}\,sin\,2\,a\,tang\,a_2$.

(O) Accurately $\qquad \dfrac{c_2}{e_1} = \dfrac{x_1}{x_2}\dfrac{c_1}{c_2}\dfrac{2\,r\,\Pi\,\cos\,a_1 - z_1\,t_1}{2\,r\,\Pi\,\cos\,a_2 - z_1\,t_2}$.

For Partial Admission.

(P) $\qquad A = (\phi\,2\,\Pi\,r\,\cos\,a - z\,t)\,c$.

(Q) Approximately $\phi = \dfrac{A}{2\,\Pi\,r\,e\,\cos\,a}$.

(R) $\qquad z = \dfrac{\phi\,2\,\Pi\,r}{P}$.

(S) Accurately $\qquad \phi = \dfrac{A + z\,t\,e}{e\,2\,\Pi\,r\,\cos\,a}$.

Effect of increasing Width of Buckets on the Performance of Axial Impulse Turbines.

By giving the buckets of a Girard wheel a greater width where the outflow takes place, it is possible to make the angle a_2 greater than would otherwise be practicable, and thus increase the efficiency by reducing the absolute residual velocity. At the same time, however, the deviations from the best speed at the outer and inner circumference also increase, and the question naturally occurs whether the advantages obtained in one direction are not neutralized by the drawbacks in another. This is easily tested graphically, and it will be found that under ordinary circumstances there is a decided net gain in efficiency. Strictly viewed, the water on either side of the mean circumference has a radial component of motion, and in consequence during its passage over the vane comes into contact with successive portions of the latter moving at different speeds, so that to some extent the same action takes place as in a radial turbine. On one side of the mean circumference this action is that of an inward-flow, on the other that of an outward-flow wheel; the relative velocity of outflow c_2 will consequently be retarded and accelerated respectively, and the residual velocity correspondingly modified. The phenomena are, however, so complicated that it is not possible to follow them with any certainty, especially as only the outer parts of the stream are rigidly constrained in a radial sense.

It is for practical purposes sufficiently accurate to take the relative velocity of outflow as unaffected by its radial tendency when measured in projection on a plane parallel with the axis and with the direction of rotation at the point in question.

On the Form of Vanes for Impulse Turbines.

Similar remarks to those made with regard to reaction wheels apply in this case, except that, so long as there is sufficient room for free deviation, the question of the sectional area of the passages does not assume quite the same form. The most essential point is, that there shall be no possibility of the buckets becoming choked, especially towards the outlet where there is the greatest risk of this occurring. Impulse turbines have in some instances proved failures owing to defective design in this respect.

The angle a_1 in impulse wheels is nearly always *negative;* this is due to the high velocity with which the water leaves the guide passages compared with reaction turbines as usually constructed. If a_1 were positive or O^o, it would generally be impossible to reconcile the two fundamental conditions of inflow without shock and outflow parallel with the axis; for the latter it is essential that the relative velocity of outflow c_2 should be nearly equal to (rather greater than) the speed of the wheel w, but with a positive value of a_1, w would be much greater than the relative velocity of inflow c_1, which is not very different from c_2.

The vane curves of impulse turbines—like those of reaction wheels—are best drawn by trial. Those for the guide passages are essentially the same as in reaction turbines, while the curves of the wheel-vanes have generally a form similar to that shown in Fig. 109. The

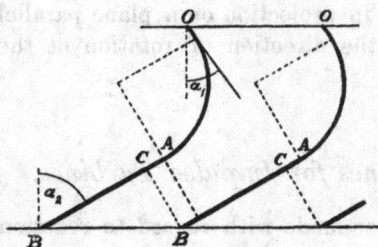

Fig. 109.

lower (outflow) end of the curves should be straight for some portion AB of their length, so that adjacent vane surfaces may run parallel for a certain distance AC to insure the water leaving at the desired angle.

A simple method of construction is to make the uppor

part of the vane curve, O A, an arc of a circle. Very great care should be taken in designing the vanes, and the section should be drawn out not only at the mean but also at the inner and outer circumferences.

In modern wheels the angle a of the guide-vanes is the same for the whole width of the passages, and similarly the relative angle of outflow a_2 is constant at all diameters. The surface of the vanes is therefore *not* helical; the form given to them in order to counteract centrifugal action is elsewhere described, but whether this is adopted or not, both a and a_2 can be made constant.

Correction of Vane Angles for different Diameters.

To partially neutralize the evils resulting from the deviation from the best speed at the outer and inner circumference of axial impulse turbines a method of construction similar to that used in the case of reaction wheels may be adopted.

The absolute velocity of outflow from the guide passages

Fig. 110.

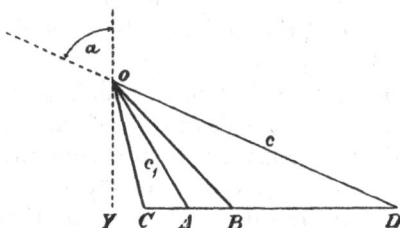

is here also to be taken as constant for the whole width, and from the speeds at the outer and inner circumferences

the appropriate values of a_1 the relative direction of inflow,
can be constructed as shown in Fig. 110, where $D\,A$,
$D\,B$, and $D\,C$ represent the speeds at the mean, inner,
and outer diameters respectively, and $O\,A$, $O\,B$, and $O\,C$
the corresponding relative velocities of inflow. So far
the process is exactly the same as for reaction turbines,

Fig. 111.

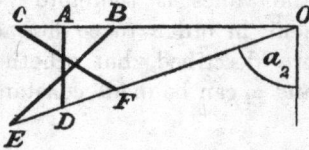

but owing to free deviation
being possible, subsequently differs. As the relative
velocity of inflow c_1 varies
for different diameters, so
also must the relative
velocity of outflow c_2, which
—neglecting the effects of friction—is greater than the
former by an amount due to the fall h_o through the wheel.

The greatest and least values of c_2 correspond to the
greatest and least values of c_1, and therefore the maximum
relative velocity of outflow takes place at the inner and
the minimum at the outer diameter. The angle a_2 being
constant for the whole width of the buckets, it is easy to
find the absolute velocity of outflow u at the inner, mean,
and outer circumferences, as illustrated by the diagram
Fig. 111; in the latter $O\,A$, $O\,B$, and $O\,C$ represent the
speeds at the mean, inner, and outer diameters respectively,
$O\,D$, $O\,E$, and $O\,F$, the corresponding relative velocities
of outflow, all having the same direction, and consequently
$A\,D$, $A\,E$, and $A\,F$ the absolute residual velocities.

It is evident that the directions of the latter cannot
all be parallel to the axis, but by the construction shown
they are made to diverge approximately to an equal extent
on either side from the desired course, so that the *mean*
direction is parallel with the axis.

When, as in the ordinary type of wheel, the edges of the
wheel-vanes are radial, it is clear that the *length* of the

curve traversed by the water at the outer circumference is *greater* than the corresponding length at the inner circumference ; on the other hand, it has been shown that the relative velocity of inflow c_1 is less for the outer than for the inner diameter. The *longer* curve is therefore passed over with the *lower* velocity, and as a consequence particles entering simultaneously at different points in the width of a bucket do not leave together, and there is a relative movement between the various parts of the stream flowing over a vane which causes internal friction and loss of energy.

Fig. 112.

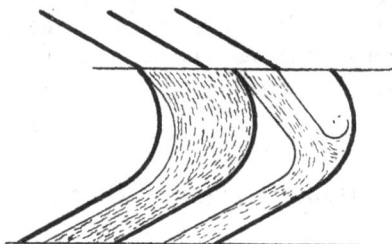

The correction of a_1 is unnecessary when the width of the buckets does not exceed $\frac{1}{9}$ to $\frac{1}{12}$ of the mean diameter. In any given case it is easy to ascertain by graphic methods the extent of the deviations at the inner and outer diameters from the assumed conditions, and if these appear considerable introduce the requisite modifications.

In impulse turbines with partial admission only, a loss of energy occurs in the wheel-buckets owing to the fact that a portion of the stream on entering a bucket does not come into contact with the upper end of the vane, but impinges on the latter much lower down, the result being a sudden instead of a gradual change of direction, and

R

consequent shock. This will be best understood by reference to Fig. 112. It is obvious that the greater the pitch of the vanes the lower down will the impact take place, and the more obtuse will be the angle between the stream and the vane. If the distance between the vanes exceeds a certain dimension a portion of the stream will pass right through the bucket without touching the vane. Even in wheels where the water is admitted over the whole circumference something similar may occur if the pitch is too great, since only those portions of the stream near to the vane will be effectively guided by the latter, while those further removed will be to a certain extent free, and cause a disturbance in the flow.

To avoid the evils just referred to, the pitch of the vanes of impulse turbines should be small, especially in those with partial admission.

Absolute Path of Water.

The method of determining the absolute path of the water in impulse turbines is exactly similar to that described for reaction wheels. The relative velocity of the fluid as it passes through the buckets increases in parallel and outward-flow wheels, but under ordinary conditions for the former, it is often sufficiently accurate to assume it as constant in constructing the curve of the absolute path.

Turbines to act both as Impulse and Reaction Wheels.

Circumstances often occur in which a variation in the quantity of water available for producing power is accompanied by a considerable alteration in the tail-water level. In order for the turbine to run clear of the water at all times, it would under these conditions be necessary to place it at a greater height above the lowest level than would be requisite if no alteration took place, the result being a loss of otherwise available fall at the very time when the best possible utilization of the power is most essential, since the tail-race level is lowest in dry seasons, when the available quantity of water is small. During floods, the increase in the quantity is generally proportionately greater than the loss of head from the rise of the lower level, so that there is a surplus of power which renders a high efficiency less essential than in the former case.

In these instances, to avoid the loss of head in question, a turbine may be so designed that in times of flood it can work partially drowned, the outflow taking place under water. It then works as a reaction wheel, since all the passages are full of water, but the proportions must be such that there is no excess of pressure over that of the atmosphere at the guide passage orifices, and consequently the velocity of flow from the latter is that due to the whole head, and as nearly as possible the same as when free deviation takes place. This can be accomplished by choosing a suitable ratio between the area of the guide passages and the area of the outflow orifices of the wheel.

For a velocity of flow equal to that occurring with free deviation, this ratio $\dfrac{A}{A_2}$ is somewhere about $\frac{1}{2}$, and with the usual values of a and a_2 is obtained by widening the buckets towards the outflow; this, of course, is exactly what is required for impulse turbines, and the dimensions must be adjusted in such a manner that the increased width at the outflow end is sufficient to allow free deviation if the wheel is acting as an impulse turbine, and at the same time give the necessary ratio $\dfrac{A}{A_2}$ when it works full.

In such cases each passage between two adjacent wheel-vanes should be so designed that the hydrostatic pressure at every cross-section of the stream is the same and equal to the atmospheric pressure. Under these conditions the stream flowing between two vanes should just *touch* the convex vane surface, and, apart from the acceleration of the relative velocity due to the fall in the wheel, the cross-sectional area of the passage should be *constant*. Allowing, however, for the acceleration of the relative velocity, the area must slightly decrease towards the outflow.

The necessary dimensions of the passages are attained generally by the use of back vanes; but some makers of turbines effect the same object by varying the *width* of the buckets, so that the wheel-casing in cross-section appears with a kind of waist. If c_x denote the relative velocity at any section, h_x the mean depth of the section below the inflow, p_x the hydrostatic pressure in that section, then

$$\frac{c_1{}^2}{2g} + p_1 + h_x = (1 + \zeta_x)\frac{c_x{}^2}{2g} + p_x.$$

Since p_1 must be equal to p_x, there follows—

$$\frac{c_x{}^2}{2g} = \left(\frac{c_1{}^2}{2g} + h_x\right)\frac{1}{1 + \zeta_x}$$

where ζ_x is the co-efficient for loss by friction in the passage as far as the section in question.

When c_x is determined, the corresponding area A_x of the passage can be calculated from

$$c_x A_x = c_1 A_1.$$

The friction may safely be neglected, and then

$$\frac{c_x}{2g} = \frac{c_1^2}{2g} + h_x.$$

A_x can be ascertained for a number of points sufficient to enable the form of the passage to be drawn.

The depth x_2 of the passage at the outflow is of course fixed by the angle a_2, the radius r, and the thickness and number of the vanes—

$$A_2 = e_2\, x_2$$

whence $$c_2 = \frac{A_2}{x_2}.$$

Assuming that the pressure both at inflow and outflow is atmospheric, the proper ratio $\frac{A}{A_2}$ is settled by the condition that the co-efficient of flow must be the same as for an impulse turbine working with the same head, hence

$$K_1 = 0.95$$

or $$\sqrt{\frac{\epsilon}{2\dfrac{A}{A_2}\, sin\ a\ sin\ a_2}} = 0.95.$$

Taking $\epsilon = 0.81$

$$\frac{A}{A_2} = \frac{0.81}{0.9025 \times 2\ sin\ a\ sin\ a_2}.$$

If $a = a_2 = 75°$, then

$$sin\ a\ sin\ a_2 = 0.9331$$

and $$\frac{A}{A_2} = \frac{0.81}{1.684} = 0.48.$$

When the difference between the heads in times of flood and under the usual conditions is considerable, then, in order to satisfy the conditions for inflow without shock, and at the same time maintain the same speed of the wheel—

$$0{\cdot}95 \sqrt{2gH} = K_1 \sqrt{2gh}$$

where H is the effective head when the wheel works as an impulse turbine, while h is the actual head when the wheel is to run full. It is obvious that K_1 only equals $0{\cdot}95$ when $H = h$.

As h is usually less than H, K_1 is generally somewhat greater than $0{\cdot}95$, and the pressure at the inflow becomes slightly less than atmospheric pressure.

With the usual variations of head occurring in practice, it is sufficiently accurate to assume a constant (atmospheric) pressure throughout the buckets.

Turbines of this class, in which the cross-section of the wheel-casing is contracted, in order to vary the width, are very liable to leakage between the guides and wheel, and this fault it is very difficult to avoid.

CHAPTER VIII.

IMPULSE TURBINES (*continued*).

Centrifugal action in parallel-flow turbines.—In reaction turbines.—Factors determining centrifugal action.—Centrifugal action in impulse turbines.—Explanation.—Constructions for correcting centrifugal action in impulse turbines.—Outwardly inclined buckets. —Inclined vanes.—Radial impulse turbines.—Distinction between action of water in axial and radial wheels.—Inward and outward flow.—The guide apparatus.—The wheel.—The number and thickness of vanes in impulse turbines.—Influence of dimensions on the efficiency of impulse turbines.—Influence of angles of vanes.—On the difference between impulse and reaction turbines.—Comparison of reaction with impulse turbines.—The Poncelet wheel.—Compound turbines.—Bearings for turbine shafts.—Friction of turbine shafts.— Hydropneumatization.—Governors.—Summary.

The Action of Centrifugal Force in Parallel-flow Turbines.

THE *so-called* effect of centrifugal force on the water passing through an axial turbine is a point on which misconceptions may easily occur and actually have arisen.

The case of reaction and impulse turbines in relation to this subject must be treated separately, and that of reaction wheels taken first.

Let *a b*, Fig. 113, in elevation and plan represent the absolute path of a filament of water during its passage through the buckets, following the outer surface of the

wheel-casing. In plan, it will be seen, the water is compelled to follow a circular path *a b*, and in consequence

Fig. 113.

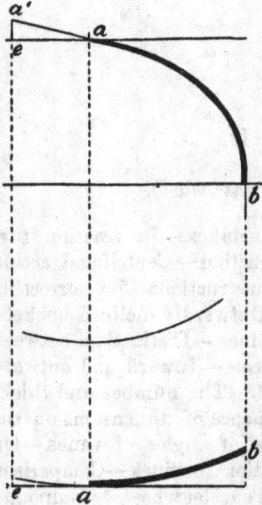

will exert a certain normal force
against the inner side of the
outer wheel-casing, due to the
deflection from its original course
e a at *a*. The amount of this
force for any particle of water is
calculated by the usual formula
for centrifugal action, and depends upon the velocity in the
circular path and the radius of
the latter. The circumferential
velocity at any point of the path
is the component of the absolute
velocity in the direction of the
tangent to the circle at that
point; for instance, the absolute
velocity at *a*, where the water
enters the wheel, is represented
by *a a*¹ (in elevation), and the horizontal projection *a e* of
this is the circular velocity. As each particle of water
advances from *a* to *b* along the absolute path, the absolute
velocity decreases, and the direction of its motion becomes
more nearly parallel to the axis of the wheel, until at *b* it
is or should be actually parallel with the axis. It is easy
to perceive that under these circumstances the horizontal
projection, and therefore the circumferential velocity of
the water, decreases in passing from *a* to *b*, until at the
latter point it is *nil*.

The centrifugal action is dependent on the *absolute
velocity* and *direction* of the water traversing the buckets,
on the distance from the axis and on the length of the
arc passed through. If, for example, the absolute path

were a straight line parallel with the axis, it is clear
there would be *no* centrifugal action, while if the water
were simply carried round in the buckets at the velocity
of the latter, without leaving the wheel, it would be a
maximum. As the outer filaments of water are forced by
the surface of the wheel-casing to follow a circular path,
so successive inner filaments are compelled by the surrounding contiguous zones of water to take a concentric
course. The resultant pressure exerted against the outer
wheel-casing is the sum of the action of all the zones of
different radii. Under ordinary conditions the centrifugal
force exerted against the outer circumference of the wheel
is only connected in a secondary manner with the velocity
of rotation, in so far as this affects the absolute velocity
of flow. The question next arises as to the effect of the
centrifugal pressure on the performance of the wheel. It
might be supposed that as a pressure between the particles
of water, increasing towards the outer circumference, is
set up, this pressure would modify in some way the
velocity of flow, and consequently the power exerted by
the motor. It is easy to show that on general grounds
this cannot be the case. In the first place, since all the
energy developed comes from that contained in the water,
it is evident that the existence of centrifugal pressure—
itself a result of the motion of the fluid—cannot produce
any *increase* of absolute velocity, which would mean a
creation of energy. In the second place, as the centrifugal
pressure is in a radial direction in which no motion occurs,
no energy is expended on work done by that pressure,
and therefore no diminution of velocity (or energy) can
result.

Centrifugal Action in Parallel-flow Impulse Turbines.

In impulse wheels where free deviation is allowed, only the outer zone of water next the circumference is compelled to take a circular path; nearer the centre the various filaments are free to follow their own course, which will naturally be in a plane, or seen in plan—the axis of the wheel being vertical to the picture plane—a straight line.

Fig. 114.

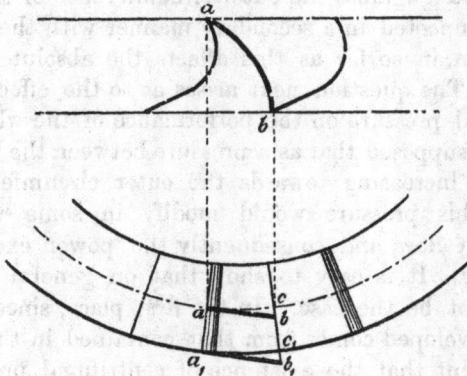

If $a\,b$, Fig. 114, represent the absolute course of a particle of fluid entering the wheel at the mean diameter at a, then—seen in plan—it will try to follow the straight line $a\,b$, instead of the circular path $a\,c$, at the mean radius; consequently, if quite unrestricted, on leaving the vane it will be further from the centre by the distance $b\,c$ than when it entered the wheel. The result of this action is that the water tends to heap itself up at the outer circumference. The extent to which this occurs depends on the absolute velocity, the form of the absolute path, and the diameter of the wheel. A particle of water entering at a_1 on the outer circumference would, if free, arrive at b_1 by the time it left the vane; it is, however, compelled to follow the cylindrical surface of the wheel-casing, and so in reality leaves at c_1. The tendency of the water to heap itself up is therefore evidently proportional to $b_1\,c_1$, and increases as the diameter diminishes. In impulse turbines the absolute velocity c with which the water enters is quite independent of the speed of rotation, but the latter influences subsequent values through its effect on the horizontal component and affects the length of the absolute path; on this length the centrifugal action in question is entirely dependent for any given diameter and absolute velocity of inflow.

With vanes of which the surfaces are helical and the edges radial, the whole centrifugal tendency is not limited to that which has just been considered, but is augmented from another cause.

Each particle of water enters the buckets in a direction which lies in a plane at right angles to the surface and upper or inflow edge of the vane (seen in plan, for a vertical axis, the direction itself is at right angles to the vane edge), but a little consideration will show, that as the particle proceeds the surface of the vane is no longer perpendicular to the plane in which that particle moves, or—what comes to the same thing—to a radial line through the point with which the particle is in contact. In consequence of this angular position of the vane-surface relatively to the course of the water as seen in plan, a radial *outward* deflection of the fluid is caused, in addition to that parallel with the circumference and similar in character. The phenomenon will be best understood by reference to Fig. 115. $a\,b$ is the upper edge of a vane,

$c\,d$ the lower edge; a particle of water entering at p
moves in the direction $p\,q$ at right angles to $a\,b$; by the
time it arrives at q, however, $a\,b$ will have travelled to
$a_1\,b_1$, and a radial line $O\,x$ through q—which touches the
vane-surface throughout the whole width of the latter—
will form the obtuse angle $p\,q\,x$ with the direction of flow
$p\,q$, so that the water traversing that part of the vane will
tend to move outwards towards x. In any perfectly con-

Fig. 115.

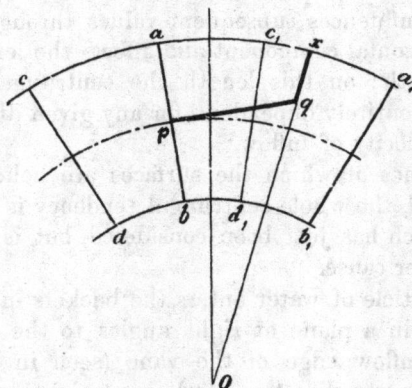

structed wheel the surfaces of the vanes should be so
formed that there is no *radial* component of the pressure
between them and the water during any part of its course.

Construction for correcting the Centrifugal Action.

When, as illustrated in Fig. 116, the upper edges of
the wheel-vanes are radial, and the absolute direction of
the water on entering is at right angles to them, or, seen
in plan, tangential to the circumference of the wheel, a

particle of fluid entering at *a* will, if unrestricted, leave
the vane at a point *b* further from the axis than *a*.   In
Fig. 116, *D D E E* represents in plan the position of a vane
when the particle of water enters at *a*, and $D_1 D_1 E_1 E_1$
that of the same vane at the moment when the particle
leaves it at *b*, by which time *a* has arrived at a_1.   The
course followed by the particle *relative* to the vane will
be $a_1 b$, and the general direction of the water at all
distances from the axis approximately parallel to this.

Fig. 116.

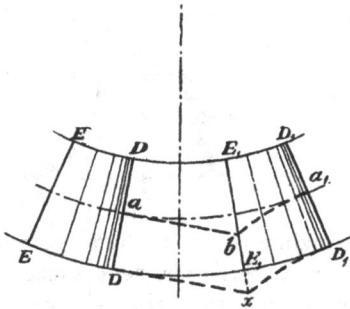

Fig. 117.

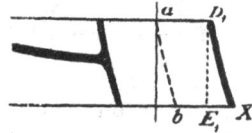

In order that the water next the outer circumference may
be able to take this course, it would be necessary to so
construct the vane and casing that a particle entering at
D_1 might leave at *x* instead of at E_1, and it is easy to see
that this requirement would be met by giving the outer
casing a larger diameter at the bottom than at the top,
somewhat as shown in section in Fig. 117.

Modern axial impulse turbines are constructed in such
a manner that in flowing over a vane the stream of water
can spread itself out so as to increase its width—generally
to about two and a half times its original dimension—and

when no regard is paid to the centrifugal action now in question, the casing has a bell-mouthed section symmetrical with respect to a vertical line (parallel to the axis) through the mean circumference, *vide* Fig. 118.

To avoid the centrifugal action when the width of the stream is increased, the lower edge of each vane should be so placed that it is bisected at the point b instead of at c by the centre line $C\,C$; this gives the unsymmetrical form of casing illustrated in Fig. 119.

The line $a_1\,b$ on the vane can be very simply constructed from a drawing of the absolute path.

Fig. 118.   Fig. 119.

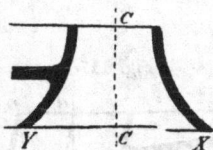
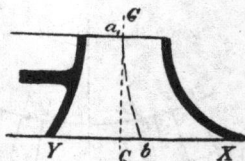

Another method of neutralizing the centrifugal action of the water is given by Meissner, which has been successfully tried in practice. This consists in giving the vane-surfaces an inclination inwards (towards the centre of the wheel) such that the relative outward tendency is counteracted by a deflection in the opposite sense, resulting from the angular position of the vane with respect to radial lines from the wheel axis. Let $A\,B$, Fig. 120, be the plan of the absolute path at the outer circumference which the water must be compelled to follow; if free, its course would be $A\,C$, and the extent to which each particle must be, so to speak, deflected inwards is represented by $A\,X$, $B\,X$ being parallel with $A\,C$ (the tangent to the circumference at A); the angle through which the deflection takes place is $X\,B\,D$, $B\,D$ being the tangent at B. The

required effect is approximately obtained by placing the
lower edge *A F* of each vane at an angle *O A F* equal to
X B D with respect to the radial

Fig. 120.

line *O A*. The upper edge *G H*
of the vane must be parallel with
the lower, so that the vane-surface
is cylindrical.[1] A result of this
is, that the length of the path
traversed by particles entering at
various points in the width of
the vanes is the same (measured
parallel with the mean circumference), and the losses arising
from internal friction between
portions of a stream moving with
different velocities are to a great
extent avoided.

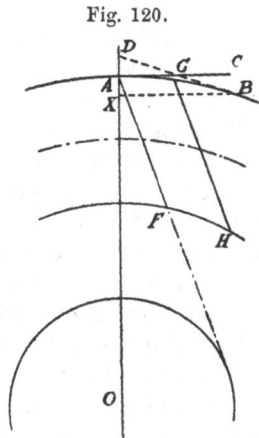

When this arrangement of the vanes of the wheel is
adopted, the guide-vanes must also be made with both
edges parallel; it is not, however, necessary that the
lower edges of the guide-vanes should be parallel with
the upper edges of the wheel-vanes, as the direction, as
seen in plan, taken by the outflowing stream is determined
by the sides of the casing. In order

Fig. 121.

to distribute more uniformly the
shocks resulting from the motion
of the wheel-vanes in front of the
guide-passage orifices, some engineers place the guide-vane edges
at an angle with those of the wheel-vanes, as shown in
the diagram, Fig. 121, where *A B* and *A B* represent the
former and *C D* and *C D* the latter.

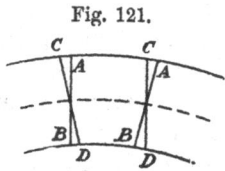

---

[1] Not necessarily circular.

Radial Impulse Turbines.

For impulse turbines the radial arrangement of vanes is generally only adopted with partial admission, and chiefly for outward-flow.

Zuppinger's tangent wheel was of the inward-flow type, designed for use with high falls and small quantities of water, but it is not now often employed, the outward-flow type being found more convenient. As a rule, turbines of this kind are made with horizontal axes, an arrangement which has some practical advantages over that with vertical axes, especially where the driving is effected by belts. Zuppinger's wheel generally has a vertical axis.

The chief distinction between the theory of axial and radial impulse turbines is that in the latter an acceleration or retardation of the relative velocity of flow occurs during the passage of the water through the buckets independently of the action of gravity. By this the relative velocity of outflow as compared with axial wheels is modified, and therefore also to some extent the relations between the vane-angles.

In other respects the construction of radial wheels is simpler, owing to the fact that the speed of rotation is constant for all points in the width of a bucket where the water enters and leaves it.

The water is admitted to the buckets of outward-flow impulse wheels with horizontal shafts at the lowest possible point of the circumference, to avoid loss of head, and enters through only a small number of guide passages, sometimes only one or two.

For Zuppinger's tangent wheels the admission is equally distributed at opposite sides, to avoid lateral pressure on the shaft and bearings.

The Guide Apparatus. The total area of the guide passages is determined exactly as for a parallel-flow wheel :

$$c = 0.95 \sqrt{2gH} \text{ to } 0.92 \sqrt{2gH}$$

$$A = \frac{Q}{0.894\,c}.$$

If ϕ denotes the proportion of the circumference over which the water enters the wheel, then approximately

$$A = \phi\; 2\; \Pi\; r_1\; e\; \cos a,$$

and from this

$$\phi = \frac{A}{2\; \Pi\; r_1\; e\; \cos a}.$$

The angle a, as before, is chosen, the value ranging between the limits previously given—for tangent wheels it is as much as 80°.

A certain ratio y between e and r_1 is selected, which in practice lies between $\frac{1}{8}$ and $\frac{1}{20}$, and then ϕ can be expressed as follows—

$$\phi = \frac{A}{2\; \Pi\; y\; r_1^2\; \cos a}.$$

Very often it is convenient to determine r_1 independently, with reference to the number of revolutions the shaft is required to make, and in that case ϕ must be calculated from the above formula, in the first instance approximately.

When the number of revolutions is not fixed, the rule $r = 1.5 \sqrt{Q}$ can be employed for tangent wheels, for outward-flow wheels $r = k \sqrt{A}$, where k is a co-efficient the value of which is given in the summary of rules and formulas.

s

The pitch P of the guide-vanes must then be selected, and the number of passages z calculated—

$$z = \frac{\phi \, 2 \, \Pi \, r_1}{P};$$

the corrected value of ϕ can finally be obtained from the accurate formula—

$$\phi = \frac{A + z \, t \, e}{e \, 2 \, \Pi \, r_1 \, \cos a},$$

where the thickness t of the vanes is empirically fixed, and the width of the passages from

$$e = y \, r_1.$$

The above formulas are the same as those used for parallel-flow wheels with partial admission, except that instead of the mean radius r appears the radius r_1 at which the water enters.

The value ϕ should never exceed $\frac{1}{4}$ of the circumference.

The Wheel. The angles a and a_2 are chosen as in the case of axial turbines. Approximately, too, the same relation between the angles obtains, namely

$$a_1 = 2 \, a - 90°.$$

This must be temporarily adopted, and c_1 and w_1 graphically determined in the now familiar way.

Having ascertained the relative velocity of inflow c_1 the relative velocity of outflow c_2 is given by formula (IVa). In this appears the depth of the buckets h_0, which, assuming that the water leaves at the lowest possible point, is for outward-flow wheels with horizontal axis about equal to the difference between the inner and outer radius, $r_2 - r_1$; this dimension is purely empiric, and varies from $\frac{1}{4}$ to $\frac{1}{8}$ of the mean radius $\frac{r_1 + r_2}{2}$.

h_o is, as previously indicated, by no means always exactly equal to the depth of the buckets, but, according to the points where the water enters and leaves the wheel, may be either greater or less. By formula (IV*a*)

$$c_2^2 = (c_1^2 + w_2^2 - w_1^2 + 2gh_o)\,\frac{1}{1 + \zeta_2'},$$

where $$\frac{1}{1 + \zeta_2} = 0.9 \text{ to } 0.95;$$

for tangent wheels with vertical axis $h_o = 0$ with respect to the middle of the buckets.

From this point on the calculation is similar to that for a parallel-flow wheel.

The ratio of the width at the outlet of the buckets to that at the inlet is

$$\frac{e_2}{e_1} = \frac{x_1}{x_2}\frac{c_1}{c_2}\left[\frac{2\,r_1\,\Pi\,cos\,a_1 - z_1\,t_1}{2\,r_2\,\Pi\,cos\,a_2 - z_1\,t_2}\right].$$

The values of this ratio adopted in practice are the same as for parallel-flow wheels.

The relation

$$sin\,a_2 = \frac{w_2}{c_2} = \frac{r_2}{r_1}\frac{w_1}{c_2}$$

gives the correct angle a_2; if this does not agree with that at first assumed, a_1 must be modified.

It should be noted that with *inward-flow* a *retardation* of the relative velocity is associated, c_2 being less than c_1.

In radial impulse wheels no correction of the angles or curves of the vanes is required; the form of the latter is therefore simpler than for axial wheels. The relative velocities of inflow and outflow are the same for all points of the width, as projected on a plane at right angles to the axis of rotation, and the particles of fluid which enter simultaneously also leave the buckets together.

For these reasons, where it is possible a radial outward-flow wheel should be adopted in preference to one with parallel-flow; as already stated, however, this type of motor is only applicable for partial admission, and for large quantities of water and low falls would assume unwieldy dimensions.

Notes on Vanes.

The guide-vanes of impulse turbines are generally of wrought-iron, or steel plate cast in with the casing, and

Fig. 122.

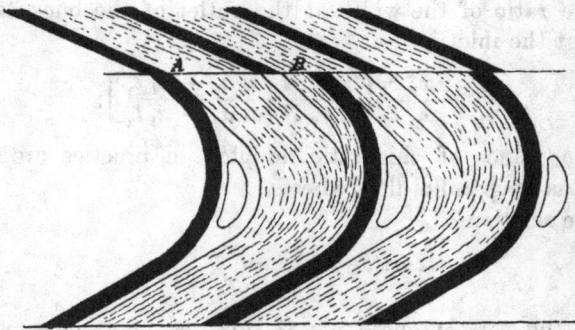

the wheel-vanes of cast-iron. It is desirable to make the vanes as thin as practicable, especially the guide-vanes, in order that they may interfere as little as possible with the flow and distribution of the water in the buckets.

Fig. 122 illustrates the nature of this interference; in practice, however, the upper edges of the wheel-vanes would be tapered off almost to a point, instead of being flat as shown. It is obvious that, as regards obstruction of the flow of water, it is advantageous to make the number of guide-vanes relatively small, but in practice there is never very much difference between the number of guide- and wheel-vanes, whichever of the views previously explained in regard to this point is adopted.

Influence of Dimensions on the Efficiency of Impulse Turbines.

For the efficiency of the wheel as distinguished from the accessory arrangements the expression was found—

$$\epsilon_1 = \frac{2 \, sin \, a \, sin \, (a \, - \, a_1)}{cos \, a_1}.$$

It was further shown that approximately

$$a_1 = 2 \, a \, - \, 90°,$$

or $\qquad a \, - \, a_1 = 90° \, - \, a \, (a_1 \text{ being negative});$

whence $\qquad sin \, (a \, - \, a_1) = cos \, a,$

and $\qquad cos \, a_1 = sin \, 2 \, a.$

Substituting these equivalents in the above formula, there follows—

$$\epsilon_1 = \frac{2 \, sin \, a \, cos \, a}{sin \, 2 \, a}.$$

The efficiency is here expressed exclusively in terms of *one* angle, and the question is—For what value of this angle is it a maximum? *sin* 2 *a* is identical in value with 2 *sin a cos a*, so that the above formula gives an efficiency of 100 per cent. whatever *a* may be. This of course is impossible, and merely a result of the approximations used in developing the expression, but it shows that, provided the conditions of inflow without shock and radial or axial outflow are observed, the value of *a* has very little influence on the efficiency.

The greater a_2, as in the case of reaction wheels, the higher must of course be the efficiency, as an increase of a_2 is accompanied by a reduced absolute residual velocity u.

On the Difference between Impulse and Reaction Turbines.

A clear understanding of the essential characteristics of a reaction as distinguished from an impulse turbine is so important for the student of the subject, that a somewhat more detailed consideration of this point is required.

It has been already explained that in reaction turbines continuity of flow [1] is a necessity, while in impulse turbines it must be avoided.

As a result of the condition of continuity the water on leaving the guide passages has a pressure p_1, which is different to the atmospheric pressure; under this pressure —neglecting the effects of leakage and other interference —it enters the wheel-buckets. When, as is now assumed, a suction-tube is employed, the height h_1 of water above the wheel is less than the total head h, and if the water issued from the guide passages at atmospheric pressure (as in an impulse turbine), the whole energy available would be represented by the velocity of outflow only, due to the head h_1, and the difference $h - h_1$ the height of the outflow orifices above the tail-water would be all *lost*. In a reaction turbine, the water generally passes through the wheel-buckets under a decreasing pressure, and in that

---

[1] It must be noted that the author uses the term "continuity of flow" to express the fact that the flowing water is always in contact with all the internal surfaces of the passages. In a general sense there must of course be continuity of flow in *all* turbines.

case the energy available is represented, not merely by the velocity, but in addition by the difference between the pressure of the water leaving the guide passages and that leaving the wheel-buckets. The available energy is therefore at the moment of leaving the guide passages made up—so to speak—partly of velocity and partly of pressure, and during the flow of the water through the wheel-buckets the energy of pressure becomes converted into energy of velocity owing to the acceleration, relative to the buckets, produced by the excess of pressure $p_1 - p_2$.

It must be distinctly understood that the hydrostatic pressure *as such* produces no effect in driving round the wheel, but only when converted into *velocity relative to the wheel.* The language of some authors on this point appears misleading, and an unnecessary distinction is drawn between the action of the fluid on the vanes in reaction and in impulse turbines.

As far as the force exerted on the vanes is concerned, the *mode* of action is precisely the same in both systems, but while in impulse turbines—apart from friction and any difference of level of the entrance and exit orifices of the buckets—there is no accelerating (or retarding) force affecting the water during its passage between the vanes, the contrary is true of reaction wheels.

The acceleration produced in the relative velocity intensifies the effort exerted by the fluid on the vanes, and, under otherwise equal conditions, increases the work done in turning the wheel.

The pressure p_1 at which the water leaves the guide passages may be either greater or less than atmospheric pressure; when it is *less* the velocity of outflow is of course greater than in the former case, and may represent an amount of energy *in excess* of that due to the total head, but under these conditions the pressure *increases* as

the fluid passes through the buckets, consequently $p_1 - p_2$
is negative, and the available energy is *less* than that due
to the velocity of outflow from the guide passages.

Where no suction-tube is used and the off-flow into the
tail-water takes place direct from the wheel, p_2 equals
atmospheric pressure, and the energy available for the
wheel is $\frac{c_2}{2g} + p_1 - a$. The *absolute* velocity u with
which the water leaves the wheel then represents the lost
energy not absorbed in the turbine. When, however, there
is a suction-tube, this loss is expressed by $\frac{u^2}{2g} + p_2 - a$,
and in many cases p_2 is less than a, and hence the energy
wasted is not so great as that due to the velocity u; this
occurs when the final velocity of off-flow c_4 under the
sluice is smaller than u, as the former must be the measure
of the loss from residual energy, and therefore

$$\frac{u^2}{2g} + p_2 - a = \frac{c_4^2}{2g}.$$

This explains the action of an enlarged suction-tube
or a Boyden diffusor in improving the efficiency of a
turbine; the principle involved may be stated by saying
that what is lost in velocity is gained in pressure. The
excess of the velocity u over c_4 is compensated by p_2 being
lower than a, and consequently a greater available energy.
It has been throughout assumed for the sake of simplicity
that there are no frictional or other resistances apart from
the work done on the wheel.

The velocities and pressures, it is scarcely necessary to
repeat, depend, under given conditions as to head of water
and position of wheel, entirely on the proportions of the
different parts of the apparatus; it is therefore not necessary
in practice to determine the pressures p_1 and p_2 as such.

Comparison of Reaction with Impulse Turbines.

In comparing one system of turbine with another it is not merely the maximum possible efficiency under the most favourable circumstances which should, for practical purposes, be the standard according to which their relative merits are estimated, but rather the average performance under the varying conditions occurring in practice, and the capacity of any motor for adapting itself with the least possible disadvantage to such variations.

Reaction turbines give a somewhat higher efficiency than impulse turbines when running at the best speed and with the full quantity of water for which they are designed, and have the further advantage that they can be used with a suction-tube and placed within certain limits at any convenient height above the tail-water.

On the other hand, all the practicable methods of regulating reaction wheels to suit varying quantities of water involve with the decrease of the latter a reduction of the efficiency, and consequently these motors are only suited for cases in which the quantity of water is tolerably constant, or at any rate does not fall below that required by the wheel when working at its best speed.

For high falls the dimensions of reaction turbines become very diminutive, and this necessitates a very large number of revolutions in order to obtain the requisite velocity at the circumference, while, when in addition to a high fall the quantity of water is small, it becomes impossible to use a reaction wheel.

The maximum efficiency obtainable with impulse turbines is not so high as in the case of reaction turbines, but—as has been previously fully explained—they can be advantageously regulated to almost any extent for varying quantities of water, and the dimensions are almost independent of the latter, so that this system of motor can be applied under nearly any conditions, either for low falls and great quantities of water, or for high falls and small quantities of water.

Owing to their great practical advantages in these respects, impulse turbines are very largely used on the continent of Europe, and have to a great extent superseded reaction turbines, since, in so very many instances where hydraulic power is available, the quantity of water is changeable with the seasons.

In America and Canada reaction turbines appear to be chiefly employed, as the practically unlimited supplies of water in the great rivers in those countries are favourable to their use.

Whether, in any given case, a reaction or an impulse turbine is appropriate, must depend on the particular circumstances—each system has its own sphere of usefulness ; the actual results obtained may be studied from the particulars of experiments given in the sequel.

Where the available *head* of water varies, a Jonval wheel may be advantageously used by dividing it into several concentric parts, and, according as the head is greater or less, using the inner or outer rings so as to obtain the appropriate speed.

A radial reaction turbine is obviously not adapted to this method of regulation for varying falls, since the velocity of all points at which the water enters is the same.

There is, of course, no hard-and-fast line at which the employment of a reaction turbine must cease and the application of an impulse wheel begin. The first Fourneyron turbine ever made, of which particulars are given elsewhere (Chap. XI.), was constructed for a high fall,

with small diameter, and to run at an enormous number
of revolutions per minute, but this is quite contrary to
present practice.

The " Poncelet " Water-wheel.

Although generally classed as a water-wheel in the
narrower sense of the term, the " Poncelet " wheel is,

Fig. 123.

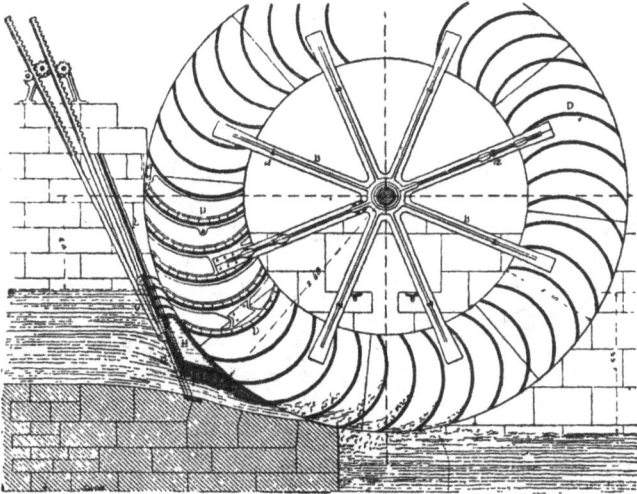

according to the definition given, in reality a turbine.
The water, *vide* Fig. 123, enters at the bottom as in an
undershot wheel, but the vanes, instead of being straight,
are curved in such a manner that the inflow takes place
without shock, and the water in consequence of its velocity
flows up a certain height along the vanes, until, brought
relatively to rest by the action of gravity, it falls again,

still following the vane surface. Both on its upward and downward course the fluid exerts pressure on the vanes as a result of the deflection caused by their curvature; to a slight extent, too, the water acts by its weight on the buckets, but this action is very insignificant in comparison with that due to the relative velocity.

The sluice under which the water is admitted to a

Fig. 124.

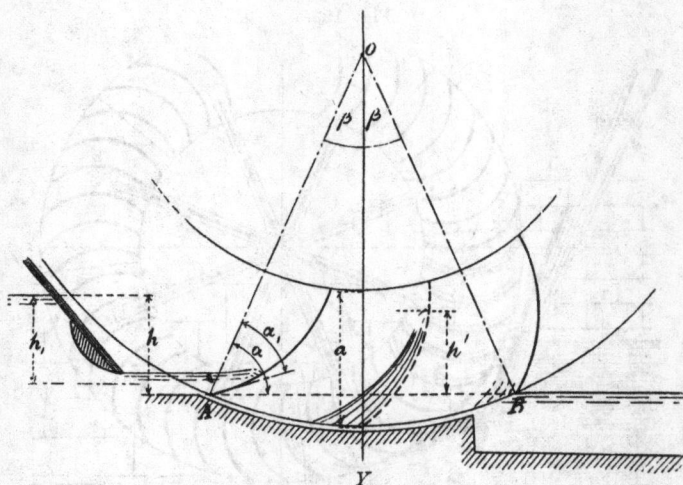

"Poncelet" wheel is generally inclined at an angle of from 40° to 60° to the horizon.

The axis, as shown by the illustration, is horizontal the diameter is comparatively great, and the velocity of rotation slow, as in ordinary water-wheels.

It is clear that, if none of the available head is to be lost, the water after attaining the highest point in the buckets should fall through exactly the same height to which it has risen before reaching again the tail-race level. This end is obtained by so choosing the point *A*, Fig. 124, at which the water enters, with reference to the speed of the wheel and velocity of the water relative to the vanes, that the fluid leaves the buckets at a point *B*, equi-distant with *A* from a vertical line *O Y* through the centre of the wheel, whence *A Y = B Y*, and *A O Y = < B O Y = β*.

The water enters the wheel from under the sluice in a horizontal direction, forming the angle *a = O A B*, with the radius *O A*, while *β = 90° − a*.

The velocity of flow from under the sluice is

$$c = 0\text{·}922 \sqrt{2gh_1},$$

where h_1 is the effective head measured to the centre of the sluice orifice.

The conditions for inflow without shock are the same as for a turbine.

If h_1 denote the height above the tail-race level to which the water rises in the buckets, then

$$\frac{c_1^2}{2g} = h_1;$$

and, neglecting friction,

$$c_2 = c_1,$$

since the water falls through the same height to which it has risen. This equation, combined with the well-known conditions for inflow without shock and radial outflow, leads to the relation

$$a_1 = \tfrac{1}{2} a.$$

It is easy to see that the depth of the buckets *a*, the difference between the outer and inner radii $r_1 − r_2$, must be at least equal to the height h^1 *plus* the depth to which the lowest point of the wheel is submerged below the tail-water.

According to Zeuner

$$r_1 \beta = 1\cdot6\, w\, \sqrt{\frac{a}{g}}\, ;$$

generally $\beta = 15°$, and consequently

$$a = 90° - \beta = 75° ; \text{ limits } 72° \text{ to } 78°.$$

If b denote the clear breadth of the wheel-buckets, and x the proportion filled by the water, then

$$Q = x\, a\, b\, w,$$
$$x = \tfrac{1}{2} \text{ to } \tfrac{2}{3}.$$

The following data may be taken as a guide in the construction of "Poncelet" wheels:

Maximum value of $b = 13$ feet.
" " $r = 10$ feet.
$P = 8$ to 10 inches; $z = 36$;
Vertical sluice opening 4 to 12 inches.

The greatest head suitable for this type of motor is about $5\frac{1}{2}$ feet.

It should be unnecessary to remark that the "Poncelet" wheel is an impulse turbine with free deviation; with it an efficiency of about 70 per cent. has been obtained.

Compound Turbines.

If a turbine is allowed to run at a much lower speed than the best, the water leaves the buckets with a very considerable absolute velocity, and there is a corresponding loss from unutilized energy. This energy might, however, be usefully employed in driving a second turbine, the water after leaving the first being deflected by a set of stationary guide-vanes to cause it to enter the second wheel at the proper angle. Both turbines could be keyed to the same shaft, and their speed would be much lower than that of a single turbine driven by the same head of water and utilizing it to the same extent. This arrangement would constitute a *compound* turbine, and it is clear that, instead of *two* wheels only, *three* or more might be employed in the same way, the speed being lower the greater the number of wheels. The only object in using a compound turbine in preference to a single one would be to reduce the speed in cases where the head was very great and a high velocity of rotation inconvenient or impracticable. The principle of the compound turbine has, in so far as the author is aware, not hitherto been applied in practice with hydraulic power, but is embodied in Mr. Parson's steam turbine, in which a very large number of wheels keyed to the same shaft are used. The velocity of rotation necessary to utilize steam efficiently in a single turbine would be impossibly high, owing to the very great velocity of flow.

With water, the losses in a compound turbine would certainly be much greater than in a single wheel, in consequence of there being a double set of guide-vanes and buckets, in *each* of which the frictional losses would be at least as great as in the single motor; notwithstanding this, it might in certain cases be desirable to adopt the compound system, in order to avoid the use of gearing and countershaft for obtaining a relatively low speed from a quick-running motor.

Bearings for Turbine Shafts.

The forms of bearings generally used for turbine shafts are the following :

(1) Subaqueous step bearing.

(2) Collar bearing above water.

(3) Overhead (Fontaine) suspension bearing.

In every case, in addition to the supporting bearing which takes the weight of the motor, there are one or more bearings or stuffing-boxes which steady the shaft laterally.

Subaqueous Step Bearings.

These require no special description, being of the usual construction, and either resting on the floor of the tail-race or supported by brackets secured to the walls of the suction-tube where one is employed. For lubrication an oil pump is necessary.

This class of bearing has the drawback that it is not easily accessible. Lignum vitæ has also been employed for step bearings, in which case no lubrication is required.

In at any rate one American turbine the shaft runs on the conical end of a cylindrical block of lignum vitæ held in a socket; the end of the shaft is of course recessed to fit the taper end of the block. (*Vide* illustration of "Risdon" turbine.)

Collar Bearings.

With these a turbine is suspended from the upper end of the shaft, which is guided lower down in a stuffing-box to insure lateral stability. In some cases the bearing rests in a spherical socket to allow greater freedom of adjustment. The same rules are applicable to collar bearings whether used for turbines or other purposes, the bearing surface being proportional to the load on the shaft.

Overhead (*Fontaine*) Suspension Bearings.

With these the load is supported on the upper end of
a spindle of steel or iron rigidly fixed in a socket of cast-
iron on the tail-race floor The turbine wheel itself is
keyed to a hollow cast-iron shaft which at its upper end
carries a kind of lantern. Through the top of this lantern,
which is formed as a boss, passes a short spindle, often
continued upwards to form the driving shaft, the lower
end of which rests on a loose steel washer, sometimes with
spherical base, running on a bearing rigidly attached to
the top of the fixed spindle previously referred to. The
position of the short spindle, which is screwed on the
outside, is adjustable by means of a large nut; a feather
key prevents it from turning relatively to the top of the
lantern. The top of the lantern is attached to the body
by bolts passing through flanges. The bearing is easily
accessible through the openings in the sides of the lantern.
The construction will be readily understood from the
illustrations, Figs. 125 and 126.

In Fig. 125, *D* is the cast-iron lantern, *G* the flanged
top in the boss *I* of which is secured the spindle or shaft
b by means of the feather key *R*; *a* the nut with flange
k fitted on the screwed portion of the spindle *b* by which
the position of the wheel is vertically adjusted; *d* a loose
washer of gun-metal between the bearing surface of the
spindle and the steel bearing-plate *e*; *f* a cup of cast-iron
containing oil for lubrication, having at its lower end a
cylindrical neck *h* fitting into a socket in the upper end
of the fixed spindle *i*. The hollow cast-iron shaft just
below the lantern is lined with a long gun-metal bush *o*
fitting the spindle *i*. The plate *e* is prevented from
turning by a pin *h*, while the cup *f* is held by a cotter *g*.
In the under side of the washer *d* are radial grooves to

T

Fig. 125.

Fig. 126.

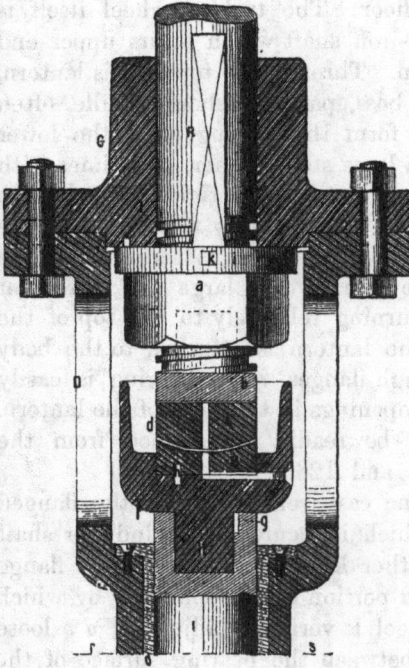

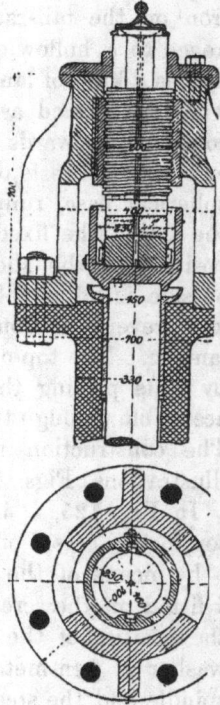

admit the oil. The lower end of the bearing spindle *b* is loose and fitted into a cylindrical recess in the end of the screwed portion; this is in order to allow the nut to be fitted on. The spindle *b* should be of steel. It is unnecessary to make the under surface of the washer *d* spherical, as shown, and this is now not generally done.

The pressure on the bearing surfaces may range from 1,100 to 1,400 pounds per square inch.

For the diameter of bearing spindles for turbines Meissner gives the following formulas:

$$\text{For } \textit{wrought-iron } d = \sqrt{\frac{P}{2,750}} + 0\cdot2 \text{ inch};$$

$$\text{For } \textit{steel } d = \sqrt{\frac{P}{7,640}} + 0\cdot12 \text{ inch};$$

where *d* is the diameter of the spindle in inches, and *P* the load on it in pounds.

Fig. 126 illustrates the construction of the overhead bearings employed for two double Jonval turbines at Schaffhausen, recently erected by Messrs. J. J. Rieter and Co. of Winterthur. A description of these will be found in Chap. XI.

The bearings in question differ somewhat in detail from that shown in Fig. 125, but the design is essentially the same.

Shaft Friction of Turbines.

Mr. J. C. Bernhard-Lehmann published in 1879 [1] the results of experiments carried out by him on the friction

[1] *Zeitschrift des Vereins Deutscher Ingenieure*, 1879, vol. xxiii. 121.

of turbine shafts, including the resistance of all the bearings and stuffing-boxes, &c.

To determine this frictional resistance a cord was passed round the driving-wheel or pulley on the turbine shaft, and to the free end weights were attached until the turbine was set slowly in motion. In this way the co-efficient of friction when the wheel was nearly at rest could be determined, and the same co-efficient was employed to calculate the power absorbed by friction at the usual speed of the motor. As the co-efficient of friction for shafting at rest is in nearly all cases greater than when it is in motion, the results thus obtained by Mr. Lehmann err if anything on the safe side.

The following is a summary of the conclusions arrived at:

(1) For Axial Turbines with vertical solid shaft, ordinary footstep, and leather collar in the guide-wheel cover, with bearings from 1 to 2·8 inches diameter—

Loss by friction, with full admission, 1·4 to 2·4 per cent.

" half " 2·3 " 3·6 "

(2) For Axial Turbines with vertical hollow shaft Fontaine's overhead bearings, and no leather collar, with bearings from 3·2 to 4·8 inches diameter—

Loss by friction, with full admission, 1·5 to 3·2 per cent.

" half " 2·3 " 5·4 "

(3) For Axial Turbines with vertical solid shaft, collar bearings, and leather collar in the guide-wheel cover, with bearings from 1·6 to 2·8 inches diameter—

Loss by friction, with full admission, 2·1 to 3·4 per cent.

" half " 2·7 " 4·7 "

(4) For Radial Outward-flow Turbines with vertical solid shaft, and step bearings without stuffing-boxes, with bearings from 1·8 to 3·6 inches diameter—

Loss by friction, with full admission, 0·8 to 1 per cent. ; with bearings from 1·8 to 2·4 inches diameter—

Loss by friction, with half admission, **0·5** to **0·9** per cent.

(5) For Radial Outward-flow Turbines with vertical hollow shaft and Fontaine's overhead bearings, with bearings from 2 to 3·2 inches diameter—

Loss by friction, with full admission, **0·9** to **1·2** per cent.

(6) For Radial Inward-flow Turbines with vertical solid shaft, step bearing with stuffing-box and leather collar in guide-wheel cover, for bearings from 2 to 2·8 inches diameter—

Loss by friction, with full admission, **0·9** to **1·1** per cent.

(7) For Radial Inward-flow Turbines with vertical solid shaft, collar bearings, and without leather collar, for bearings from 2·4 to 3·2 inches diameter—

Loss by friction, with full admission, **1·3** to **1·7** per cent.

(8) For Radial Outward-flow Impulse Turbines with horizontal axis and ordinary bearings, with bearings from 1·6 to 3·6 inches diameter—

Loss by friction, **1** to **1·6** per cent.

A few experiments with bearings of lignum vitæ showed that with this material the friction was considerably higher than with metal.

All the experiments were undertaken with the motors in their ordinary condition and with the usual lubrication, while the weight of the water in the wheel, and, for axial turbines, its downward pressure on the vanes, was taken into account in calculating the total friction.

The pressure on a parallel-flow turbine tending to increase the shaft friction is made up of two factors, (1) the excess of pressure $(p_1 - p_2)$ at the inflow acting on the horizontally measured area of the buckets, and (2) the force resulting from the deflection of the stream $\left(Y = \frac{1}{g} \ (c \ cos \ a \ - \ c_2 \ cos \ a_2) \ \text{for the unit of weight} \right)$.

In radial-flow turbines this source of increased friction is absent.

Hydropneumatization.

To enable impulse turbines to be used with very low heads, which would otherwise be impracticable on account of the very small depth of water left above the guide-apparatus, Girard devised a process to which he gave the above somewhat cumbrous title.

This process consists in enclosing the turbine in a casing dipping below the surface of the tail-water and pumping in air until the level is sufficiently depressed to allow the necessary clearance for the wheel, while keeping the lower edge of the latter at or below the original tail-race level. In this way it is possible to maintain a depth of water over the turbine which would otherwise be reduced by the required clearance. There is no actual gain of head by this process, as the water issues from the guide orifices against a pressure which exceeds that of the atmosphere in proportion to the clearance obtained; the advantage achieved consists solely in the avoidance of the disturbances arising when the depth of fluid over the wheel is very small, which cause air to be drawn in with the water and interfere with its proper action. When once the tail-water level has been depressed, only sufficient air has to be further supplied to compensate for leakage.

Hydropneumatization has not been adopted in practice to any considerable extent, as it is complicated and expensive, and the power necessary to replace the air lost by leakage is very appreciable.

Girard Turbine with Draft Tube.

In Vol. III. of the *Annales du Conservatoire des Arts et Metiers* (1891, p. 291), a description is given of an interesting installation, designed with a view to combining the advantages of the draft tube with those of the impulse turbine.

It was necessary, owing to the special circumstances of the case, to place the turbine at a considerable height above the tail-water level, and this entailed the use of the draft tube. In order to employ this in connection with an impulse wheel, the column of water in the tube is allowed to sink a few inches below the lower edge of the wheel, and is maintained at that level by means of a regulating device consisting of a float in connection with a small air-valve. When the water rises too far in the tube, the valve is opened and admits air above the column, slightly increasing the pressure there, so that the column sinks again.

By this means the wheel works as an *impulse wheel in a partial vacuum,* and of course the velocity of flow from the guide passages corresponds to the increase of head due to the vacuum, and compensates for the height at which the turbine is placed above the tail-water.

The principle involved is really the same as that of hydropneumatization, only the air pressure is *less* than that of the atmosphere, and the tail-water level is raised instead of being depressed.

Governors.

Centrifugal governors when used with turbines have to be of the *indirect acting* class, that is, the governor does not itself work the regulating mechanism, but merely put it into gear with a shaft driven by the motor itself, so that the latter exerts the force required for regulation. This indirect action is rendered necessary by the very considerable power which has generally to be exerted to move the regulating mechanism of turbines. In some of these the governor acts on reversing clutches which it throws in and out of gear; in others a balanced valve is employed which admits water under pressure to either side of a piston actuating the regulating mechanism of the turbine.

The residual velocity of the water leaving the wheel has also been utilized for governing turbines by acting on an auxiliary small wheel.

When turbines are used for driving dynamos, a good governing mechanism is of great importance, while a reduction of the efficiency at part gate is of less moment than the maintenance of a constant speed.

A balanced cylindrical sluice-gate is in such instances probably the best form of regulator, used in conjunction with an inward- or outward-flow radial turbine, subdivided by partitions perpendicular to the axis.

As previously explained, a sluice on the draft tube of a Jonval turbine does not act very rapidly, and when employed in connection with a governor the construction must be such that a comparatively small movement of the sluice causes a considerable alteration of the outflow orifice.

	Large quantity of water Q and low fall H where $\frac{Q}{z}$ greater than 16 sq. ft.	Medium quantity of water and medium fall where $\frac{Q}{z}$ greater than 2 and less than 16 sq. ft.	Small quantity of water and high fall where $\frac{Q}{z}$ less than 2 sq. ft.
Q = Quantity of water in cubic feet per second	Given.		
h = Total available head in feet	Given.		
E_{hp} = Available horse-power $= Q \times h \times 0.1135$.			
a = Mean angle of outflow from guide-vanes Assumed	70° to 68°	73° to 70°	75° to 73°.
a_2 = Mean angle of outflow from wheel-vanes Assumed	70° to 66°	74½° to 73°	77° to 74°.
$\frac{A}{A_2} = \left\{ \begin{array}{l} \text{Ratio of area of guide passages to outflow} \\ \text{area of wheel-buckets, approximately} \end{array} \right\} = \frac{c_2 \cos a}{c_2 \cos a_2}$	1·5 to 0·5, usually about 1	1·5 to 0·5, usually about 1	1·5 to 0·5, usually about 1
$\frac{c_1}{c_2} = \left\{ \begin{array}{l} \text{Ratio of width of buckets at inflow to} \\ \text{width at outflow} \end{array} \right\} = 1$, but ranges to 0·75.			
c = Velocity of flow from guide passages $= K_1 \, 8.025 \sqrt{h}$	K_1 from Table I.	K_1 from Table I.	K_1 from Table I.
K_1 = Co-efficient of velocity $= \sqrt{\dfrac{1}{2\dfrac{A}{A_2} \sin a \sin a_2 + \left(1 + \left(\dfrac{A}{A_2}\right)^2 \left(c_2 + \cos^2 a_2\right)\right)}} = \sqrt{\dfrac{a}{2\dfrac{A}{A_2} \sin a \sin a_2}} = \sqrt{\dfrac{a \cos a_2}{\frac{A}{A_2} \sin (a + a_2) \sin a_2}}$, where a may be taken as 0·81.		For ordinary turbines $K_1 = 0.97$.	
w = Velocity of wheel at mean diameter $= K_2 \, 8.025 \sqrt{h}$	can be determined graphically.		
$K_2 = K_1 \dfrac{A}{A_2} \sin a_2 = \dfrac{K_1 \sin (a + a_2)}{\cos a_1}$. For ordinary turbines $K_2 = 6.64$.			
$\tan g \, a_1$ = Tangent of mean angle of inflow $= \dfrac{A}{A_2} \dfrac{\sin a_2}{\cos a} - \tan g \, a$			
a_1 can be most easily determined graphically, vide p. 135, Fig. 61.			
A = Effective area of guide passage $= \dfrac{Q}{c}$.			
r = Mean radius of wheel $= k \sqrt{A}$	$\sqrt{A}$ to 1·25 $\sqrt{A}$	1·25 $\sqrt{A}$ to 1·5 $\sqrt{A}$	1·5 $\sqrt{A}$ to 2 $\sqrt{A}$.
p = Pitch of guide-vanes	10 to 12 inches	$\dfrac{r}{3.75}$ to $\dfrac{r}{4.5}$	4½ to 6 inches.
s = Number of guide-vanes $= \dfrac{2 \pi r}{p}$.			
z = Number of wheel-vanes $= s + 1$ or $s + 2$. N.B.—s and z_1 should never be exactly the same.			
t = Thickness of guide-vanes } near ends	$\frac{1}{2}''$ to $\frac{1}{4}$ for cast-iron.		
t_2 = Thickness of wheel-vanes } near ends	$\frac{1}{8}$ to $\frac{3}{8}$ for wrought-iron.		
In the following formulæ the values of t and t_2 introduced should be those at the ends; where the ends are rounded off or taper, it is safest to take the full thickness without allowing for this.			
e = Width of guide passage $= \dfrac{A}{2 \pi r \cos a - \pi t - z_1 t_1 \frac{\cos a}{\cos a_1}}$	$\dfrac{r}{1.25}$ to $\dfrac{r}{3}$	$\dfrac{r}{2}$ to $\dfrac{r}{3.5}$	$\dfrac{r}{2.5}$ to $\dfrac{r}{3}$
x = Depth of guide passage perpendicular to direction of flow measured from drawing			
x_1 = Depth obstructed by upper end of one wheel-vane, measured from drawing.			
e = Width of guide passage $= \dfrac{A}{s \, x - x_1 \, z_1}$, where x and x_1 can be measured from drawing.			
A_2 = Outflow area of wheel-buckets $= \left(\dfrac{A}{A_2}\right) A$	$\left(\dfrac{A}{A_2}\right)$ as determined above.		
c_1 = Width of wheel-buckets at inflow $= c + \frac{1}{16}''$ to $c + \frac{1}{8}''$.			
t_2 = Thickness of wheel-vanes at outflow	Same limits as for t_1.		
c_2 = Corrected width of buckets at outflow $= \dfrac{A_2}{z \pi r \cos a_2 - z_1 t_2}$			
$e_2 = \dfrac{A_2}{z_1 x_2}$, where x_2 can be measured from drawing.			
x_2 = Depth of outflow passage from bucket perpendicular to direction of flow measured from drawing.			
If e_2 as calculated differs too much from e_2, the thickness or number or both of the vanes can be modified.			
h_v = Depth of guides and wheel measured parallel with the axis, generally alike	$\dfrac{r}{3}$ to $\dfrac{r}{6}$	$\dfrac{r}{4}$ to $\dfrac{r}{6}$	$\dfrac{r}{2.5}$ to $\dfrac{r}{6}$

II.—RADIAL-FLOW REACTION TURBINES.

	Radial and mixed inward Flow.	Radial Outward Flow.
Q = Quantity of water in cubic feet per second .... Given	2·4 to 262 and over.	2·5 to 350 and over.
h = Total available head in feet .... Given	3 to 80	3 to 26*.
a = Mean angle of outflow from guide-vanes .... Assumed	50° to 66°	75° to 66° and less.
a_1 = Mean angle of outflow from wheel-vanes .... Assumed	80° to 66°	30° to 70° and less.
$\dfrac{A}{A_2}$ = Ratio of area of guide passages to outflow area		
of wheel-buckets = approximately $\dfrac{c_1}{c_2}\dfrac{\cos a}{\cos a_1}\dfrac{r_1}{r_2}$ .... Assumed	1·6 to 0·6 In mixed-flow wheels r_2 is the mean value.	0·5 to 1.
$\dfrac{r_1}{r_2}$ = Ratio of radius at inflow to radius at outflow .... Assumed	2 to $\frac{4}{3}$	$\frac{3}{4}$ to $\frac{4}{5}$.
$\dfrac{b_1}{b_2}$ = Ratio of width of buckets where water enters to width where it leaves, calculated approximately $\dfrac{A}{A_2}\dfrac{\cos a_1}{\cos a}\dfrac{r_2}{r_1}$ to insure reasonable proportions	Where, as often happens in American wheels, the outflow takes place along a curved orifice, c_2 must be measured parallel with the curve at right angles to direction of flow.	
c = Velocity of flow from guide passage .... $= K_1\, 8·025\sqrt{h}$	K_1 from Table No. I.	K^1 from Table No. I.
K_1 = Co-efficient of velocity $= \sqrt{\dfrac{1}{2\dfrac{A}{A_2}\dfrac{r_1}{r_2}\sin a \sin a_1 + c_1 + \left(\dfrac{A}{A_2}\right)^2 (c_1 + \cos x_2)}} = \sqrt{\dfrac{c}{2\dfrac{A}{A_2}\dfrac{r_1}{r_2}\sin a \sin a_1}} = \sqrt{\dfrac{c\,\cos a_1}{2\sin(a+a_1)\sin a}}$ for ordinary turbines $K_1 = 0·725$ to 0·84.		
w_1 = Velocity of wheel at diameter where water enters $= K_2\, 8·025\sqrt{h}$.		
K_2 = Co-efficient of speed .... $= K_1\dfrac{r_1}{r_2}\dfrac{A}{A_2}\sin a_2 = K_1\dfrac{\sin(a+a_1)}{\cos a_1}$, for ordinary turbines $K_2 = 0·64$.		
$\tan g\, a_1$ = Tangent of mean angle of inflow .... $= \dfrac{r_1}{r_2}\dfrac{A}{A_2}\dfrac{\sin a_2}{\cos a} - \tan g\, a$ or a_1 can be more conveniently determined graphically, vide p. 135, Fig. 61.		
A = Effective area of guide passages .... $= \dfrac{Q}{c}$.		
r_1 = Radius of wheel where water enters .... $= k\sqrt{A}$	0·75 $\sqrt{A}$ to 1·75 $\sqrt{A}$.	1·5 $\sqrt{A}$ to 2 $\sqrt{A}$.
r_2 = Radius of wheel where water leaves .... $= \left(\dfrac{r_2}{r_1}\right)r_1$		
P = Pitch of guide-vanes	12 to 4½ inches.	$\dfrac{r}{16}$ to $\dfrac{r}{6}$.
z = Number of guide-vanes .... $= \dfrac{2\,r_1\,\Pi}{P}$		
z_1 = Number of wheel-vanes.	... to 0·7 z.	1·2 z to 1·8 z.
t = Thickness of guide-vanes } near ends t_1 = Thickness of wheel-vanes }	Same as for Jonval turbines; in American wheels often nearly knife-edge.	
s = Width of guide passages $= \dfrac{A}{0·9\left(2\,\Pi\,r_1\cos a - z\,t - z_1\,t_1\dfrac{\cos a}{\cos a_1}\right)}$	Co-efficient of contraction 0·9	Co-efficient of contraction 0·9.
A_2 = Outflow area of wheel-buckets $= \left(\dfrac{A_2}{A}\right)A$.		
c_2 = Width of buckets at outflow $= \dfrac{A_2}{(3\,\Pi\,r_2\cos a_2 - z_1\,t_1)\,(0·9\text{ to }1)}$ When measurements can be taken from a drawing and s = Effective depth of guide passage perpendicular to direction of flow, s_1 = Depth obstructed by one wheel-vane, s_2 = Effective depth of outflow passage from bucket measured perpendicular to direction of flow ; s = Width of guide passages $= \dfrac{A}{0·9\,(s\,z - z_1\,t_1)}$	Whether a co-efficient in contraction is required depends on the construction of the buckets. If the outflow is axial and vanes parallel none is required.	Co-efficient of contraction is not generally required, but is in some cases necessary ; vide p. 125.
c_2 = Width of buckets at outflow $= \dfrac{A_2}{z_1\,s_2}$		

That method of estimating the areas A and A_2 should be adopted which gives the largest values of s and c_2, as being on the safe side. Where there is a considerable space between guide and wheel it appears probable, from certain experiments, that the most correct method for inward or mixed-flow turbines is to take the outflow areas as proportional to the shortest measured distance between the vanes. (Vide Chapter V.)

* The first Fourneyron turbine at St. Blaise was constructed for fall of 108 metres = about 354 feet, but this is exceptional.

III.—PARALLEL-FLOW GIRARD TURBINES, with Full Admission.

	Formula			
Q = Quantity of water in cubic feet per second	Given	176 to 530 cubic feet	53 to 176 cubic feet	35 to 53 cubic feet.
h = Total available head in feet	Given	1·96 to 7 feet	5 to 26 feet	26 to 40 feet.
a = Mean angle of outflow from guide-vanes	Assumed	60° to 66°	66° to 72°	72° to 75°.
a_1 = Mean angle of outflow from wheel-vanes	Assumed	62° to 68°	65° to 73°	74° to 77°.
H = Effective head for producing velocity of flow	$h - h_f$ (Losses before entering guide-passages).			
h_f = Height of orifices of guide passages over tail-water level	About 1 foot.			
L = Losses before entering guide passages	$= \zeta \frac{c^2}{2g}$, where ζ is made up of various co-efficients of friction, area-ration, &c.			
c = Velocity of outflow from guide passages	$= 0.95 \sqrt{2gH}$			
A = Area of outflow from guide passages	$= \dfrac{Q}{0.89 c}$			
r = Mean radius of wheel to centre of buckets		1·25 to 1·5 $\sqrt{A}$	1·5 to 1·75 $\sqrt{A}$	1·75 to 2 $\sqrt{A}$.
P = Pitch of guide-vanes at mean diameter	$= \dfrac{2\Pi r}{P}$	$\frac{r}{10}$ to $\frac{r}{11}$	$\frac{r}{9}$ to $\frac{r}{10}$	$\frac{r}{8}$ to $\frac{r}{9}$
s = Number of vanes in guide apparatus		60 to 66	54 to 60	45 to 54.
t = Thickness of guide-vanes at outflow	of wrought iron or steel	$\frac{1}{4}''$ to $\frac{3}{8}''$	$\frac{3}{8}''$ to $\frac{1}{2}''$	$\frac{1}{2}''$ to $\frac{3}{4}''$.
e = Clear width of guide passages	$= \dfrac{A}{2\Pi r \cos a - st}$	$\frac{r}{4}$ to $\frac{r}{4\cdot5}$	$\frac{r}{4\cdot5}$ to $\frac{r}{5\cdot5}$	$\frac{r}{5\cdot5}$ to $\frac{r}{6}$
$y = \frac{c}{r}$ assumed ; $r = \sqrt{\dfrac{A}{2\Pi y \cos a}}$		$\frac{1}{4}$ to $\frac{1}{4\cdot5}$	$\frac{1}{4\cdot5}$ to $\frac{1}{5\cdot5}$	$\frac{1}{5\cdot5}$ to $\frac{1}{6}$
a_1 = assumed ; $r = \dfrac{A}{2\Pi y \cos a}$	$2a - 90°$ .... negative	30° to 42°	42° to 54°	54° to 60°.
u = Mean angle of rotation at mean radius or by graphical construction	$= c \dfrac{\sin(a - a_1)}{\cos a_1}$			
c_1 = Relative velocity of inflow	$= c \dfrac{\cos a}{\cos a_1}$			
h_1 = Depth of wheel	$= c \dfrac{\cos a}{\cos a_1}$	$\frac{r}{5}$ to $\frac{r}{5\cdot5}$	$\frac{r}{5}$ to $\frac{r}{5\cdot5}$	$\frac{r}{4}$ to $\frac{r}{5}$
c_2 = Relative velocity of outflow	$= \sqrt{c_1^2 + \frac{2gh_1}{1 + \zeta_1}} = 0.95$ to $0.97 \sqrt{c_1^2 + 64\cdot4\,h_1}$			
$\sin a_1 = \frac{w}{c_2}$, c_2 determined accurately from this, and if the agreement with the assumed value is not sufficiently good, c_2 corrected.	Graphical construction of a_1 more convenient.			
$\left(\frac{A}{c}\right)$ = Width of buckets at outflow	approximately $\dfrac{z_1}{z_2} \sin 2\,dm\,a_1$	2·2 to 2·5	2·5 to 3	2·7 to 3·5
z_1 = Proportion of depth of bucket filled at inflow	$= 0.8$ to 1.			
z_2 = Proportion of depth of bucket filled at outflow	$= 0.5$ to 0.75.			
s_1 = Number of wheel-vanes	= either rather less or rather greater than s.	56 to 60	50 to 56	40 to 50.
$\left(\frac{A}{c_2}\right)$ accurate formula	of cast iron. $= \dfrac{z_1}{z_2}\left(\dfrac{2\Pi r \cos a_1 - s_1 t_1}{}\right)$			
t_1 = Thickness of wheel-vanes at outlet	Generally the same as t.	$\frac{1}{4}''$ to $\frac{1}{2}''$	$\frac{1}{4}''$ to $\frac{3}{8}''$	$\frac{1}{4}''$ to $\frac{3}{8}''$.
e_1 = Width of buckets at inflow	$= e + \frac{1}{4}''$ to $+ \frac{3}{4}''$.			
h_2 = Depth of buckets measured parallel with axis		$\frac{r}{5}$ to $\frac{r}{5\cdot5}$	$\frac{r}{5}$ to $\frac{r}{5\cdot5}$	$\frac{r}{4}$ to $\frac{r}{5}$
h_3 = Depth of guide apparatus parallel with axis		$\frac{1}{8}$ to $\frac{1}{4}\,h_f$	$\frac{1}{4}$ to $\frac{1}{2}\,h_f$	$\frac{1}{3}$ to $\frac{1}{2}\,h_f$.

IV.—RADIAL OUTWARD-FLOW GIRARD TURBINES, *with Partial Admission.*

Q = Quantity of water in cubic feet per second	Given	0·175 to 85 cubic feet.
h = Total available head in feet	Given	25 to 1600 feet.
a = Mean angle of outflow from guide-vanes	Assumed	65° to 75°.
a_2 = Mean angle of outflow from wheel-vanes	Assumed	68° to 76°.
H = Effective head for producing velocity of flow	$= h - h_2 -$ (Losses before entering guide passages).	
h_2 = Height of orifices of guide passages above tail-water level	Assumed	about 1 foot.
L = Losses before entering guide passages	$= \zeta \dfrac{c^2}{2}$, where ζ is made up of various co-efficients of friction, area-ratios, &c.	
c = Velocity of flow from guide passages	$= 0.92$ to $0.95 \sqrt{2gH}$.	
A = Area of outflow from guide passages	$= \dfrac{Q}{0.89\, c}$.	
$y = \dfrac{e}{r_1}$ = Ratio of width of guide passages to inner radius	Assumed	$\dfrac{1}{7}$ to $\dfrac{1}{10}$.
r_1 = Radius of wheel at inflow	$= 1.3\sqrt{A}$ to $13\sqrt{A}$.	
When the number of revolutions is fixed, r_1 must be calculated after w_1 the speed of the inner circumference has been determined.		
P = Pitch of guide-vanes at inner diameter	Assumed	$1\frac{1}{2}''$ to 8".
ϕ = Proportion of the circumference over which the water enters the wheel, not including thickness of vanes }	$= \dfrac{A}{2\pi y r_1^2 \cos a}$	
s = Number of guide passages	$= \dfrac{\phi\, 2\pi r_1}{P}$.	
t = Thickness of guide-vanes at outflow	Assumed wrought iron	$\frac{1}{16}''$ to $\frac{3}{8}''$
ϕ *corrected value including thickness of vanes*	$= \dfrac{A + s t e}{2\pi r_1 e \cos a}$	
P *corrected value*	$= \dfrac{\phi\, 2\pi r_1}{s}$	
e = Clear width of guide passages	$= y r_1$	
a_1 = Mean angle of inflow	$= 2a - 90°$.	
w_1 = Velocity of rotation at inner radius r_1	$= c\, \dfrac{\sin(a - a_1)}{\cos a_1}$.	
c_1 = Relative velocity of inflow	$= c\, \dfrac{\cos a}{\cos a_1}$.	
w_1 and c_1 are best determined graphically.		
h_u = Radial depth of buckets = $r_2 - r_1$	Assumed	$\dfrac{1}{3.5}$ to $\dfrac{1}{7} r_1$.
$r_2 = r_1 + h_u$ = Radius of wheel at outflow. $\quad w_2 = \dfrac{r_2}{r_1} w_1$.		
c_2 = Relative velocity of outflow	$= \sqrt{\dfrac{c_1^2 + w_2^2 - w_1^2 + 2gh_u}{1 + \zeta_2}}$	$\dfrac{1}{1+\zeta_2} = 0.9$ to 0.95.
$\sin a_2 = \dfrac{w_2}{c_2}$. $\quad a_2$ determined accurately from this, and if the agreement with the assumed value is not sufficiently good, a_2 corrected.		
z_1 = Proportion of depth of bucket filled at inflow		0·8 to 1.
z_2 = Proportion of depth of bucket filled at outflow		0·5 to 0·75.
P_1 = Pitch of wheel-vanes; less than or equal to that of guide-vanes		$1\frac{1}{2}''$ to 3".
z_1 = Number of wheel-vanes	$= \dfrac{2\pi r_1}{P_1}$	
t_1 = Thickness of wheel-vanes at inflow	$\frac{1}{16}''$ to $\frac{3}{8}''$	
t_2 = Thickness of wheel-vanes at outflow	$\frac{1}{16}''$ to $\frac{3}{8}''$	
$\dfrac{s_2}{s_1}$ = $\dfrac{\text{Width of buckets at outflow}}{\text{Width of buckets at inflow}}$	$= \dfrac{z_1 c_1}{z_2 c_2}\left(\dfrac{2\pi r_1 \cos a_1 - z_1 t_1}{2\pi r_1 \cos a_2 - z_2 t_2}\right)$.	2·5 to 4.
Radial depth of guide apparatus	$= \frac{1}{5}$ to $\frac{2}{3}(r_2 - r_1)$.	

CHAPTER IX.

SUMMARY OF RULES AND FORMULAS AND NUMERICAL EXAMPLES.

Rules and formulas:—Reaction turbines: parallel-flow; radial flow.—Impulse turbines: parallel-flow; radial-flow.—Examples: Jonval wheel No. 1; Jonval wheel No. 2; inward-flow wheel; axial Girard turbine No. 1; turbine to work both as impulse and reaction wheel, axial No. 2; radial outward-flow impulse wheel No. 3.

NUMERICAL EXAMPLES.

Reaction Turbines.

(1) *Axial Turbines No.* 1.

AN axial turbine is required to utilize a fall of 13·5
feet and a quantity of water 210 cubic feet per second.

Data:—$h = $ **13·5** feet. $Q = $ **210** cubic feet.

Weight of water per second $G = Q\lambda = 13,109·25$ lbs.

Available horse-power

$$E_{hp} = \frac{13,109·25 \times 13·5 \times 60}{33,000} = \text{say } \mathbf{322}.$$

In Jonval turbines the width of the buckets is the same
throughout.   In the variety known as the Henschel wheel
the width where the water leaves the buckets is greater
than where it enters.

It is assumed in the present instance that the width is
to be constant, or nearly so; in that case, as has been
already shown, the ratio $\dfrac{A}{A_2}$ is fixed as soon as the angles
a and a_2 are chosen.

Assume $a = $ **75°** and $a_2 = $ **80°**

$$\frac{A}{A_2} = \frac{\cos a}{\cos a_2} = \frac{0·259}{0·174} = 1·488, \text{ say } \mathbf{1·5}.$$

From (viii), p. 128, $\dfrac{e_1}{e_2} = \dfrac{A}{A_2}\dfrac{r_2}{r_1}\dfrac{\cos a_2}{\cos a}$; for axial turbines

$\dfrac{r_2}{r_1} = 1$, hence

$\quad \dfrac{e_1}{c_2} = 1\cdot5\ \dfrac{1}{1\cdot488} = \mathbf{1\cdot008}$, practically 1, as assumed.

The velocity of flow c must next be determined :—

$$c = K_1 \sqrt{2gh} = K_1 \sqrt{2 \times 32\cdot2 \times 13\cdot5} = K_1\,29\cdot483.$$

On referring to Table I. it will be found that for

$\dfrac{r_1}{r_2} = 1$, $a_2 = 80°$, $a = 75°$, and $\dfrac{A}{A_2} = 1\cdot5$, the correspond-

ing value of $K_1 = 0\cdot535$, hence

$\quad c = 0\cdot535 \times 29\cdot483 = \mathbf{15\cdot773}$ feet per second.

The mean velocity of rotation w can now be calculated by (v)

$$w = c\,\dfrac{A}{A_2}\,sin\ a_2;$$

therefore $w = c\ 1\cdot5 \times sin\ 80° = 1\cdot5 \times 0\cdot985 \times 15\cdot946,$
$\qquad\qquad w = 1\cdot477 \times 15\cdot773 = \mathbf{23\cdot297}$ feet per second,

or $\qquad K_2 = \dfrac{w}{\sqrt{2gh}} = \dfrac{23\cdot297}{29\cdot483} = \mathbf{0\cdot8}$.nearly.

w may also be graphically determined, by the construction so frequently referred to, as one side of a right-angled triangle of which the hypothenuse c_2 and one of the angles $a_2 = 80°$ is given, *vide* Fig. 60 (p. 134).

$c_2 = c\,\dfrac{A}{A_2} = 15\cdot773 \times 1\cdot5 = \mathbf{23\cdot659}$ feet per second.

It will be instructive to compare the values so far

obtained with those resulting from other preliminary assumptions, for instance,

$$a_2 = 75° \text{ and } a = 75°.$$

In this case $\dfrac{A}{A_2} = 1$ (since $cos\ a_2 = cos\ a$); from Table I.,

$$K_1 = 0.665, \text{ hence}$$

$$c = 0.665 \times 29.483 = 19.606,$$

and $K_2 = 0.665 \times 1 \times sin\ 75° = 0.665 \times 0.966 = 0.642$; therefore $w = 0.642 \times 29.483 = 18.928$ feet per second.

From this it is evident that a very much lower speed is necessary in the second case than in the first. The head and quantity of water assumed are the same as for the "Goeggingen" turbine in Table A, for which the best experimental results are there given; on reference to these it will be seen that the experimental value of K_2 agrees very well with that last calculated, for $a = 75°$ and $a_2 = 75°$. The angles of the vanes for the turbine in question are not given, but from the correspondence in the values of K_2 it is probable that a and a_2 are somewhere near $75°$; for the sake of further comparison the calculation will be continued with the figures obtained on this assumption instead of those first proposed.

The next step is to ascertain the angle a_1; this when designing can be done most conveniently by the graphic method, *vide* Figs. 61 and 62 (p. 135).

By formula (ix) *tang* $a_1 = tang\ a_2 - tang\ a$, since in this instance a_2 and a are equal; *tang* $a_1 = 0$, and therefore $a_1 = 0$, that is to say, the relative direction of entrance is parallel with the axis of the turbine.

The effective area of the guide passages

$$A = \frac{Q}{c} = \frac{210}{19.606} = \text{say } 10.7 \text{ square feet.}$$

Had the first assumptions been adhered to for which $c =$ 15·773, it is clear that the area A must have been greater and the wheel altogether larger.

For medium falls and quantities of water the empirical formula may be used—

$$r = 1·25 \text{ to } 1·5 \sqrt{A},$$

say $r = 1·25 \sqrt{10·7} = 1·25 \times 3·3 = \mathbf{4·125}$ feet,

or $d = \mathbf{8·25}$ feet, the mean diameter.

The Jonval turbine in the table referred to has a mean diameter of 8·036 feet; for the turbine under consideration, $d = 8$ feet may be assumed, or $r = 4$ feet.

The number of guide- and wheel-vanes is respectively 36 and 38 for the wheel actually constructed, and the same numbers may be adopted for the subsequent calculations—

$$z = \mathbf{36}; \; z_1 = \mathbf{38}.$$

Assuming the vanes to be of cast-iron, they may be taken as having a thickness of $\frac{1}{2}$ inch at the extremities, $t = \frac{1}{2}$ in., $t_1 = \frac{1}{2}$ in. From the preceding data the width e of the guide passages can now be found from

$$e = \frac{A}{2\, r\, \Pi\, \cos a - z\, t - z_1\, t_1\, \dfrac{\cos a}{\cos a_1}}$$

t and t_1 must of course be expressed as fractions of a foot, as the foot is the unit employed for other measurements;

$$t = t_1 = \tfrac{1}{24} \text{ foot.}$$

$$e = \frac{10·7}{25·13 \times 0·259 - \dfrac{36}{24} - \dfrac{38}{24} \times \dfrac{0·259}{1}} = \frac{10·7}{4·6}.$$

$$e = \mathbf{2·326} \text{ feet} = \text{say } \mathbf{2} \text{ feet } \mathbf{4} \text{ inches.}$$

The turbine of which the particulars are given in the table has a width of only 1 foot $5\frac{3}{4}$ inches; it is evident from this that the angles a and a_2 must be less than has been assumed; it will be noted on reference to Table I. that for equal values of a and a_2 with a given ratio $\dfrac{A}{A_2}$, the co-efficient K_1 is very nearly the same in all cases, so that provided $a = a_2$ the velocity of flow would remain very nearly unchanged; for $a = a_2 = \mathbf{70°}$, for instance, $K = \mathbf{0\cdot672}$ and $K_2 = \mathbf{0\cdot632}$.

On going through the calculations again with these values it will be found that

$$e = \mathbf{1\cdot65} \text{ feet } = 1 \text{ foot } \mathbf{7\cdot8} \text{ inches.}$$

This is not very different from the width of the Jonval turbine in question, and if the vanes were of wrought-iron, and therefore thinner, the agreement would be closer.

The calculated efficiency with $a = a_2 = 70°$ is **80** per cent., that actually obtained **83** per cent. including friction.

This proves that the co-efficients of resistance ζ_1 and ζ_2 for this wheel were in reality lower than has been assumed, and accounts for the best speed as proved by experiment being somewhat greater than that determined by calculation, in the ratio of $\mathbf{0\cdot642}$ to $\mathbf{0\cdot632}$ approximately.

It has been assumed throughout that $e_1 = e_2$. To allow for inaccuracies in erection and vibration it is usual in practice to make the width e_1 of the wheel slightly greater than that e of the guide passages; this clearance ranges from $\frac{1}{8}$ to $\frac{3}{8}$ inch.

To insure agreement with the preliminary assumptions as far as possible, e_2 may be calculated from

$$e_2 = \frac{A_2}{2\, r\, \Pi\, \cos a_2 - z_1\, t_2}$$

(where in the present case $A_2 = A$), and its value will be found slightly less than that of e, if the thickness of the ends of the vanes where the water leaves the wheel is the same as where it enters; substituting numerical values

$$e_2 = \frac{10 \cdot 7}{25 \cdot 13 \times 0 \cdot 259 - \frac{38}{24}} = \textbf{2·172 feet,}$$

$$e_2 = \textbf{2 feet 2}\,\tfrac{1}{16}\ \textbf{inches.}$$

If, however, e_2 is made exactly the same as e, then the assumed condition $A = A_2$ may be complied with by altering the thickness t_2 of the vanes where the water leaves them.

Reaction Turbines.

(1) *Axial Turbines No. 2.*

A Jonval turbine is to be designed for a fall of **15** feet and to utilize **136** cubic feet of water per second.

Assumed value of $a = \textbf{73° 30′.}$

,,   ,, $a_2 = \textbf{74° 30′.}$

$$\frac{A}{A_2} = \frac{\cos a}{\cos a_2} = \frac{0 \cdot 284}{0 \cdot 267} = \textbf{1·064.}$$

Velocity of flow $c = \sqrt{\dfrac{2gh}{2 \times 1 \cdot 064 \times 0 \cdot 959 \times 0 \cdot 964 + \overline{0 \cdot 125 + 1 \cdot 132 \times 0 \cdot 271}}}$·

$K_1 = \textbf{0·646.}$

$c = 0 \cdot 646 \sqrt{2g15} = 0 \cdot 646 \times 31 = \textbf{20·026 feet per} $ second, say **20** feet per second.

Velocity of rotation $w = K_2 \sqrt{2gh}.$

$K_2 = 0 \cdot 646 \times 1 \cdot 064 \times 0 \cdot 964 = \textbf{0·662.}$

$w = 0\!\cdot\!662 \times 31 = \mathbf{20\!\cdot\!52}$ feet per second.

$tang\ a_1 = 3\!\cdot\!606 - 3\!\cdot\!376 = 0\!\cdot\!230.$

$a_1 = \mathbf{13°}.$

$A = \dfrac{136}{20} = \mathbf{6\!\cdot\!8}$ square feet.

$r = 1\!\cdot\!25 \sqrt{A}$ (since $\dfrac{Q}{c}$ less than 16 sq. ft.).

$r = 1\!\cdot\!25 \sqrt{6\!\cdot\!8} = 1\!\cdot\!25 \times 2\!\cdot\!6 = \mathbf{3\!\cdot\!25} = \mathbf{3}$ ft. **3** in., or diameter $= \mathbf{6}$ feet **6** inches.

$P = \dfrac{r}{4\!\cdot\!5} = \dfrac{39}{4\!\cdot\!5} = $ say **9** inches.

$z = \dfrac{3\!\cdot\!1416 \times 6\ \text{ft. } 6\ \text{in.}}{9\ \text{in.}} = \dfrac{245''}{9''} = \mathbf{27}.$

$z_1 = z + 1 = \mathbf{28}.$

$t = \tfrac{3}{8}'' = \mathbf{0\!\cdot\!03}$ ft.

$t_1 = \tfrac{3}{8}'' = \mathbf{0\!\cdot\!03}$ ft.

$e = \dfrac{6\!\cdot\!8}{20\!\cdot\!42 \times 0\!\cdot\!284 - 27 \times 0\!\cdot\!03 - 28 \times 0\!\cdot\!03 \times \dfrac{0\!\cdot\!284}{0\!\cdot\!974}}.$

$e = \dfrac{6\!\cdot\!8}{5\!\cdot\!8 - 0\!\cdot\!81 - 0\!\cdot\!24} = \dfrac{6\!\cdot\!8}{4\!\cdot\!695} = \mathbf{1\!\cdot\!235}$ ft.

$e = \mathbf{1}$ ft. $\mathbf{2\tfrac{13}{16}}$ in.

$A_2 = \dfrac{1}{1\!\cdot\!064} \times 6\!\cdot\!8 = \mathbf{6\!\cdot\!39}$ sq. ft.

$e_2 = \dfrac{6\!\cdot\!39}{5\!\cdot\!45 - 0\!\cdot\!84} = \mathbf{1\!\cdot\!386}$ ft. $= \mathbf{1}$ ft. $\mathbf{4\tfrac{5}{8}}$ in.

$e_1 = e + \tfrac{3}{16} = \mathbf{1}$ ft. **3** in.

The depth of wheel $h_o = \dfrac{r}{5} = \dfrac{39''}{5} = $ say **8''**. It is very usual to make the depth h_o of the guides the same as that of the wheel, and in this case $h^1 = 8''$. Some engineers make h^1 $\tfrac{2}{3}$ to $\tfrac{3}{4}$ h_o. To some extent the depth of the guides and wheel must depend on the pitch of the vanes;

with an exceptionally large pitch in proportion to the diameter, a greater depth than usual would be necessary in order to insure the water being sufficiently guided between the vanes and given the required direction of flow.

Reaction Turbines.

Inward Mixed-Flow Turbine.

A fall of 30 feet, with a quantity of water equal to 120 cubic feet per second, is to be used for driving a turbine of the inward mixed-flow class.

The water—it is assumed—enters radially and leaves axially.

Owing to the nature of the ground and surroundings the water has to be conducted to the wheel through a wrought-iron pipe having a length l of about 50 feet, consequently a certain amount of loss by friction is incurred before the water enters the turbine.

This loss may be approximately calculated by the well-known formula for pipe friction when the diameter of the pipe has been determined. The velocity c_o in the pipe may be taken as 5 feet per second, and the area of the pipe must, in that case, be

$$A = \frac{120}{5} = 24 \text{ square feet,}$$

whence the diameter

$$d = \text{say } 5\cdot5 \text{ feet.}$$

The loss of head from friction is given by the formula

$$L = 0\cdot025 \; \frac{l}{d} \; \frac{c_o^2}{2g},$$

and when the numerical values above given are substituted—

$$L = 0.025 \frac{50}{5 \cdot 5} \frac{25}{64 \cdot 4} = \mathbf{0 \cdot 088} \text{ foot.}$$

This is so small a proportion of the total head—only $0 \cdot 29$
per cent.—that in calculating the velocity of flow it may
safely be neglected. If there are no sharp bends or other
sources of loss in the pipe (and this in the present instance
is assumed to be the case), the co-efficient of velocity K_1
may be taken from Table No. I., or calculated from one of
the formulas given—that in which the hydraulic efficiency
occurs is the best for practical purposes, a value of ϵ being
employed which has been proved by experience to be safe
for a well-designed wheel.

Should the losses before entering the wheel be considerable, say 5 per cent., then it would be necessary to
take into account their effect on the velocity of flow c.
This can most easily be accomplished by using a value
of the efficiency less by 5 per cent. than that otherwise
employed in determining the velocity of flow; instead
of 81 per cent., for instance, only 76 per cent.

Returning now to the design of a mixed-flow turbine
for the given fall and quantity of water, the angles a and
a_2 and the ratios $\frac{A}{A_2}$ and $\frac{r_1}{r_2}$ may be assumed as follows:

$$a = \mathbf{75^\circ}; \; a_2 = \mathbf{70^\circ};$$

$$\frac{A}{A_2} = \mathbf{0 \cdot 75}; \; \frac{r_1}{r_2} = \tfrac{3}{2}.$$

The angle a_2 of outflow from the buckets is taken as
greater than a, the angle of outflow from the guides, in
order to avoid a very great width of bucket.

It must be understood that the assumed values apply
to the mean angle of each stream of water issuing from a
passage.

Referring to Table No. I., the value of K_1 corresponding to the preceding assumptions will be found to be

$$K_1 = 0\cdot653.$$

The theoretical velocity due to the head is

$$V = \sqrt{2gh} = 43\cdot953,$$
whence $\qquad c = K_1\, V = 28\cdot70.$

The effective area of the guide passages,

$$A = \frac{Q}{c} = \frac{120}{28\cdot70} = \text{say } 4\cdot2 \text{ square feet.}$$

Approximately $\dfrac{e_1}{e_2} = \dfrac{r_1}{r_2}\dfrac{A}{A_2}\dfrac{\cos a_2}{\cos a} = 0\cdot66$;

this is a perfectly practicable ratio, so that the calculations may be proceeded with on the above basis.

The outer radius of the wheel

$$r_1 = 1\cdot25\sqrt{A} = 1\cdot25 \times 2\cdot049,$$
or say $\qquad r_1 = 2\cdot5$ feet.
$$r_2 = \tfrac{2}{3}\, r_1 = 1\tfrac{2}{3} \text{ foot.}$$

The pitch of the guide-vanes at the radius r_1 may be taken as

$$P = \frac{r_1}{4} = \frac{2\cdot5}{4} = 0\cdot62,$$
say $\qquad P = 0\cdot625$ foot $= 7\tfrac{1}{2}$ inches.

Number of guide-vanes

$$z = \frac{2\,\Pi\,r_1}{P} = 25.$$

True pitch $P = 0\cdot6284$ foot $= 7\cdot540$ inches.
Number of wheel-vanes

$$z_1 = 0\cdot7\,z = \text{say } 18;$$

this is perhaps rather a small value of z_1, resulting in a somewhat large pitch for the wheel-vanes, and it will be better to make

$$z_1 = 20.$$

The relative angle of inflow a_1 can be determined graphically or by the formula

$$tang \ a_1 = \frac{A}{A_2} \frac{r_1}{r_2} \frac{sin \ a_2}{cos \ a} - tang \ a,$$

whence $tang \ a_1 = 4 \cdot 0826 - 3 \cdot 7320 = 0 \cdot 3506$,
and $a_1 = 19° \ 20'$.

It will be found on drawing out the vanes that, for the same *mean* angles a and a_2, the depth available for the passage of the stream flowing between two guide-vanes is least when it is taken as the shortest distance x between them (measured as explained in Chapter V., p. 122), and as this method of gauging the outflow areas agrees best with practical results, it will be here adopted.

Taking the thickness of the guide-vanes as about $\frac{3}{8}$ inch, it will be found on drawing out the curves that the depth of each stream is

$$x = 1\tfrac{3}{8} \text{ inch} = 0 \cdot 1145 \text{ foot},$$

and the depth s_1 obstructed by each passing wheel-vane

$$s_1 = \tfrac{3}{32} \text{ inch} = 0 \cdot 0078 \text{ foot}.$$

The width of the guide passages must therefore be

$$c = \frac{A}{0 \cdot 9 \ (25 \times 0 \cdot 1145 - 18 \times 0 \cdot 0078)},$$

$$c = \frac{4 \cdot 2}{0 \cdot 9 \times 2 \cdot 724} = 1 \cdot 713 \text{ foot},$$

a co-efficient $0 \cdot 9$ being allowed for contraction.

As the water leaves axially, the wheel-vanes can be made parallel at the outlet, so that no contraction need be allowed for. The distance x_2 between the vanes is, as measured from a drawing,

$$x_2 = 1 \cdot 91 \text{ inch} = 0 \cdot 1591 \text{ foot}.$$

The area $A_2 = \frac{4}{3} A = $ **5·6** square feet,

whence $e_2 = \dfrac{5·6}{20 \times 0·1591} = $ **1·757** foot.

If the outflow is entirely axial, this width of buckets would make the greatest diameter at the outflow $3·333 + 1·757 = 5·087$ feet, that is, larger than the diameter at the inflow, and although there are some American turbines constructed in this way, it is in the author's opinion very undesirable, as either that portion of the water entering at the lower edge of the buckets has to make a very sharp turn to fill the outflow orifices, or, what is more probable, the outflow orifices are only partially filled, and then an element of uncertainty is introduced into the design.

Under these circumstances it will be preferable to make the outflow inclined, with both a radial and axial component, but more nearly radial than axial. The same amount of contraction may then be allowed for as at the inflow. On setting out the vanes again to suit this modification in the design, the depth x_2 will be found as

$$x_2 = 1\tfrac{13}{16} \text{ inch} = \textbf{0·1514 foot,}$$

and

$$e_2 = \frac{5·6}{0·9 \times 20 \times 0·1514} = \textbf{2·06 feet.}$$

The co-efficient of speed

$$K_2 = K_1 \frac{r_1}{r_2} \frac{A}{A_2} \, sin \, a_2 = K_1 \frac{sin \, (a + a_1)}{cos \, a_1},$$
$$K_2 = \textbf{0·690.}$$

The two methods of calculation afford a check on the accuracy of the calculation or construction of the angles.

It is not necessary to calculate K_2, as w_1 the velocity of rotation can be obtained by graphical methods, and if the construction is correct the inflow should be without shock and the outflow radial. The velocity of rotation (by whichever method determined) is

$$w_1 = 0.690 \ V = \textbf{30·327} \text{ feet per second};$$

hence the number of revolutions per minute,

$$n = \frac{30·327 \times 60}{15·71} = \textbf{115·8},$$

or say $n = \textbf{115}$ revolutions.

The available power,

$$E_{hp} = 30 \times 120 \times 0·1135 = \textbf{408·6} \text{ horse-power},$$

and assuming the efficiency to be 75 per cent., the work done,

$$W_{hp} = \textbf{306·45}.$$

The total work done by the motor per second may be theoretically calculated by the formula

$$W = \frac{G}{g} \ c^2 \frac{r_1}{r_2} \frac{A}{A_2} \ sin \ a \ sin \ a_2 \text{ (foot lbs.)},$$

whence $W = 195,742·3$ foot lbs. $= \textbf{355·9}$ horse-power, and the hydraulic efficiency

$$\epsilon = \frac{355·9}{408·6} = 0·87 \text{ or } \textbf{87} \text{ per cent.};$$

allowing 3 per cent. off this for friction, the actual efficiency of the wheel would be

$$\epsilon_e = \textbf{84} \text{ per cent.}$$

In a wheel of the radial mixed-flow type, it must not be forgotten that the guide- and wheel-*vane angles* are much greater than the mean angles of outflow from the guides and buckets respectively; in the present case the vane angle for the guides, as measured from a drawing, is

$$a = \textbf{82}°,$$

while that for the buckets is

$$a_2 = \textbf{83}°.$$

For the quantity of water and head given it would be better to give the turbine a larger diameter than that assumed, and thus reduce the width of the guide passages and buckets. It would then be possible to make the outflow axial, or nearly so, without any abrupt change in the direction of flow, and at the same time avoid the necessity of allowing for contraction in the bucket orifices.

By making the angles a and a_2 *less* than has been assumed the dimensions of the motor could be still further reduced, and although this would involve some increase in the residual velocity and of the unutilized energy due to it, there would probably be no loss of efficiency, as the form of the guide and bucket passages would be more favourable, the changes in sectional area being less abrupt than with the greater angles.

Impulse Turbine.

(1) Axial No. 1.

A Girard turbine is required to utilize a quantity of water of 176 cubic feet per second, with a fall of 10 feet; the wheel must be of the parallel-flow type with vertical shaft.

$$Q = 176 \, ; \; h = 10.$$
$$a \text{ assumed} = 64°.$$
$$a_2 \quad \text{,,} \quad = 66°.$$
$$h_2 \quad \text{,,} \quad = 1 \text{ foot.}$$

As the fall is moderate the water may be conducted to the wheel through a casing or tube of sufficiently large area to render the loss in it by friction, before the water enters the guide apparatus, practically negligible, so that $L = O$.

The effective head H is therefore $h - h_2 = 10 - 1 = 9$, and the velocity of flow from the guide passages

$$c = 0\text{·}95\sqrt{2g\ 9} = \mathbf{22\text{·}871} \text{ feet per second.}$$

The area of outflow of the guide passages

$$A = \frac{Q}{0\text{·}89\ c} = \frac{176}{20\text{·}355} = \mathbf{8\text{·}646} \text{ square feet.}$$

The mean radius of the wheel may be taken as

$$* \ r = 1\text{·}4\sqrt{A} = 1\text{·}4 \times 2\text{·}94 = \mathbf{4\text{·}116} \text{ feet;}$$

consequently the diameter $= 8\text{·}236$, say **8** feet **3** inches.

The pitch of the guide-vanes

$$P = \frac{r}{10} = \text{ say } \mathbf{5} \text{ inches;}$$

hence $\qquad z = \dfrac{2\ \Pi\ r}{P} = \mathbf{62.}$

The turbine No. 7 in Table B, actually constructed for the assumed conditions, has 64 guide-vanes, and as within reasonable limits z is a matter of choice, to facilitate further comparison it may be given the same value.

The thickness of the guide-vanes, which should be of wrought-iron or steel, may be assumed as

$$t = \tfrac{5}{16} \text{ inch} = \mathbf{0\text{·}026} \text{ foot.}$$

The clear width of the guide passages

$$e = \frac{A}{2\ \Pi\ r \cos a - z\ t} = \frac{8\text{·}646}{11\text{·}353 - 64 \times 0\text{·}026}.$$

$$e = \mathbf{0\text{·}892} \text{ feet} = \mathbf{10\text{·}7} \text{ inches.}$$

$a_1 = \text{(approximately) } 2\ a - 90° = 128° - 90° = \mathbf{38°}\ (-)$

---

* As the rule for determining r is empirical, the factor—ranging from 1·25 to 1·5—has been chosen so as to give a result agreeing with turbine No. 7 in Table B.

$$w = c \; \frac{sin \; (a + a_1)}{cos \; a_1} = \; 22 \cdot 871 \; \times \; 0 \cdot 556 = \mathbf{12 \cdot 716} \text{ feet}$$

per second, should in designing be determined graphically.

$$c_1 = w = \mathbf{12 \cdot 716}.$$

$$h_o = \frac{r}{5} = \text{say } 10'' = 0 \cdot 833 \text{ feet}.$$

$$c_2 = 0 \cdot 96 \sqrt{(12 \cdot 716)^2 + 64 \cdot 4 \times 0 \cdot 833}$$
$$= 0 \cdot 96 \sqrt{215 \cdot 33}.$$

$$c^2 = \mathbf{14 \cdot 086} \text{ feet per second}.$$

True value of a_2 for vertical outflow calculated by

$$sin \; a_2 = \frac{w}{c_2} = \frac{12 \cdot 716}{14 \cdot 086} = \mathbf{0 \cdot 903} \, ;$$

a_2 can be conveniently ascertained graphically

$$a_2 = \mathbf{64° \; 30'}.$$

The assumed value of a_2 was 66°, and if vertical outflow is to be rigidly adhered to, a_1 must be revised. For practical purposes, however, it is sufficiently accurate. The outflow merely deviates very slightly, almost inappreciably, from the vertical direction; the originally assumed values may therefore be retained. For the turbine No. 7 in the table, $a_1 = 40°$ instead of 38°. The velocity w of this wheel has the ratio $0 \cdot 500$ to the theoretical velocity of flow $\sqrt{2gh}$; it will be interesting to compare this with the result obtained above. It was found that

$$w = 0 \cdot 556 \; c = \mathbf{12 \cdot 716}.$$

The theoretical velocity $V = \sqrt{64 \cdot 4 \times 10} = \mathbf{25 \cdot 375}$,

hence $\qquad \dfrac{w}{V} = \dfrac{12 \cdot 716}{25 \cdot 375} = \mathbf{0 \cdot 501} \, ;$

from this it will be seen that the agreement is very close.

For z_1, as for z, the number given for turbine No. 7 in the table may be assumed, $z_1 = \mathbf{60}.$

t_1 and t_2 can be taken as equal to t,

$$t_1 = t_2 = t = \tfrac{5}{16}'' = \mathbf{0·026} \text{ foot.}$$

The ratio of the width at outflow to that at inflow, should be calculated by the accurate formula

$$\frac{e_2}{e_1} = \frac{x_1}{x_2} \frac{c_1}{c_2} \left(\frac{2\,\Pi\,r\,\cos a_1 - z_1\,t_1}{2\,\Pi\,r\,\cos a_2 - z_1\,t_2} \right);$$

$x_1 = \mathbf{0·9}$ and $x_2 = \mathbf{0·7}$, both assumed.

It should be noted that $\dfrac{e_2}{e_1}$ depends only on the ratio of $x_1 : x_2$. The absolute value of x_1 is fixed by the number and thickness of the vanes, and must be settled graphically.

$$\frac{e_2}{e_1} = \frac{0·9}{0·7} \frac{12·716}{14·086} \frac{25·92 \times 0·788 - 60 \times 0·026}{25·92 \times 0·407 - 60 \times 0·026}.$$

$$\frac{e_2}{e_1} = 1·3 \times 0·903 \times 2·146 = \mathbf{2·5}.$$

This is the same ratio as that given in the table for turbine No. 7.

Depth of guides $h^1 = \tfrac{3}{4} h_o = \tfrac{3}{4} 10'' = \mathbf{7\tfrac{1}{2}''}$.

Impulse Turbine.

(1) *Axial No. 2.*

A turbine has to be constructed to utilize a fall of water subject to the following conditions:—

During nine months of the year, on the average, the available head is **11** feet and the quantity of water **106** cubic feet per second; for the rest of the year (in times of flood) the head is **9** feet and the quantity of water **114** cubic feet.

It will be seen by comparing the products of the head and quantity of water in the two cases, that the power available is *less* during times of flood with the smaller head and greater quantity of water; as, however, this state of affairs only lasts for three months out of the twelve, it is desirable that the turbine should be designed chiefly with a view to suit the normal conditions, but at the same time to give a reasonably good efficiency in times of flood. The diminution of the head in the latter period is almost entirely due to the rise of the water-level in the tail-race, amounting to 2 feet. Unless, therefore, the wheel is suspended at least 2 feet above the normal tail-water level, it must work *drowned* in times of flood, but an impulse turbine would give poor results under such conditions. On the other hand, to place the wheel 2 feet above the lower surface of the water means the loss of over 18 per cent. of the available power during three-quarters of the year. The best solution of the difficulty consists in so designing the turbine that at ordinary times it acts as an impulse wheel, with the usual clearance above the water surface, while in times of flood it works as a reaction turbine, and may, without detriment to its efficiency, be submerged.

In the first place, it is evident, the guide passages and wheel-buckets must be of sufficient area to allow for the flow of the maximum quantity of water under the minimum fall, and therefore with the minimum velocity; in other respects the proportions must be determined as for an impulse turbine with free deviation.

As the wheel has to run at all times at a constant speed, matters must, if possible, be so arranged that the *best speed* is very nearly the same whether the motor works with free deviation or with reaction. As no long pipes are necessary for leading the water to the turbine, and the latter can be placed in a chamber or reservoir of relatively large sectional area, the velocity of flow from the guide passages, when free deviation is permitted, may be taken

as $0.95\sqrt{2gh}$, and in order that the conditions as regards
speed may be the same with reaction as with free deviation,
the co-efficient of flow K_1 must be so chosen that the
following equation is satisfied :—

$$0.95 \sqrt{(h - h_2)\, 2g} = K_1 \sqrt{2gh_1},$$

where h is the maximum and h_1 the minimum head, while
h_2 denotes the height above the lower water-level of the
outflow orifices of the guide passages.

Taking as usual $h_2 = 1$ foot, there follows

$$0.95\sqrt{2g\ 10} = K_1 \sqrt{2g\ 9}, \qquad \text{and since } 2g$$

is a factor common to both sides of the equation,

$$0.95 \sqrt{10} = K_1 \sqrt{9},$$

whence

$$K_1 = 0.95 \sqrt{\frac{10}{9}} = 0.95 \times 1.054,$$

$$K_1 = 1.001, \text{ or say } 1;$$

that is, the velocity of flow when reaction is used must
equal the velocity due to the head.

To attain this the turbine must be constructed with a
low value of the ratio $\dfrac{A}{A_2}$, which can easily be determined
as follows.

By the well-known formula—

$$K_1 = \sqrt{\frac{\epsilon}{2\dfrac{A}{A_2}\, sin\ a\ sin\ a_2}}\ ;$$

a and a_2 may be chosen, say $a = 64^c$ and $a_2 = 68°$. The
efficiency ϵ, in order to be on the safe side, should not,
in a turbine of this class, be taken as more than 75 per

cent., which, when friction is deducted, is equivalent to a mechanical efficiency of 70 to 71 per cent. From the above equation, when the assumed values are substituted, the ratio $\dfrac{A}{A_2}$ can be calculated.

$$\frac{A}{A_2} = \frac{\epsilon}{K_1^2\,2\,sin\,a\,sin\,a_2}$$

$$\text{or } \frac{A}{A_2} = \frac{0.75}{2 \times 0.899 \times 0.927} = \frac{0.75}{1.667},$$

$$\frac{A}{A_2} = \mathbf{0.45}.$$

In a turbine with these proportions, it may be noticed, the pressure at the guide-passage orifices will be slightly *less* than that of the atmosphere, as otherwise the velocity of flow could not equal that due to the head.

The velocity of flow

$$c = 0.95\sqrt{64.4 \times 10} = 1.001\sqrt{64.4 \times 9},$$

$$c = \mathbf{24.099}, \text{ say } \mathbf{24} \text{ feet per second.}$$

The area of the guide passages A must be calculated for the largest quantity of water likely to pass through the wheel, 114 cubic feet, hence

$$A = \frac{114}{0.89 \times 24} = \mathbf{5.337} \text{ square feet.}$$

In determining A, the formula has been used which is applicable to Girard turbines, in which the obstruction of the wheel-vanes passing in front of the guide passages is allowed for by the co-efficient 0.89 instead of the method given for reaction wheels; the former is somewhat more simple than the latter method, and gives results which err on the safe side.

The mean radius of the wheel

$$r = 1.5\sqrt{A} = 1.5 \times 2.31 = \mathbf{3.46},$$

say $r = 3.5$ feet; $d = 7$ feet.

The pitch of the guide-vanes may be taken as—

$$P = \frac{r}{9} = \text{say } 4.5 \text{ inches,}$$

and $z = \dfrac{2\,r\,\Pi}{P} = \dfrac{263.9 \text{ inches}}{4.5} = \text{say } 58.$

The thickness of the guide-vanes of wrought-iron
$$t = \tfrac{5}{16} \text{ inch} = 0.0259 \text{ foot.}$$

The width of the guide passages
$$e = \frac{5.34}{21.99 \times 0.438 - 58 \times 0.0259} = \frac{5.34}{8.133},$$

$$e = 0.656 \text{ foot} = 7\tfrac{13}{16} \text{ inches.}$$

$$a_1 = 2 \times 64° - 90° = 38° \; (-).$$

The mean velocity of rotation—

$$w = 24 \times \frac{\sin(64° - 38°)}{\cos 38°} = 24 \times \frac{\sin 26°}{\cos 38°} = 24 \times \frac{0.438}{0.788}$$

$$w = 24 \times 0.556 = 13.34 \text{ feet per second.}$$

c_1 the relative velocity of inflow $= w$.

The depth of the wheel h_o for a turbine intended to work, as in this case, both as an impulse and as a reaction turbine, should be somewhat greater than for an ordinary impulse turbine, since back vanes are used, and it will be found in drawing out the form of these that a greater length than usual is required in order to avoid too sudden curvature, and to make the lower ends parallel both on the back and front side.

It may therefore be assumed that

$$h_o = \frac{r}{4} = \text{say } 10 \text{ inches} = 0.833 \text{ foot,}$$

instead of $\dfrac{r}{5}$, as would be the case for an impulse wheel under ordinary circumstances.

The relative velocity of outflow from the buckets c_2 can now be calculated.

$$c_2 = 0.96 \sqrt{177.9 + 64.4 \times 0.833} = 0.96 \sqrt{231.54},$$

$$c_2 = 14.60 \text{ feet per second.}$$

The exact value of a_2 for vertical outflow would be given by—

$$sin\ a_2 = \frac{13.34}{14.60} = 0.914,$$

whence $\qquad\qquad a_2 = 66°;$

this differs by only 2° from the assumed value, so that the latter may be allowed to remain unaltered.

The ratio of the width of the buckets at the outflow to that at the inflow $\frac{e_2}{e_1}$ is determined for the wheel when working submerged by the ratio $\frac{A}{A_2}$ in conjunction with the other dimensions. On the other hand, when the action is that of an impulse turbine and the wheel runs clear of the tail-water, the width e_2 of the buckets at the outflow must be sufficiently great to prevent any possibility of the buckets being choked, that is, the proportion x_2 of the outflow orifice filled must be less than 1; it remains to be ascertained whether these two conditions are compatible with each other.

The relative velocity of outflow when the wheel works submerged is

$$c_2 = c\ \frac{A}{A_2} = 24 \times 0.45 = 10.80 \text{ feet};$$

with free deviation it was found to be

$$c_2 = 14.60 \text{ feet.}$$

It is evident from this that if in the first case the water just *fills* the orifice, in the second, where the velocity is greater, it will *not* quite fill it, so that clearance between the convex surface of the vane and the water is insured. It is consequently only necessary to proportion the widths of the buckets according to the requirements of the turbine when working submerged.

When the section through a guide passage is drawn out, the depth x can be measured, also the depth s_1 of the portion of the passage obstructed by one wheel-vane; similarly, the depth x_2 of the outflow orifices of the buckets; then, as before demonstrated—

$$A = c\,(z\,x - z_1\,s_1)$$

and

$$A_2 = e_2\,z_1\,x_2\,;$$

whence

$$\frac{e_2}{c} = \frac{A_2}{A}\,\frac{z\,x - z_1\,s_1}{z_1\,x_2}\,.$$

Assuming $t_1 = t = \tfrac{5}{16}''$ and $z_1 = 60$ it will be found that

$$\frac{e_2}{e_1} = \mathbf{2 \cdot 4}.$$

Impulse Turbines.

(2) *Radial Outward-Flow No. 3.*

A head of water $h = \mathbf{590}$ feet with a supply $Q = \mathbf{16}$ cubic feet per second is to be utilized. The head is measured to the orifices of the guide passages. In this case—the fall being very great and the quantity of water small—an outward-flow radial impulse wheel with horizontal shaft is the most suitable type of motor. In Table B the particulars of a turbine designed by Messrs. J. J. Rieter and Co. for similar conditions, and working at Terni in Italy, will be found, and may be compared with the results of calculation.

The available power is

$$E = \frac{590 \times 16 \times 62 \cdot 425}{550} = \mathbf{1071 \cdot 4} \text{ horse-power.}$$

x

$$a \text{ assumed } = \mathbf{70}°,$$
$$a_2 \quad ,, \quad = \mathbf{70}°.$$

It may be supposed for the sake of simplicity that the pipe conducting the water to the turbine is of such dimensions that the loss in it by friction is very small and is covered by using a lower value of the co-efficient with which $\sqrt{2gh}$ has to be multiplied to determine c. It is very easy to calculate the losses in the pipe when the dimensions are known; with high falls like that in question, the velocity of flow in the pipe may be 5 or 6 feet per second. For a velocity of 5 feet per second the area of the pipe must be $\frac{16}{5} = 3\cdot2$ square feet and the diameter say 2·1 feet.

If the length of the pipe be twice the head, or say 1,600 feet, it will be found on calculation that the total loss from friction is equivalent to a head of about 5·1 feet, or less than 0·9 per cent. of the total fall; for a sharp bend with a mean radius of curvature equal to the diameter of the pipe, the loss of head would be—for the same velocity—about 0·8 of a foot. These figures make it clear that a co-efficient 0·92 will amply cover all losses incurred up to the moment the water leaves the guide orifices.

$$c = 0\cdot92\sqrt{2g \ 590} = \mathbf{179\cdot33},$$
say $c = \mathbf{179}$ feet per second.

Area of guide passages $A = \dfrac{Q}{0\cdot89 \ c} = \dfrac{16}{159}.$

$$A = \mathbf{0\cdot1} \text{ square foot.}$$

For the inner radius r_1 a large value may be assumed in order to avoid too great a number of revolutions :

$$r_1 = 13 \sqrt{A} = 13 \times 0\cdot316 = \mathbf{4\cdot125},$$
or $r_1 = \mathbf{4}$ feet $1\frac{1}{2}$ inch.

The pitch P assumed = 3 inches = 0·25 foot.

Ratio $\dfrac{e}{r_1} = y = \frac{1}{10}$ assumed.

$$e = 0\cdot4125 \text{ foot,}$$

or say $\qquad e = 5 \text{ inches} = 0\cdot417 \text{ foot.}$

Proportion of circumference, exclusive of thickness of vanes, over which the water is admitted

$$\phi = \frac{A}{2\,\Pi\,r_1\,e\,\cos\,a} = \frac{0\cdot1}{25\cdot920 \times 0\cdot417 \times 0\cdot342},$$

$$\phi = 0\cdot027.$$

Number of vanes in guide apparatus

$$z = \frac{2\,\Pi\,r_1\,\phi}{P} = \frac{25\cdot920 \times 0\cdot027}{0\cdot25},$$

$$z = 2\cdot8, \text{ say } 3.$$

On reference to the dimensions of the turbine for similar conditions given in Table B it will be seen that there are only *two* guide passages; this is a matter of minor importance in a turbine in this class. Provided the water is sufficiently guided to leave the orifices in the required direction, the resistance is reduced by diminishing the number of vanes. The pitch in such cases is greater than the pitch of the wheel-vanes, and consequently the proportion of the depth of the buckets occupied by water at *the inflow* must be taken as $x_1 = 1$.

Adhering to the calculated number of guide passages $z = 3$, the *corrected* value of ϕ can now be determined after fixing the thickness of the vanes; this latter for steel vanes let into the guide-casing may be taken as $t = \frac{3}{16}$ inch $= 0\cdot0156$ foot.

$$\phi = \frac{A + z\,t\,e}{2\,\Pi\,r_1\,e\,\cos\,a} = \frac{0\cdot1 + 3 \times 0\cdot417 \times 0\cdot0156}{3\cdot708},$$

$$\phi = 0\cdot032.$$

From this the corrected value of the pitch

$$P = \frac{2\,\Pi\,r_1\,\phi}{z} = \frac{25\cdot920 \times 0\cdot032}{3} = 0\cdot276,$$

or $P = 3\frac{5}{16}$ inches.

This of course can be more simply ascertained by graphical construction in a way requiring no description.

$$a_1 = 2\,a - 90° = 50°,$$

a_1 being of course negative, as is always the case in impulse wheels.

The velocity of the inner circumference of the wheel

$$w_1 = c\,\frac{sin\,(a - a_1)}{cos\,a_1} = c\,\frac{sin\,20°}{cos\,50°},$$

$$w_1 = \frac{0.342}{0.643}\,c = \mathbf{0.532}\,c,$$

or $0.532 \times 0.92\,\sqrt{2gh} = 0.489\,\sqrt{2gh}$;

$$w_1 = \mathbf{95.228} \text{ feet per second,}$$

and also $c_1 = \mathbf{95.228}$ feet per second.

The depth of the buckets $h_o = r_2 - r_1$ may be taken as $h_o = \frac{1}{5}\,r_1 = \mathbf{0.825}.$

As the flow of the water through the buckets is *radially outward*, an increase of the relative velocity takes place over and above that due to the additional fall h_o, and hence

$$c_2 = 0.95\,\sqrt{c_1{}^2 + w_2{}^2 - w_1{}^2 + 2gh_o},$$

or $c_2 = 0.95 \times 114.502 = \mathbf{108.777},$

the velocity of the outer circumference being

$$w_2 = \frac{r_2}{r_1}\,w_1 = \frac{r_1 + h_o}{r}\,w_1 = \tfrac{6}{5}\,w_1,$$

$$w_2 = 1.2 \times 95.228 = \mathbf{114.27}.$$

The correct value of a_2 should be given by the relation $sin\,a_2 = \dfrac{w_2}{c_2}$ (for radial outflow). As c_2 is less than w_2, it is evident that the outflow cannot be radial with the assumed angles; a_1 must therefore be modified.

By way of trial, let it be assumed that

$$a_1 = \mathbf{54°},$$

then

$$w_1 = c\,\frac{sin\ 16°}{cos\ 54°} = c \times \frac{0\text{·}276}{0\text{·}588} = 0\text{·}470\ c,$$

or $w_1 = 0\text{·}432\sqrt{2gh} = \mathbf{84\text{·}13}$ feet per second; $(K_2 = 0\text{·}432)$.

$$c_1 = c\,\frac{cos\ 70°}{cos\ 54°} = c\,\frac{0\text{·}342}{0\text{·}588} = \mathbf{0\text{·}581}\ c,$$

$$c_1 = \mathbf{104} \text{ feet per second,}$$

$$w_2 = \tfrac{9}{5}\,w_1 = \mathbf{100\text{·}9} \text{ feet per second,}$$

and

$$c_2 = 0\text{·}95\sqrt{c_1^2 + w_2^2 - w_1^2 + 2gh_o},$$

$$c_2 = \mathbf{112\text{·}29}.$$

The number of revolutions per minute will be

$$n = \frac{w_1 \times 60}{2\ \Pi\ r_1} = \text{say } \mathbf{195}$$

$$sin\ a_2 = \frac{w_2}{c_2} = \frac{100\text{·}9}{112\text{·}29} = \mathbf{0\text{·}898}$$

whence

$$a_2 = \mathbf{64°}.$$

Referring to the Table B, the angle a_1 for the turbine before mentioned will be found to have the value 54° assumed last, and the angle a_2 calculated from this differs by 6° from the corresponding angle in the turbine as actually made. The proportion $\frac{r_2}{r_1}$ is a little different to that adopted in the preceding calculations, and this modifies slightly the results. The effect of these differences on the performance of the turbine is not great; the calculated velocity of rotation agrees very well with that adopted by the makers, as may be seen by comparing the values of K_2.

The ratio of the width of the buckets at the outflow to that at the inflow has now to be determined.

$$\frac{e_2}{c_1} = \frac{x_1}{x_2}\frac{c_1}{c_2}\left(\frac{2\ \Pi\ r_1\ cos\ a_1 - z_1\ t_1}{2\ \Pi\ r_2\ cos\ a_2 - z_1\ t_2}\right).$$

Assuming $P = 3''$, the number of vanes z_1 in the wheel is

$$z_1 = \frac{2\ \Pi\ r_1}{P} = \frac{25\cdot920}{0\cdot25} = \text{say } \mathbf{100},$$

whence the correct value of $P = \mathbf{0\cdot259}$, or

$$P = \mathbf{3\cdot111} \text{ inches.}$$

The thickness of the vanes at the inflow and outflow may be taken as the same—

$$t_1 = t_2 = \tfrac{1}{2} \text{ inch} = \mathbf{0\cdot0416}.$$

By substituting these values in the preceding formula there follows :

$$\frac{e_2}{e_1} = \frac{x^1}{x_2}\frac{104}{112\cdot29}\left(\frac{25\cdot920 \times 0\cdot588 - 0\cdot0416 \times 100}{31\cdot102 \times 0\cdot438 - 0\cdot0416 \times 100}\right).$$

$$x_1 = \mathbf{1}, \text{ and } x_2 \text{ can be assumed as}$$
$$x_2 = \mathbf{0\cdot5},$$

whence $\dfrac{e_2}{e_1} = 2 \times 0\cdot926 \times 1\cdot083 = \mathbf{2\cdot17},$

say $\dfrac{e_2}{e_1} = \mathbf{2\cdot2}.$

$$c_1 = e + \tfrac{1}{2} = \mathbf{5\tfrac{1}{2}} \text{ inches,}$$
$$e_2 = 2\cdot2\ e_1 = \text{say } \mathbf{12} \text{ inches.}$$

With the greater value of a_2 adopted in the turbine of Messrs. Rieter and Co. a smaller residual velocity is obtained, and although the outflow is not exactly radial, it only deviates by a few degrees from the radial direction. The formulas employed, although not mathematically accurate under such conditions, give results which are close enough to the truth for practical purposes, and the gain in efficiency is very appreciable. A strict agreement might be obtained by choosing a somewhat smaller value of a_1—between 54° and 50°—but it is hardly worth while to repeat the calculation for this purpose.

As is evident from the formula for the ratio $\frac{e_2}{e_1}$, an increase in the angle a_2 results in a greater value of that ratio, and for $a_2 = 70°$—other factors remaining as before— $\frac{e_2}{e_1} = 2\cdot94$. It is possible to carry the increase of a_2 so far that the advantages obtained are neutralized by the excessive axial components of the motion of the streams leaving the buckets, but this does not happen with the proportions adopted in practice.

Finally, for the sake of comparison, the work done can be calculated from the formula

$$W_{hp} = \frac{M \, w_1 \, c \, \sin \, a}{550},$$

$$M = \frac{\text{Weight of water per second}}{32\cdot2} = \mathbf{31\cdot018},$$

$$W_{hp} = \frac{31\cdot018 \times 84\cdot13 \times 179 \times 0\cdot94}{550},$$

$$W_{hp} = \mathbf{798} \text{ horse-power.}$$

This gives a hydraulic efficiency

$$\epsilon = \frac{W}{E} = \frac{798}{1071} = \mathbf{0\cdot745},$$

or say $\epsilon = \mathbf{74\frac{1}{2}}$ per cent.

In consequence of the greater value of the angle a_2 in the wheel as actually made, the loss by unutilized energy is about 2·9 per cent. less than under the conditions assumed in the calculations, and this brings the hydraulic efficiency up to **77·4** per cent. Allowing 2 per cent. for shaft friction, the actual efficiency would thus be 75·4 per cent. The makers guarantee 75 per cent.

For this type of wheel the power absorbed in friction does not in practice exceed 1·6 per cent., so that in reality the allowance of 2 per cent. is more than sufficient, and leaves a certain margin beyond that already included in the co-efficients used in determining the velocities of flow.

CHAPTER X.

MEASUREMENT OF THE QUANTITY OF FLOWING WATER.

Enumeration of methods.—Direct measurement.—Measurement by current meters.—Measurement by floats.—Measurement over weirs: theoretical discharge, actual discharge, influences affecting discharge, Braschmann's formula, Donkin and Salter's experiments and table, Weisbach's formula, Francis's formula, Fteley and Stearns's experiments, comparisons.—Comparative table of results of experiments, &c., on discharge over weirs.—Measurement by meters.—Measurement of water levels: hook gauge, positions for measuring levels.—New method of gauging the flow of water; the "Venturi" meter.

In determining the available energy of a fall of water the most important—and at the same time most difficult—measurement to be made is that of the quantity of water passing in a given time. Errors amounting to many per cent. may arise from carelessness or a defective method in determining this factor of the power of a falling stream.

The following are the principal methods which have been employed for ascertaining the quantity or volume of water flowing through any given section of a stream or channel:—

(1) Direct measurement of volume in a calibrated collecting tank or reservoir;

(2) Measurement of velocity of flow in any section of known dimensions by means of current meters;

(3) Measurement of velocity of flow by floats;
(4) Measurement over weirs or tumbling bays;
(5) Measurement by various forms of meter.

(1) *Direct Measurement in a calibrated collecting Tank.*

It is very rarely that this method for the measurement of water is practicable for any but the very smallest hydraulic motor; when applicable, it is by far the most reliable, being independent of experimental co-efficients or the errors of current or other meters, and requiring only average care in determining fixed dimensions.

The water as it leaves the motor is all collected in a tank or reservoir of known capacity, which it is most convenient to calibrate beforehand. In some instances it may be practicable to use several tanks in succession, care being taken that in transferring the inflow from one tank to another no loss of fluid is incurred.

Sometimes with small quantities of water it is possible to use two comparatively small tanks or vessels, which are alternately filled and emptied.

(2) *Measurement by Current Meters.*

With this method the velocity of flow of the water in a channel of known section is measured by current meters, and the quantity of water calculated from it. To arrive at accurate results it is necessary to subdivide a section of the stream, taken at right angles to the general direction of flow, into a number of parts, prefer-ably of equal area, and to observe the velocity, as indi-cated by the current meters, in each of these parts for some time. From the observed velocities in the different parts the mean velocity for the whole section is determined.

The current meters used are those in which a screw-shaped fan is made to revolve by the action of the moving water; in some cases these are connected with electrical stopping and starting apparatus.

It is preferable to use several meters simultaneously as checks on each other, and all of them, previous to the experiments, should be carefully tested to ascertain their degrees of accuracy.

The axis of each current meter should be parallel to the general direction of flow in which the velocity of the stream is to be measured, as, if free to adjust itself to the direction of flow at the particular point where it is placed, the recorded velocity may be too great, since *locally* the water may have a motion deviating in direction from that in which the velocity is required, and in that case only the *component* of the local velocity in the general direction of flow must be taken into account.

If carefully carried out there is no doubt that this method gives very reliable results; for very large quantities of water it is often the only one possible, with the exception of that in which floats are used, to be subsequently described.

In each portion of the section of the stream several observations should be taken at intervals. During the trial of the Jonval turbine at Goeggingen, of which an account is elsewhere given, *eight* separate observations, each lasting *one minute*, were made with each of the instruments employed.

As the surface of the water forms the boundary of the top of the uppermost row of subdivisions of the channel section, the area of these divisions of course varies somewhat during the experiments in consequence of changes in the water level, and where accuracy is desired the latter must be observed and allowed for.

(3) *Measurement by Floats.*

This consists in observing the speed with which a float is carried down the stream to be measured, and inferring the velocity of the latter from the former. It depends for its accuracy on a knowledge of the relation existing between the velocity of the float and the mean velocity of the water in the particular part of the stream traversed by the float.

As in the previous method, the channel should be divided into several parts, in each of which observations must be made, and the velocity should be ascertained at three different depths at least. For this purpose a body of slightly greater specific gravity than the fluid, and presenting a considerable surface to the action of the stream, is suspended at the desired depth from another body floating on the surface, of such a form as to offer much less resistance to the stream than the first, so that without sensible error the velocity with which the instrument is carried along by the current is that of the submerged body and of the stream at the particular depth below the surface at which it is placed. A hollow metallic ball supported from a small ball or cylinder of wood has been employed for this kind of float.

Another form of float is that consisting of two bodies of equal size and shape connected by a thin rod; one of these swims just below the surface, the other at any desired depth, regulated by the connecting rod. The velocity with which the apparatus moves is taken as being the mean of the velocities of the water at the depths at which the two bodies float. In this way the mean between the surface velocity and that at any depth can be ascertained for several positions of the lower body. It is not necessary that the upper body should be close to the surface; it may be made to swim at any required depth below, but must then be kept in place by being connected to a light float of small resistance at the surface.

Another form of measuring float is a cylindrical wooden staff weighted so as to swim in a nearly vertical position. Its velocity gives the mean velocity of the water throughout the depth occupied by the staff. For weighting this kind of float a metal cap containing shot, attached to the lower end, is employed.

(4) *Measurement of Discharge of Water over Weirs or Tumbling Bays.*

This method of water-gauging is perhaps that which has been most frequently employed to measure the quantity of water consumed by hydraulic motors, chiefly because in many instances it was the most convenient method practicable.

The stream to be measured is dammed by a weir or other suitable obstruction, and all the water compelled to flow through a rectangular aperture open at the top, cut out from the upper edge of the weir, and generally known as a Tumbling Bay. Very frequently the water is allowed to fall over the whole length of the weir, in which case the sides of the canal or stream form the ends of the bay, and should be vertical and parallel for some distance above the weir (up-stream).

If h denote the height of the water-level above the lower edge of the bay, measured some distance above the weir where there is no appreciable gradient on the surface of the water, and L the length of the bay, then the theoretical quantity of water which should be discharged over the bay is

$$Q = L \, \tfrac{2}{3} \, \sqrt{2gh^3}.$$

Experiment has proved that in reality the quantity discharged is *less* than this, so that the theoretical result has to be multiplied by some co-efficient C_1 less than unity to arrive at the true quantity.

A large number of investigations have been carried out under various conditions to determine the co-efficient, and assuming that all the observations are correct, the accuracy of the results obtained depends upon whether the co-efficient employed is applicable to the particular conditions under which the experiment is made. Very great care should therefore be taken to select a co-efficient determined under circumstances corresponding as closely as possible to those in which the measurement is to be made.

It has been found that the co-efficient C_1 varies with the following dimensions and conditions :—

Length of bay;

Head over bay ;

Width of canal of approach ;

Nature and thickness of edges of bay ;

Presence or absence of thin edges at ends of bay (full width of canal or otherwise);

Distance from bottom of bay to bottom of canal.

With a given thickness and shape of the edges of the bay, the co-efficient C_1 varies chiefly with the *ratio* of the length of the bay to the width of the canal and with the *head* over the bay ; the distance from the bottom of bay to the bottom of the canal appears to have little influence on the results.

When the length of the bay is less than the width of

the canal of approach, so that the bay has thin edges at the ends, it is said to have *end contractions*, since the thin ends cause a contraction of the stream.

Braschmann has constructed a formula for determining the value of the co-efficient, *including the factor* ⅔, in which the ratio of the length L of the bay to the width B of the canal of approach and the head h over the bay are taken into account; it is as follows :—

$$Q = \left(0\cdot3838 + 0\cdot0386 \ \frac{L}{B} + \frac{0\cdot00174}{h}\right) L \ \sqrt{2gh^3},$$

all dimensions being in *feet*.

This formula is very generally used in Germany, and in many cases in which the author has compared it with experimental results shows a very fair agreement with the latter, in others a considerable discrepancy.

A formula of this kind must be used with discretion, and only applied within the limits for which it has been constructed.

The following table has been chiefly compiled from that given by Messrs. Donkin and Salter in their paper "On the Measurement of Water over Weirs," published in the *Minutes of Proceedings of the Institution of Civil Engineers*, vol. lxxxiii., 1885-86; the arrangement, however, is somewhat different, being according to the length of the bays to facilitate reference for practical purposes. Some slight additional data have been added, and two columns containing respectively the ratio of the length of the bay to the width of the canal of approach and the ratio of the length of the bay to the head.

The values of the co-efficients given are in most cases those resulting from the mean curves, also taken from Messrs. Donkin and Salter's paper.

Weisbach gives the following formula as embodying

the results of the experiments of Poncelet and Lesbros on the flow of water over tumbling bays :—

$$Q = \mu\, L\, h\, \sqrt{2gh},$$

where $\qquad\qquad \mu = \mu_o\, (1 + 1{\cdot}718\, n^4),$

and $\qquad\qquad n = \dfrac{L\, h}{F_o},$

F_o being the sectional area of the canal of approach and μ_o the co-efficient of Poncelet and Lesbros for a bay $0{\cdot}2$ mètre ($7{\cdot}874$ inches) long.

Francis, whose data are generally employed in America, gives the formula

$$Q = 3{\cdot}33\, L\, \sqrt{h^3},$$

equal to $\qquad\qquad Q = 0{\cdot}416\, L\, \sqrt{2gh^3},$

when there are no end contractions,

and $\qquad\qquad Q = 3{\cdot}33\, (L - 0{\cdot}1\, n\, h)\, \sqrt{h^3},$

when end contraction takes place, where n is the number of end contractions, generally 2.

Braschmann's formula agrees very well with Francis's results, but it will be seen on reference to the latter, in the case in which the weir has a length equal to the width of the canal of approach without end contractions, that the co-efficient is practically constant. The same is true of the experiments of Fteley and Stearns with the longer weir. With the shorter weir, 5 feet long, the results obtained by the same experimenters, with heads under 6 inches, show that the discharge is influenced by the head ; above 6 inches the co-efficient is nearly constant.

According to Braschmann's formula, the co-efficient varies in all cases with the head.

COMPARATIVE TABLE OF RESULTS OF EXPE-

Name of experimenter.	Length of bay of weir L	Depth over weir h	Width of canal of approach B		Co-efficient in formula $Q = C L \sqrt{2gh^3}$ C
	Inches.	Inches.	Ft.	In.	
Lesbros.	0·7874	3	12	1	0·435
,,	,,	6	,,		0·434
,,	,,	12	,,		0·433
,,	,,	18	,,		0·431
,,	,,	23½	,,		0·425
Castel.	1·181	1½	2	5⅛	0·4198
,,	,,	2	,,		0·4194
,,	,,	3	,,		0·4186
,,	,,	4	,,		0·4182
,,	,,	5	,,		0·4182
,,	,,	6	,,		0·4185
,,	,,	7	,,		0·4186
,,	,,	8	,,		0·4190
Donkin and Salter.	1·5	⅓	1	6	0·4256
,,	,,	¾	,,		0·421
,,	,,	1¼	,,		0·4174
,,	,,	1½	,,		0·4156
,,	,,	2	,,		0·4146
,,	,,	2¾	,,		0·4122
,,	,,	3	,,		0·412
Prof. Kennedy.	1·75	⅓	1	5	0·4666
,,	,,	¾	,,		0·435
,,	,,	1	,,		0·4256
,,	,,	1½	,,		0·414
,,	,,	2	,,		0·4053
Castel.	1·968	1½	2	5⅛	0·4090
,,	,,	2	,,		0·4086
,,	,,	3	,,		0·4082
,,	,,	4	,,		0·4080
..	..	5	..		0·4080

·RIMENTS ON FLOW OF WATER OVER WEIR.

Thickness of crest of bay or weir t	Distance from crest of bay to bottom of canal d		Distance back from weir at which head was measured l		Ratio of length of bay to width of canal $\dfrac{L}{B}$	Ratio of length of bay to head $\dfrac{L}{h}$
	Ft.	In.	Ft.	In.		
Very thin		$2\frac{1}{8}$	11	6	0·00543	0·2624
metal.		,,		,,	,,	0·1312
,,		,,		,,	,,	0·0656
,,		,,		,,	,,	0·0437
,,		,,		,,	,,	0·0335
$\frac{1}{16}$		$6\frac{3}{4}$	1	$7\frac{1}{4}$	0·0405	0·787
,,		,,		,,	,,	0·590
,,		,,		,,	,,	0·393
,,		,,		,,	,,	0·295
,,		,,		,,	,,	0·236
,,		,,		,,	,,	0·197
,,		,,		,,	,,	0·169
,,		,,		,,	,,	0·148
$\frac{1}{16}$		7	1	6	0·0833	3
,,		,,		,,	,,	2
,,		,,		,,	,,	1·2
,,		,,		,,	,,	1
,,		,,		,,	,,	0·75
,,		,,		,,	,,	0·545
,,		,,		,,	,,	0·5
$\frac{1}{16}$		7	1	9	0·1029	3·5
,,		,,		,,	,,	2·333
,,		,,		,,	,,	1·75
,,		,,		,,	,,	1·166
,,		,,		,,	,,	0·875
$\frac{1}{16}$		$6\frac{3}{4}$	1	$7\frac{1}{4}$	0·0679	1·312
,,		,,		,,	,,	0·984
,,		,,		,,	,,	0·656
,,		,,		,,	,,	0·492
,,		,,		,,	,,	0·393

Y

Name of experimenter.	Length of bay or weir L	Depth over weir h	Width of canal of approach B		Co-efficient in formula $Q = C\,L\sqrt{2gh^3}$ C
	Inches.	Inches.	Ft.	In.	
Castel.	1·968	6	2	$5\frac{1}{8}$	0·4082
,,	,,	7	,,		0·4086
,,	,,	8	,,		0·4090
,,	,,	9	,,		0·410
Smeaton.	3	$\frac{1}{4}$	Not given.		0·442
,,	,,	1	,,		0·424
,,	,,	2	,,		0·412
,,	,,	3	,,		0·388
Bidone.	3·048	3·47	2	1·31	0·3994
,,	,,	6·66	,,		0·4113
Castel.	3·937	$1\frac{1}{4}$	2	$5\frac{1}{8}$	0·411
,,	,,	$1\frac{1}{2}$	,,		0·403
,,	,,	2	,,		0·398
,,	,,	3	,,		0·395
,,	,,	4	,,		0·394
,,	,,	6	,,		0·395
,,	,,	8	,,		0·396
,,	,,	$9\frac{1}{2}$	,,		0·396
Smeaton.	6	$\frac{1}{2}$	Not given.		0·445
,,	,,	1	,,		0·431
,,	,,	2	,,		0·425
,,	,,	3	,,		0·420
,,	,,	4	,,		0·412
,,	,,	5	,,		0·402
,,	,,	6	,,		0·390
Bidone.	6·721	3,97	2	1·31	0·4003
Castel.	7·874	$1\frac{1}{4}$	2	$5\frac{1}{8}$	0·417
,,	,,	2	,,		0·407
,,	,,	3	,,		0·397
,,	,,	4	,,		0·395
,,	,,	6	,,		0·395
,,	,,	7	,,		0·397
,,	,,	8	,,		0·397

Thickness of crest of bay or weir t	Distance from crest of bay to bottom of canal d		Distance back from weir at which head was measured l		Ratio of length of bay to width of canal $\frac{L}{B}$	Ratio of length of bay to head $\frac{L}{h}$
	Ft.	In.	Ft.	In.		
$\frac{1}{16}$	6¾		1	7¼	0·0679	0·328
,,	,,		,,		,,	0·281
,,	,,		,,		,,	0·246
,,	,,		,,		,,	0·218
Not given.	Not given.		Not given.		—	12
,,	,,		,,		—	3
,,	,,		,,		—	1·5
,,	,,		,,		—	1
1·33	5·76		1	7¼	0·1204	0·878
,,	,,		,,		,,	0·457
$\frac{1}{16}$ approx.	6¾		1	7¼	0·1351	3·149
,,	,,		,,		,,	2·624
,,	,,		,,		,,	1·968
,,	,,		,,		,,	1·312
,,	,,		,,		,,	0·984
,,	,,		,,		,,	0·656
,,	,,		,,		,,	0·492
,,	,,		,,		,,	0·414
Not given.	Not given.		Not given.		—	12
,,	,,		,,		—	6
,,	,,		,,		—	3
,,	,,		,,		—	2
,,	,,		,,		—	1·5
,,	,,		,,		—	1·2
,,	,,		,,		—	1
1·33	5·76		1	7¼	0·2655	1·693
$\frac{1}{16}$	6¾		1	7¼	0·2703	6·299
,,	,,		,,		,,	3·937
,,	,,		,,		,,	2·625
,,	,,		,,		,,	1·968
,,	,,		,,		,,	1·312
,,	,,		,,		,,	1·125
,,	,,		,,		,,	0·984

Name of experimenter.	Length of bay or weir L	Depth over weir h	Width of canal of approach B		Co-efficient in formula $Q = C L \sqrt{2gh^3}$ C
	Inches.	Inches.	Ft.	In.	
Lesbros.	7·874	$1\frac{1}{4}$	12	1	0·410
,,	,,	2	,,		0·403
,,	,,	3	,,		0·397
,,	,,	4	,,		0·395
,,	,,	5	,,		0·394
,,	,,	6	,,		0·393
,,	,,	7	,,		0·392
Castel.	15·748	$1\frac{3}{8}$	2	$5\frac{1}{8}$	0·423
,,	,,	2	,,		0·417
,,	,,	3	,,		0·414
,,	,,	4	,,		0·414
,,	,,	5	,,		0·415
,,	23·622	$1\frac{1}{4}$	,,		0·435
,,	,,	$1\frac{1}{2}$	,,		0·430
,,	,,	2	,,		0·429
,,	·,,	3	,,		0·430
,,	,,	4	,,		0·429
Boileau.	34·4	2	2	11·43	0·420
,,	,,	4	,,		0·429
,,	,,	6	,,		0·437
,,	,,	8	,,		0·444
Blackwell.	36	1	30	0	0·450
,,	,,	2	,,		0·446
,,	,,	3	,,		0·453
,,	,,	4	,,		0·411
,,	,,	5	,,		0·401
Fteley and Stearns.	60	$1\frac{1}{4}$	5	0	0·456
,,	,,	$1\frac{1}{2}$	,,		0·443
,,	,,	$1\frac{3}{4}$	,,		0·431
,,	,,	2	,,		0·428
,,	,,	$2\frac{1}{2}$	,,		0·424
,,	,,	3	,,		0·422

Thickness of crest of bay or weir t	Distance from crest of bay to bottom of canal d		Distance back from weir at which head was measured l		Ratio of length of bay to width of canal $\frac{L}{B}$	Ratio of length of bay to head $\frac{L}{h}$
	Ft.	In.	Ft.	In.		
Knife edge.	1	$9\frac{1}{4}$	11	6	0·0543	6·299
,,	,,		,,		,,	3·937
,,	,,		,,		,,	2·625
,,	,,		,,		,,	1·968
,,	,,		,,		,,	1·575
,,	,,		,,		,,	1·312
,,	,,		,,		,,	1·125
$\frac{1}{16}$	6$\frac{3}{4}$		1	$7\frac{1}{4}$	0·5406	11·453
,,	,,		,,		,,	7·874
,,	,,		,,		,,	5·249
,,	,,		,,		,,	3·937
,,	,,		,,		,,	3·149
,,	,,		,,		0·8109	18·897
,,	,,		,,		,,	15·748
,,	,,		,,		,,	11·811
,,	,,		,,		,,	7·874
,,	,,		,,		,,	5·905
Sharp edge.	1	1·381	Head measured by glass gauge.		0·9709	17·2
,,	,,		,,		,,	8·6
,,	,,		,,		,,	5·733
,,	,,		,,		,,	4·3
$\frac{1}{16}$	Not given.		Not given.		0·1000	36·0
,,	,,		,,		,,	18·0
,,	,,		,,		,,	12·0
,,	,,		,,		,,	9·0
,,	,,		,,		,,	7·2
0·0792	3·17 feet.		6	0	1·000	48·0
,,	,,		,,		,,	40·0
,,	,,		,,		,,	34·289
,,	,,		,,		,,	30
,,	,,		,,		,,	24
,,	,,		,,		,,	20

Name of experimenter.	Length of bay or weir L	Depth over weir h	Width of canal of approach B		Co-efficient in formula $Q = C L \sqrt{2gh^3}$ C
	Inches.	Inches.	Ft.	In.	
Fteley and Stearns.	60	4	5	0	0·419
,,	,,	5	,,		0·418
,,	,,	6	,,		0·4174
,,	,,	7	,,		0·4170
,,	,,	8	,,		0·4166
,,	,,	9	,,		0·4170
,,	,,	10	,,		0·4174
Deacon.	96	4·248	—		0·4504
,,	,,	6·000	—		0·4343
,,	,,	6·756	—		0·4167
Francis.	120	7	14	0	0·4146
,,	,,	8	,,		0·4146
,,	,,	9	,,		0·414
,,	,,	10	,,		0·414
,,	,,	12	,,		0·414
,,	,,	14	,,		0·4146
,,	,,	16	,,		0·4148
,,	,,	18	,,		0·4143
Francis.	168	9	14	0	0·4162
,,	,,	10	,,		0·4163
,,	,,	11	,,		0·416
,,	,,	12	,,		0·416
Fteley and Stearns.	228	6	19	0	0·412
,,	,,	8	,,		0·4114
,,	,,	10	,,		0·411
,,	,,	12	,,		0·4106
,,	,,	14	,,		0·4110
,,	,,	16	,,		0·4111
,,	,,	18	,,		0·4110
..	..	19½	..		0·4103

Thickness of crest of bay or weir t	Distance from crest of bay to bottom of canal d	Distance back from weir at which head was measured l		Ratio of length of bay to width of canal $\dfrac{L}{B}$	Ratio of length of bay to head $\dfrac{L}{h}$
		Ft.	In.		
0·0792	3·17 feet.	6	0	1·000	15
,,	,,	,,		,,	12
,,	,,	,,		,,	10·000
,,	,,	,,		,,	8·571
,,	,,	,,		,,	7·500
,,	,,	,,		,,	6·666
,,	,,	,,		,,	6·000
$\frac{3}{32}$	—	—		,,	22·599
,,	—	—		,,	16·000
,,	—	—		,,	14·209
$\frac{1}{4}$	5 0	6	0(?)	0·7143	17·143
,,	,,	,,		,,	15·000
,,	,,	,,		,,	13·333
,,	,,	,,		,,	12·000
,,	,,	,,		,,	10·000
,,	,,	,,		,,	8·571
,,	,,	,,		,,	7·500
,,	,,	,,		,,	6·666
$\frac{1}{4}$	5 0	6	0(?)	1·000	18·666
,,	,,	,,		,,	16·800
,,	,,	,,		,,	15·273
,,	,,	,,		,,	14·000
$\frac{1}{4}$	6·55	6	0	1·000	38·000
,,	,,	,,		,,	28·500
,,	,,	,,		,,	22·800
,,	,,	,,		,,	19·000
,,	,,	,,		,,	16·285
,,	,,	,,		,,	14·250
,,	,,	,,		,,	12·666
,,	,,	,,		,,	11·692

In order that Francis's formula may give reliable results, its author states that the edges of the weir (or bay) presented to the current must be sharp; if bevelled or rounded off in any perceptible degree a material effect will be produced on the discharge. It is essential also, that the stream should touch the orifice only at these edges, after passing which it should be discharged through the air.

The velocity of the water approaching the weir should be very slow, and to this end the depth of the bottom of the canal below the weir-crest should not (according to Francis) be less than *one-third* of the length of the weir, otherwise the velocity of approach must be taken into account, as it tends to increase the quantity of water discharged.

With a thick weir-crest the co-efficient of discharge is greater than with a thin one. Experiments with weirs formed of planks 1 inch thick and over were made by Blackwell, but as the majority of other experiments of this kind have been carried out with thin bays, it has not been considered necessary to record the former here.

The most recent experiments on this subject, as far as the author is aware, are those carried out by Bazin, and recorded in the *Annales des Ponts et Chaussées* (October 1888, p. 446).

(5) *Measurement by Meters.*

The gauging of water by meters is only applicable to comparatively small quantities, and is rarely practicable in the case of hydraulic motors. It depends for its accuracy on the construction of the particular form of meter used, which in every case should be carefully tested, and the error, if any, ascertained before it is employed.

Measurement of Water Levels.

For ascertaining the level of the surface of the water, for the purpose of determining either the head acting on a turbine or the depth over a weir, the hook gauge is the best instrument to employ. It consists of a metal rod bent up at one end to form a hook with a sharp point, and adjustable by means of a nut working on a thread cut in the long shank of the hook, the readings being taken by a micrometer. The hook is used point upward, the point being at first submerged, and the position noted on the scale at which it emerges from the surface of the water; this can be done with great accuracy, much exceeding that obtainable by the reverse process, that of observing the position in which a point, at first above the water, comes into contact with it. The hook must, of course, be raised and lowered vertically, and should be enclosed in a box or casing, open at the top, to which the water is admitted only through a comparatively small hole in the bottom in order to protect the surface of the fluid in the immediate neighbourhood of the gauge from disturbances.

In ascertaining the upper level of the water driving a turbine, the measurement should *not* be taken *directly over the motor,* but in the head-race immediately before the water enters the turbine chamber or wheel pit, at a point where the surface is as nearly as possible undisturbed. Over the motor, especially when the entrance to the chamber is not central, there is often a depression of the surface caused by a whirling motion or vortex, and if observed here the head would have too small a value.

The "Venturi" Water Meter.

A new method of measuring the quantity of water passing through a pipe has recently been proposed and tried by Mr. Clemens Herschel, the engineer of the Holyoke Water-power Company.[1]

This method is based upon the fact, investigated by Venturi, that when water flows through a pipe of which the section is contracted and subsequently gradually increased, the pressure in the smallest section is much less than in the largest on either side of the contraction, and may with suitable proportions sink below the atmospheric pressure, so that it can be measured by a vacuum gauge. The velocity in the smallest section is *theoretically* that due to the effective head corresponding to the difference between the pressure in the largest section before the contraction and that in the smallest section *plus* the influence of the velocity in the largest section, generally very slight. To obtain the *actual* velocity the theoretical quantity has to be multiplied with an experimental co-efficient. It is evident that when the latter is accurately known, together with the sectional area of the smallest part of the pipe, the true velocity of flow can be calculated from the observed difference of pressure above referred to.

Mr. Clemens Herschel's experiments were directed to determining the value of the co-efficient. Two sizes of "Venturi" ajutages were tried of exactly similar geometric form, one inserted in a pipe of 1 foot diameter, the other in a pipe of 9 feet diameter. The ajutage (or "venturi," as it has been christened by Mr. Herschel) consists of a pipe which converges and then diverges, and constitutes

[1] *The Venturi Water Meter*, by Clemens Herschel, M.Am.Soc.C.E., *Transactions of the American Society of Civil Engineers*, vol. xvii., Nov. 1888.

with the apparatus for recording the pressure the actual meter.

The diameter at the smallest section was one-third the diameter of the large ends of the conical portions, the angle of the cones being small. The length of the smaller cone was in each case about 1·68 diameters, that of the larger 7·37 diameters —referred to the greater diameter. The apparatus was placed so that the *shorter* cone was at the up-stream end. In the trials with the 1-foot tube an annular air-chamber surrounded the "venturi" at the smallest diameter, communicating by four accurately drilled holes, ¼ inch in diameter, with the interior of the pipe. The interior of the "venturi" at the small part was carefully polished, leaving the edges of the holes sharp and square.

For the 9-foot pipe the "venturi" had *eight separate* air-chambers, one for each ¼-inch hole communicating with the interior. These several air-chambers had each a suction-pipe attached, while only a single suction-pipe was used with the annular chamber of the 1-foot "venturi."

The trials with the smaller size of "venturi" were made first, and in the subsequent experiments with the 9-foot meter certain defects which had made themselves apparent were remedied, so that the second series of trials may be taken as the more accurate.

Experiments were made with velocities of flow ranging from 2 to 50 feet per second, while the depression at the "venturi" ranged from a very small amount to about 40 feet of water column.

In the case of the 9-foot tube the deviation of any single experiment from the resultant mean was only ½ per cent. for the whole range of velocities, and if both series of experiments are taken, the deviation did not exceed 3 per cent.

By limiting the use of the meter to velocities greater

than 9 feet per second through the "venturi," or about 1 foot per second through the corresponding pipe, the range of variation becomes even less than that above stated.

The discharge during the experiments always took place under water. The quantity of water discharged was measured in a collecting tank in the case of the 1-foot tube, and by a weir for the 9-foot tube.

The value of the co-efficient C to be used in the formula—

$$\text{Velocity} = C \sqrt{2gh},$$

where h is the effective head on the "venturi," was found in most cases to be near 0·99, but varied slightly with the velocity.

If p_1 denote the pressure in the largest section of the meter on the up-stream side of the contraction, p_2 the pressure in the smallest section, A_1 and A_2 respectively the areas of the largest and smallest sections, then by the hydrodynamic equation—

$$p_1 + \left(\frac{A_2}{A_1}\right)^2 \frac{c^2}{2g} = p_2 + \frac{c^2}{2g}.$$

—where c is the velocity of flow through the smallest section—and thence

$$c = \sqrt{\frac{(p_1 - p_2)\, 2g}{1 - \left(\frac{A_2}{A_1}\right)^2}},$$

where p_1 and p_2 are expressed as *head*, and therefore

$$p_1 - p_2 = h.$$

Since $\dfrac{A_2}{A_1} = \tfrac{1}{9}, \ 1 - \left(\dfrac{A_2}{A_1}\right)^2 = \dfrac{80}{81},$

and theoretically $c = 1\cdot0062 \sqrt{2gh}.$

Figures 127 and 128 show the construction of the "Venturi" meter in the 9-foot pipe.

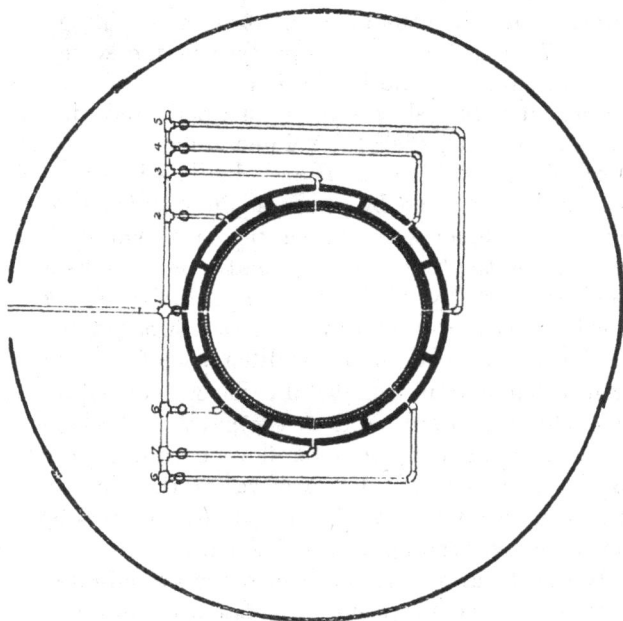

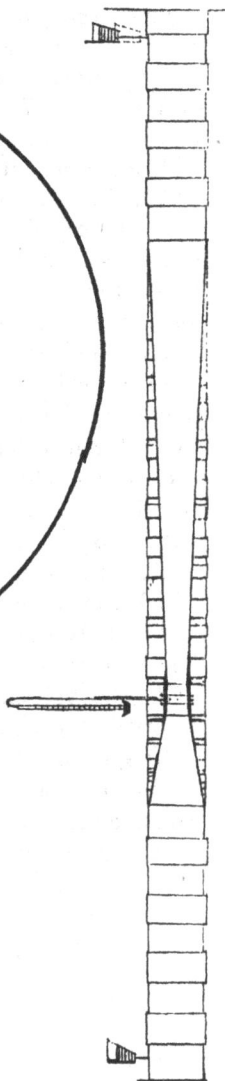

Fig. 128.

Since the publication of the first edition of this work, the "Venturi" meter has come into practical use in the United States. At the Columbian Exposition at Chicago (1893) a 36″ "Venturi" meter was employed for measuring the main water supply to the Exhibition.

The meter is now furnished with an automatic recorder, which shows by ordinary dials the volume of flow.

As described by Mr. Clemens Herschel, "The Recorder is connected with the tube by pressure pipes, which lead to it from the chamber surrounding the up-stream end, and the throat of the tube. It is operated in part by a weight and in part by clockwork. The difference of pressure or head at the entrance and at the throat of the meter is balanced in the Recorder by difference of level in two columns of mercury in cylindrical receivers, one within the other. The inner carries a float, the position of which is indicative of the quantity of water flowing through the tube. By its rise and fall the float varies the time of contact between an integrating drum and the counters by which the successive readings are registered.

"Usually the drum revolves once in every ten minutes, and at each revolution the counter registers, on ordinary dials, the volume flowing for that period of time. The interval of time may be shortened if desired."

The apparatus may be placed in any convenient position within one thousand feet of the tube, and there may also be an electric device by which the record can be made at any distance from the meter.

CHAPTER XI.

DESCRIPTIONS OF AND EXPERIMENTS WITH TURBINES.

General remarks.—Rittinger's experiments with Jonval turbines.
—Tables of results.—Haenel's turbine.—Tables of results.—Curves.
—Jonval turbines at Goeggingen.—Turbines at Olching.—Francis's
experiments; Tremont and Boott turbines.—Analysis of results.
—Holyoke testing station.—"Collins" turbine; with and without suction-tube. — "Boyden" turbine. — "Hercules" turbine. —
Francis's experiments with "Humphrey" turbine.—"Little Giant"
turbine.—Girard turbines at Varzin.—Outward-flow impulse turbines at Terni.—Turbine at Immenstadt.—Turbines at Schaffhausen.—The "Pelton" wheel.—Small impulse turbines for high
falls.—Fourneyron turbine at St. Blaise.—Niagara Installation.—
Francis's experiments with a "Swain" turbine.—Table of experimental and calculated results, A.—Tables of dimensions of turbines
actually made, B and C.—The best speed of turbines as tested by
experiment.—Causes of differences between calculated and experimental results.—Fliegner's experiments.—General remarks.

ALTHOUGH a large number of experiments and tests of
turbines have been carried out at various times, only a
comparatively small proportion of the published results
obtained are of any use for the purpose of accurate comparison and study. In this respect, some of the older
experiments are more valuable than those of more recent
date, owing to their greater thoroughness and the more
general conditions under which they were performed, more

especially as regards the range of speeds. In many instances the dimensions given are incomplete, so that a comparison of observed with calculated results is not possible; generally it is the angles of the vanes which are omitted. In all tests with turbines, the two chief quantities measured are the available power and the power actually developed. The latter can be easily determined with tolerable accuracy by means of the brake when ordinary precautions are taken, but for the former it is necessary to ascertain the quantity of water used by the motor, and this is an operation requiring more than average care and nicety of observation if reliable results are to be obtained.

In the following, those experiments have been selected for description of which the fullest account of the water- and power-measurements is available, and which appear from the character of the latter to afford the best internal evidence of their trustworthiness. In general the most complete record obtainable of the dimensions of the turbines employed is given, and will be found, together with the relative proportions and various particulars as to speed and velocity of flow for the maximum efficiency, in Table A, as well as in the detailed account of the experiments in question.

In Table A are also stated the results obtained by calculation from observed quantities. From the *observed* quantity of water and *measured effective area* of guide passages the velocity of flow has been determined, and the corresponding speed of rotation calculated for comparison with that experimentally obtained; the theoretical co-efficient K_1 for the velocity of flow has been calculated from the observed efficiency by the formula—

$$K_1 = \sqrt{\dfrac{\epsilon}{2\dfrac{A}{A_2}\sin a\,\sin a_2}}\ (vide \text{ p. 109)};$$

this affords the means of testing the agreement between theory and practice.

In the case of the older turbines, such as Rittinger's, the efficiencies of which are considerably below those now attained by the best wheels, the values of the co-efficients ζ and ζ_1 as previously given are not applicable, and if c is calculated by formula (vi A) with those values, the agreement between the observed and calculated results will not be good. For this reason the above expression for K_1 containing the efficiency, which can be directly noted, has been chosen as affording an accurate test of the correctness of theoretical assumptions.

The *measured* area A it must be remembered is not the *effective* area, but to arrive at the latter an allowance must be made for the obstruction of the orifices by the passing wheel-vanes.

In some cases this has only been estimated, as the thickness of the wheel-vanes at the edges is not given.

Rittinger's Experiments with Jonval Turbines.

Results of these experiments have been several times quoted in illustration of various points in the theory of reaction turbines. They are of special interest for several reasons: in the first place, each series was made with turbines of approximately the same power and working with the same head, but differing in some of their proportions; in the second place, they extended over a considerable range of speed, and last, but not least, the quantity of water was actually measured in a collecting

z

reservoir, so that there is no doubt as to the accuracy of this part of the observations, which in ordinary tests is most open to question.

Three different series of experiments at various times were made by Mr. Rittinger with eight turbines in all, and the leading results of these are given in the sequel, to some extent re-arranged, together with the principal dimensions and proportions.

All these experimental turbines of Rittinger's were as compared with present practice somewhat abnormal in their design, chiefly in the dimensions of the vane-angles; in most of them the angle a_1 was negative, and several had back-vanes. It is probably partly on this account that the efficiency did not prove higher. In turbines Nos. 1, 4, 5, 6, 7 and 8, the angle a_1 was negative, while in turbines 2 and 3 the same angle had a very considerable positive value, in all cases very different to that now usually adopted.

Experiments were made with all the guide passages fully open, and also with part of them closed, but only the former are here given, as the others have not so much interest for the purpose of comparison with theoretical results.

For comparison with the observed speed, the mean velocity of rotation has been calculated for three assumptions, (1) that the inflow takes place without shock, (2) that the outflow is vertical, and (3) that the best speed is represented by $\frac{1}{2}$ (c_2 sin a_2 + c sin a), an expression, it will be remembered, obtained on the hypothesis that the velocity of flow is constant for all speeds, and therefore from a theoretical point of view inaccurate.

It will be seen that the calculated speeds for inflow without shock and vertical outflow with some exceptions agree very fairly; this proves that the turbines have been designed in accordance with the generally accepted principles.

The correspondence between the speed thus fixed and that obtained by experiment is, however, not very close, but is generally better than between the latter and the velocity as calculated on assumption (3).

That the agreement is not closer is due to the unusual dimensions of the vane-angles; as long as the latter are within the limits usual in ordinary construction, theory and practice harmonize very well together, but from causes explained in the chapter "On the Best Speed of Turbines," this concord is not so complete under abnormal conditions, as then the relative direction of inflow accompanied by least resistance does not coincide with the vane-angle.

The results in the following tables are given in the original Austrian measures, which do not differ very greatly from English standards.

1 Vienna foot   = 1·03710 English foot.
1 Vienna sq. foot   = 1·07557 English sq. foot.
1 Vienna cubic foot = 1·11546 English cubic foot.

Results of Rittinger's Experiments.

Turbine No. 1.

a = 78° 0′. Angle of guide-vanes at outflow.
a_1 = −61° 0′. Angle of wheel-vanes at inflow.
a_2 = 64° 0′. Angle of wheel-vanes at outflow.
r = 10·833 Vienna inches : Mean radius of wheel.
z = 12 : Number of guide-vanes.
z_1 = 24 : Number of wheel-vanes.
e_1 — 4·333 Vienna inches : Width of wheel-buckets at top.
e_2 = 4·333 Vienna inches : Width of wheel-buckets at bottom.

Depth of guide passages at outflow, 15 Vienna lines.
Depth of wheel-buckets at outflow, 18 Vienna lines.

N.B.—1 *Vienna inch* = 12 *lines.*

$$\frac{A}{A_2} = 0\cdot522. \quad A = 0\cdot452 \times 0\cdot9.$$

(0·9 is a co-efficient to allow for obstruction by wheel vanes.)

	h	Q	n	ε_e	Remarks.
	Vienna feet.	Vienna cubic ft.			
1	6·25	4·98	160	0	
2	6·21	5·25	147	0·326	
3	6·19	5·33	144	0·395	
4	6·18	5·40	140	0·455	
5	6·17	5·48	136	0·509	
6	6·16	5 57	131	0 553	
7	6·15	5·66	126	0·590	Low degree of reaction.
8	6·14	5·74	122	0·626	
9	6·13	5·83	116	0·646	
10	6·12	5·92	110	0·659	
11	6·11	6·02	105	0·671	
12	**6·10**	**6·07**	**102**	**0·673**	
13	**6·09**	**6·12**	**99**	**0·673**	15·044 = c.
14	6·08	6·16	96	0·672	
15	6·08	6·22	93	0·667	
16	6·07	6·32	87	0·656	
17	6·06	6·43	80	0·631	

Feet per second

Best speed as observed . . . = 9·358
(1) Speed for inflow without shock . . = 9·056
(2) Speed for vertical outflow . . . = 7·060
(3) Speed calculated by

$$w = \tfrac{1}{2}(c_2 \sin a_2 + c \sin a) \quad . \quad . = 10\cdot866$$

Mean of (1) and (2) . . . . = 1·058

Turbine No. 2.

$a\ \ = 73°\ 0'$

$a_1 = 54°\ 0'.$

$a_2 = 74°\ 0'.$

$r\ \ = 10·833$ Vienna inches.

$z\ \ = 12.$

$z_1 = 24.$

$c_1 = 4·333$ Vienna inches.

$c_2 = 4·333$ Vienna inches.

Depth of guide passages at outflow, 17 Vienna lines.

Depth of wheel-buckets at outflow, 6 Vienna lines.

$\dfrac{A}{A_2} = 1·274.$ $A = 0·511 \times 0·9$ Vienna square feet.

	h	Q	n	z_e	Remarks.
	Vienna feet.	Vienna cubic ft.			
1	6·25	5·25	242	0	
2	6·25	5·12	201	0·452	
3	6·25	5·13	189	0·533	
4	6·25	5·11	177	0·601	
5	6·25	5·08	164	0·655	High degree of reaction.
6	6·25	5·06	150	0·687	
7	6·25	5·06	143	0·695	
8	**6·25**	**5·04**	**135**	**0·697**	**10·955** $= c.$
9	6·25	5·02	127	0·695	
10	6·25	5·01	119	0·688	
11	6·25	4·99	102	0·650	
12	6·25	4·97	84	0·586	

Feet per second.

Best speed as observed . . . $= 12·762$

(1) Speed for inflow without shock . . $= 14·899$

(2) Speed for vertical outflow . . . $= 13·400$

(3) Speed calculated by

$\qquad w = \frac{1}{2}\,(c_2\ sin\ a_2 + c\ sin\ a)$ . . $= 11·943$

Mean of (1) and (2) . . . $= 14·149$

Turbine No. 3.

$a = 68° 0'.$
$a_1 = 65° 0'.$
$a_2 = 75° 0'.$
$r = 10.833$ Vienna inches.
$z = 12.$
$z_1 = 12.$
$e_1 = 4.333$ Vienna inches.
$e_2 = 4.333$ Vienna inches.

Depth of guide passages at outflow, 24 Vienna lines.
Depth of wheel-buckets at outflow, 13 Vienna lines.

$\dfrac{A}{A_2} = 1.661.$ $A = 0.722 \times 0.9$ Vienna square feet.

	h	Q	n		Remarks.
	Vienna feet.	Vienna cubic ft.			
1	5 95	6·97	268	0	
2	6·01	6·66	224	0·405	
3	6·03	6·60	211	0·479	
4	6·05	6·55	198	0·543	Very high degree of
5	6·06	6·49	183	0·589	reaction.
6	6·08	6·43	168	0·622	
7	6·09	6·40	160	0·631	
8	**6·10**	**6·38**	**151**	**0·632**	9·811 = *c.*
9	6·10	6·35	142	0·630	
10	6·11	6·32	132	0·618	
11	6·13	6 24	113	0·587	
12	6·15	6·19	91	0·319	

Feet per second.
Best speed as observed   .   .   = 14·272
(1) Speed for inflow without shock   .   . = 16·933
(2) Speed for vertical outflow   .   .   = 15·746

Feet per second.

(3) Speed calculated by
$$w = \tfrac{1}{2} (c_2 \sin a_2 + c \sin a) \qquad . = 12{\cdot}418$$
Mean of (1) and (2) . . . $= 16{\cdot}339$

With respect to Rittinger's turbines Nos. 1, 2, and 3, with which the first series of trials was carried out at Blankso, it should be noticed that they were chiefly designed to test the influence of varying degrees of reaction on the performance of the motors. The results showed that turbine No. 3, with the highest degree of reaction, gave the lowest efficiency, but in the experiments made with part of the guide passages closed, it also proved to be less sensitive in the matter of efficiency to the influence of this method of regulation.

The closing of half the guide passages reduced the efficiency of turbine No. 1 by 13 per cent., of turbine No. 2 by 19 per cent., and of turbine No. 3 by only 10 per cent.

A deviation from the *best speed* of 10 per cent. reduced the efficiency

of turbine No. 1 by 1 to 2 per cent.

„ No. 2 by $\tfrac{1}{2}$ per cent.

„ No. 3 by $\tfrac{1}{3}$ per cent.

The ratio of the best speed to the speed without a load was

for turbine No. 1 as 1 to 1·61,

„ No. 2 as 1 to 1·79,

„ No. 3 as 1 to 1·77.

The quantity of water consumed with turbine No. 1 increased as the speed diminished, owing to the negative value of the angle of inflow a_1, while with turbines Nos. 2 and 3, in which a_1 was positive, the contrary occurred. This is in accordance with theory.

Turbine No. 4.

$a = 78°$.

$a_1 = -63$.

$a_2 = 65°$.

$r = 8.333$ Vienna inches.

$z = 12$.

$z_1 = 24$.

$c_1 = 3.333$ Vienna inches.

$e_2 = 3.333$ Vienna inches.

Depth of guide passages at outflow, 8·17 Vienna lines.

Depth of wheel-buckets at outflow, 8·8 Vienna lines.

Depth of wheel-buckets at inflow,   9·33 Vienna lines.

$\dfrac{A}{A_2} = 0.418$. $A = 0.1887 \times 0.9$ Vienna square feet.

	h	Q	n	ε_e	Remarks.
	Vienna feet.	Vienna cubic ft.			
1	9·34	3·06	289	0	Turbine with ordinary
2	9·22	3·78	186	0·693	bucket form.
3	9·20	3·89	171	0·711	
4	**9·20**	**3·94**	**164**	**0·715**	**23·20 = c.**
5	9·19	4·00	156	0·711	
6	9·17	4·05	141	0·705	Gradually decreasing
7	9·16	4·17	126	0·674	depth of passages between
8	9·14	4·30	112	—	wheel-vanes.

Feet per second

Best speed as observed   .   .   . = 11·922

(1) Speed for inflow without shock   . = 13·224

(2) Speed for vertical outflow   .   . = 8·779

(3) Speed calculated by

$w = \frac{1}{2} (c_2 \sin a_2 + c \sin a)$.   . = 15·738

Mean of (1) and (2)   .   .   . = 11·001

Turbine No. 5.

$a = 76° 0'.$

$a_1 = -61° 0.$

$a_2 = 65° 0'.$

$r = 8\cdot333$ Vienna inches.

$e_1 = 3\cdot333$ Vienna inches. $z = 12.$

$c_2 = 3\cdot333$ Vienna inches. $z_1 = 24.$

Depth of guide passages at outflow, $9\cdot5$ Vienna lines.

Depth of wheel-buckets at outflow, $8\cdot5$ Vienna lines.

Depth of wheel-buckets at inflow, $9\cdot5$ Vienna lines.

$\dfrac{A}{A_2} = 0\cdot504.$ $A = 0\cdot220 \times 0\cdot9$ Vienna square feet.

	h	Q	n	ε_o	Remarks.
	Vienna feet.	Vienna cubic ft.			
1	9·18	3·03	289	0	Passages between wheel-
2	9·14	3·60	184	0·623	vanes of equal *horizontal*
3	9·14	3·66	176	0·636	depth throughout.
4	9·13	3·71	167	0·641	
5	**9·13**	**3·77**	**159**	**0·645**	**19·040** $= c.$
6	9·12	3·82	150	0·639	
7	9·12	3·95	143	0·618	
8	9·11	4·08	115	0·575	

Feet per second.

Best speed as observed . . . $= 11\cdot471$

(1) Speed for inflow without shock . $= 11\cdot464$

(2) Speed for vertical outflow . . $= 8\cdot697$

(3) Speed calculated by

$w = \frac{1}{2} (c_2 \sin a_2 + c \sin a)$. $= 13\cdot581$

Mean of (1) and (2) . . . $= 10\cdot080$

The experiments with turbines Nos. 4 and 5 were

made at Mariazell and intended to test the effect on their performance of different methods of constructing the buckets.

In turbine No. 4 the buckets were of the form usual in turbines of the type in question, with negative angle a_1; in No. 5 the wheel-vanes were so designed that the clear distance between them *measured parallel with the direction of rotation* was constant throughout.

Turbine No. 4 gave a maximum efficiency of $71\frac{1}{2}$ per cent., while No. 5 developed only $64\frac{1}{2}$ per cent.

With only $\frac{3}{4}$ of the guide passages open, the efficiency sank to about 40 per cent. in the case of No. 4, and 37 per cent. with $\frac{1}{2}$ the passages closed; under the latter conditions turbine No. 5 had an efficiency of 35 per cent.

The ratio of the best speed to the speed without a load was

for turbine No. 4 as 1 to 1·76,
„　　No. 5 as 1 to 1·81.

Turbine No. 6.

$a = 75°\ 20'.$

$a_1 = -\ 59°\ 40'.$

$a_2 = 63°\ 50'.$

$r\ =\ 0·4856$ Vienna feet.

$e_1 = 0·195$ Vienna feet.　$z = 12.$

$e_2 = 0·195$ Vienna feet.　$z_1 = 24.$

Depth of guide passages at outflow, 7·2 Vienna lines.

Depth of wheel-buckets at outflow, 6·0 Vienna lines.

Depth of wheel-buckets at inflow,　7·2 Vienna lines.

$\dfrac{A}{A} = 0·54.$　$A = 0·1174 \times 09$ Vienna square feet.

	h	Q	n	ε_e	Remarks.
	Vienna feet.	Vienna cubic ft.			
1	17·61	2·41	583	0	
2	17·58	2·48	540	0·231	
3	17·55	2·55	493	0·410	
4	17·52	2·62	447	0·544	
5	17·49	2·69	400	0·633	
6	17·46	2·75	354	0·687	
7	17·45	2·79	331	0 697	
8	**17·43**	**2·82**	**308**	**0·700**	**26·689** = c.
9	17·42	2·86	285	0·694	
10	17·40	2·89	262	0·679	
11	17·37	2·96	216	0·626	

Feet per second.

Best speed as observed . . . $= 15·645$
(1) Speed for inflow without shock . . $= 14·252$
(2) Speed for vertical outflow . . . $= 12·934$
(3) Speed calculated by
$$w = \tfrac{1}{2}\,(c_2 \sin a_2 + c \sin a) \quad . \quad = 19·375$$
Mean of (1) and (2) . . . $= 13·593$

Turbine No. 7.

$a = 75° 50'$.
$a_1 = -66° 0'$.
$a_2 = 61° 0'$.
$r = 0·4856$ Vienna feet.
$e_1 = 0·195$ Vienna feet. $z = 12$.
$e_2 = 0·195$ Vienna feet. $z_1 = 20$.
Depth of guide passages at outflow, 6·9 Vienna lines.
Depth of wheel-buckets at outflow, 8·6 Vienna lines.
Depth of wheel-buckets at inflow, 6·9 Vienna lines.

$\dfrac{A}{A_2} = 0·433$. $A = 0·1117 \times 0·9$ Vienna square feet.

	h	Q	n	ε_e	Remarks.
	Vienna feet.	Vienna cubic ft.			
1	17·45	2·84	430	0·485	
2	17·41	2·92	392	0·574	Ordinary buckets.
3	17·37	3·00	357	0·638	
4	17·33	3·08	320	0·670	
5	17·30	3·12	302	0·678	31·03 = c.
6	17·28	3·16	283	0·676	
7	17·24	3·24	247	0·659	
8	17·20	3·32	210	0·616	

Feet per second

Best speed as observed . . . = 15·352
(1) Speed for inflow without shock . . = 13·034
(2) Speed for vertical outflow . . . = 11·756
(3) Speed calculated by
$$w = \tfrac{1}{2} (c_2 \sin a_2 + c \sin a) \quad . \quad . = 20\cdot927$$
Mean of (1) and (2) . . . . = 12·365

Turbine No. 8.

The dimensions of this turbine were the same as those of No. 7, the only difference being in the form of the buckets.

	h	Q	n	ε_e	Remarks.
	Vienna feet.	Vienna cubic ft.			
1	17·49	2·92	440	0·481	Bucket with passages
2	17·36	3·00	403	0·576	of nearly equal
3	17·32	3·08	367	0·641	depth throughout.
4	17·29	3·18	332	0·679	
5	17·27	3·20	315	0·691	
6	17·25	3·24	297	0·691	32·222 = c.
7	17·24	3·28	279	0·690	
8	17·22	3·32	261	0·680	
9	17·19	3·40	225	0·645	

Feet per second.

Best speed as observed . . . $= 15\cdot100$

(1) Speed for inflow without shock . . $= 13\cdot533$

(2) Speed for vertical outflow . . . $= 12\cdot205$

(3) Speed calculated by

$\qquad w = \frac{1}{2}\ (c_2\ sin\ a_2 + c\ sin\ a)$ . . $= 21\cdot730$

Mean of (1) and (2) . . . . $= 12\cdot869$

If no allowance be made in calculating the effective value of A for the obstruction of the guide passages by the passing wheel-vanes, then all of Rittinger's turbines have been designed for simultaneous inflow without shock and vertical outflow. Rittinger in designing the wheels appears to have neglected the obstruction caused by the wheel-vanes; this seems to be confirmed by the fact that turbines 4 and 5 were designed for a consumption of water per second of respectively 4·77 and 5·01 cubic feet (Austrian), whereas at the best speed the experiments showed that only 3·94 and 3·77 cubic feet were actually used, indicating probably an insufficient area of the guide orifices.

Turbines 6, 7, and 8 were also tested at Blankso. No. 6 had ordinary wheel-vanes, while Nos. 7 and 8, in other respects exactly alike, differed in the construction of the vanes. No. 7 was made with ordinary vanes and passages of decreasing depth towards the outflow, while No. 8 had passages of nearly equal depth throughout, measured perpendicular to the direction of flow—the object being to determine which method of design gave the better results.

The difference in efficiency of the three wheels was very slight.

The ratio of the best speed to the speed without a load was

for turbine No. 6 as 1 : 1·90.

An inspection of the experimental and calculated results obtained with Rittinger's turbines, as given in a compendious form in Table A for the best speeds, shows very plainly the effect of varying ratios $\dfrac{A}{A_2}$ on the velocity of flow and speed.

The agreement between the theoretical and experimental values of K_1 is on the whole satisfactory, especially when it is considered that the effective area of outflow from the guide passages, from which the experimental velocity of flow must be calculated, is somewhat uncertain.

For turbines 1, 5, 6, 7, and 8, the speed for inflow without shock agrees most closely with the observed best speed; for turbine 4 the mean of the speeds for inflow without shock and vertical outflow gives the closest approximation to the experimental value. In all of these wheels the relative angle of inflow a_1 is negative; and for 1, 5, 6, 7, and 8, the observed speed is *greater* than that corresponding to inflow without shock.

With turbines 2 and 3 the actual speed agrees best with that calculated for vertical outflow. Each of these wheels has a very unusually large *positive* angle a_1 and exceptionally high ratios $\dfrac{A}{A_2}$. The observed speed is *less* than that calculated for inflow without shock.

In the absence of available data as to the thickness of the wheel-vanes at the inlet, it has been assumed that the obstruction of the guide orifices by the passing wheel-vanes reduces their area by 10 per cent., so that to arrive at the effective value of A, the measured area A^1 has been multiplied by 0·9. This would be rather too small a co-efficient for wheels of large or moderate size, but for small turbines like those in question is probably not too low, since in a small motor the thickness of the vanes is relatively to the area of the passages greater than in a large one.

Haenel's Experiments with an Axial Turbine.

About the year 1861 Edward Haenel, who is generally credited with having constructed the first turbines with back-vanes, carried out a series of trials with a turbine of his own design, which were very complete and instructive in their results.

Haenel, who was at that time manager of the Gräflich Stolberg'schen Maschinenfabrik at Magdeburg, made the experiments in question in conjunction with Mr. J. C. Bernhard-Lehmann, the present manager of the Maschinenfabrik von Queva at Erfurt.

Description of the Turbine.

The turbine with which the trials were made was one of eight, all exactly alike, constructed to drive a mill at Rothenburg on the Saale. Fig. 130 (p. 362) shows a vertical section of this motor.

The guide apparatus consisted of an inner casing carrying the guide-vanes, and an outer casing connected with the supporting-frame. The vanes, thirty-two in number, were fastened by rivets to projections on the inner casing. The thickness of twenty-four of these vanes was $\frac{1}{4}$ inch; that of the remaining eight, $\frac{3}{8}$ inch. To the ends of the latter were attached radial wrought-iron bars projecting beyond the outer circumference of the vanes and fitting into corresponding recesses in the outer casing; these bars carried the weight of the guide-wheel and regulating apparatus. Surrounding the outer casing was an annular channel, closed at the top by a wrought-iron ring, and divided by radial ribs into thirty-two compartments, each communicating with a guide passage by means of a hole drilled through the casing. The wrought-iron ring had an equal number of corresponding holes, in each of which was inserted a wrought-iron pipe reaching above the head-water level and communicating with the atmosphere. A boss with four arms was connected by set screws to the inner casing of the guide-wheel, and the spaces between the arms filled in by sectors of iron plate.

The boss contained a bearing for the turbine-shaft consisting of three blocks of lignum-vitæ held in cast-iron frames; these were adjustable by means of wedges attached to rods projecting above the upper water-level, so that they could be conveniently manipulated.

The wheel itself consisted of a cast-iron ring with four arms, wrought-iron vanes riveted on, and an outer casing of sheet-iron.

The ring was cast with projecting lugs to which the arms were riveted. There were thirty-two back-vanes constructed of iron $\frac{1}{4}$ inch thick.

The surfaces of all the vanes both of the guide apparatus and wheel were polished in order to reduce friction as much as possible.

The entrance of the water to the guide passages was controlled by a scroll-regulator consisting of two conical rollers diametrically opposite, carrying bands of gutta-percha, as elsewhere described. The regulator was worked by a spur-wheel and pinion, the latter keyed to a vertical shaft carried up above the head-water level. The gutta-percha bands were furnished with strips of wrought-iron on the outer circumference, to give stiffness and resist the pressure of the water; these strips were $\frac{1}{8}$ inch thick, $\frac{7}{8}$ inch wide, and placed $\frac{1}{8}$ inch apart, each being attached by four copper rivets to the band, which was itself $\frac{1}{4}$ inch thick.

The turbine-shaft was of cast-iron, hollow, with the main-bearing above water, the weight being carried by a wrought-iron spindle fixed in a socket resting on the foundation.

The passages between the wheel-vanes were so designed as to be of equal section throughout, measured normally to the direction of flow of the water, so that the relative velocity of flow in the buckets would remain constant when the wheel worked under water. To attain this end, the buckets were made wider at the outflow than at the inflow, and back-vanes, as already stated, were employed.

The following were the leading dimensions and proportions of the turbine :—

Outer diameter of guide and wheel at
 inflow . . . . . $= 5$ ft. $7\frac{1}{2}$ in.
Inner diameter of guide and wheel at
 inflow . . . . . $= 4$ ft. $4\frac{1}{2}$ in.
Mean diameter of guide and wheel at
 inflow . . . . . $d = 5$ ft. 0 in.
Outer diameter of wheel at outflow . $= 6$ ft. 3 in.
Inner diameter of wheel at outflow . $= 3$ ft. 9 in.
Width of wheel-buckets at inflow . $c_1 = 1$ ft. 3 in.
Width of wheel-buckets at outflow . $e_2 = 2$ ft. 6 in.
Depth of wheel . . . $h_o = 1$ ft. 0 in.
Angle of outflow from guide-vanes . $a = 67° 30'$.
Angle of inflow of wheel-vanes . $a_1 = 45° 0'$.
Angle of outflow of wheel-vanes on the
 concave side . . . . $a_2 = 63° 40'$.

A A

Angle of outflow of wheel-vanes on the
  convex side . . . . . $a_2 = 67°\ 0'.$
Measured area of outflow from
  guide passages . . . $A^1 = 3\cdot28824$ sq. ft.
Effective area of outflow from
  guide passages . . . $A = 2\cdot877984$ sq. ft.
Effective area of outflow from
  wheel-passages . . . $A_2 = 6\cdot0155$ sq. ft.
$$\frac{A}{A_2} = 0\cdot47843 \text{ sq. ft.}$$

All dimensions given are in *Prussian* feet and inches.
1 lineal Prussian foot $= 1\cdot02972$ English foot.
1 square Prussian foot $= 1\cdot06032$ English square foot.
1 cubic Prussian foot $= 1\cdot09183$ English cubic foot.

Measurement of Power developed.

The effective work done by the turbine was measured
with the ordinary "Prony" brake, applied to a brake-
pulley fitted on to the spur-wheel keyed to the turbine-
shaft, and made in halves bolted together.

The brake-lever was connected by a hinged rod with
a bell-crank lever, to one arm of which the weights were
suspended. One of the brake-blocks prolonged to form
the brake-lever was suspended by a chain to the roof.
The brake-pulley had an outer diameter of $65\frac{3}{4}$ inches
and a breadth of $6\frac{7}{8}$ inches, with two flanges $\frac{1}{4}$ inch
deep. Water was the lubricant employed.

To determine the work actually developed by the
wheel, that portion of it spent in overcoming shaft-
friction had to be estimated and added to the value obtained by brake-measurement. For this purpose the
weight of the different parts carried by the main-bearing

was ascertained, and the frictional resistance calculated with the co-efficient 0·075 given by Morin for footstep bearings. The diameter of the bearing-spindle was 3 inches, consequently the effective radius at which the resistance must be taken as acting was 1 inch. The downward pressure exerted by the water during its passage through the buckets was neglected in determining the work of friction, as being very small compared with the weight of the turbine. The total weight on the bearing was taken as 6,932 lbs. (Prussian).

The ratio of the actual work done by the wheel (work measured by brake *plus* work required to overcome friction) to the available energy gives the hydraulic efficiency ϵ, as distinguished from the actual efficiency ϵ_e.

Water Measurements.

In determining the quantity of water used by the turbine, a float was employed for measuring the velocity of the fluid in the head-race. The latter was of exactly rectangular section, with horizontal floor, and 8 feet 11 inches wide.

The float was formed of two wooden discs, 3 feet in diameter and 2 inches thick, connected by twelve round bars of wood equidistant from each other in a circle of 32 inches diameter, the diameter of each bar being $1\frac{1}{4}$ inch; the total depth of the float was 2 feet 6 inches.

Detailed particulars of the manner in which the measurements were carried out are not given in the original account of the trials.

The time was observed occupied by the float in traversing a distance of 35 feet after it had attained a uniform motion.

The water-levels were measured in close proximity to

the turbine. The level of the tail-water was indicated by means of a float inserted in a pipe $1\frac{1}{2}$ inch diameter, dipping into the tail-race and passing up through the floor of the head-race. To the float was attached by one end a cord passing over a roller, and carrying at its other end a pointer, which indicated the water-level on a scale. The level of the head-water was read off on another scale at the grating through which the water passed into the turbine-chamber.

During the experiments no depression or vortex motion of the water over the wheel was noticeable. By means of a sluice the lower water-level could be varied so as to alter the head.

Observations were made at speeds varying from 0 to the speed at which the wheel ran without a load; the ratio to the latter of the best speed, as shown by the first series of experiments, was, with all the guide passages open, 0.482; with twenty passages, 0.507. It must, however, be pointed out that the head was not in these cases quite the same when the wheel ran without a load as when working at the best speed, so that the comparison is not strictly accurate; it will be seen that the ratio is in both instances very nearly $\frac{1}{2}$.

Comparison of Experimental and Theoretical Results.

For inflow without shock the ratio of the best velocity at the mean diameter to the velocity of flow should be

$$\frac{w}{c} = \frac{\sin (a + a_1)}{\cos a_1},$$

a_1 being in this case negative.

For vertical outflow the same ratio should be expressed by

$$\frac{w}{c} = \frac{A}{A_2} sin\ a_2,$$

and if the turbine has been designed according to the principles stated, both values should be identical.

Substituting the particular values for Haenel's turbine of a and a_1, a_2 and $\frac{A}{A_2}$ in the two formulas, it will be found that

(1) $$\frac{sin\ (a+a_1)}{cos\ a_1} = 0\cdot5412$$

and

(2) $$\frac{A}{A_2}\ sin\ a_2 = 0\cdot4344\ ;$$

in calculating the latter quantity, the mean value of the angles of outflow on the concave and convex sides of each vane has been used.

It is clear from the above results that inflow without shock and vertical outflow do not occur at the same speed.

The actual ratio of the best speed as ascertained by experiment to the velocity of flow, with all the guide passages open, is in one case

$$0\cdot51489,$$

in another $\qquad 0\cdot57724,$

with different heads; the mean of these two ratios is $0\cdot54606$, which agrees very well with the theoretical ratio for inflow without shock. A variation of 7 or 8 per cent. above or below the best speed has a very slight influence on the efficiency, so that for all practical purposes the values obtained on the assumption of inflow without shock are sufficiently close.

It is noticeable that with turbines which have not been designed for simultaneous vertical outflow and inflow without shock, the latter condition usually agrees best with experiment. In such cases, some writers advocate the adoption of a mean between the values calculated on the two assumptions; for Haenel's turbine this would give

$$\frac{w}{c} = \frac{0\cdot5412 + 0\cdot4344}{2} = 0\cdot4878,$$

which is sufficiently near one experimental ratio, $0\cdot51489$, but does not agree very well with the other, $0\cdot57724$.

The ratio to the theoretical velocity, due to the head over the guide-orifices, of the velocity of flow as calculated from the *observed* efficiency by the formula

$$K_1 = \sqrt{\frac{\epsilon}{2\,\frac{A}{A_2}\,sin\,a\,sin\,a_2}},$$

is for the efficiency $0\cdot6949$, (the best result recorded in Table 3,) $K_1 = 0\cdot930$; for the highest efficiency given in Table 2, K_1 is practically the same. Comparing this with the experimental values of K_1, it will be seen that in the first case

$$K_1 = 0\cdot8484,$$

and in the second

$$K_1 = 1\cdot1035.$$

The last of these values indicates a pressure *below* that of the atmosphere in the clearance space between guide passages and wheel, since the velocity of flow is greater than that theoretically due to the head.

That the co-efficient K_1 should vary so considerably for the same wheel is explained by the probable supposition that the actual effective sections of the stream passing through the turbine were not at all times those represented by the dimensions of the wheel, but were altered by the formation of eddies or *dead* water within the passages and buckets, not participating in the flow, but, so to speak, modifying the shape of the walls of the channels.

During the experiments recorded in Table 2, the immersion of the wheel was only about 0·5 foot, while the results given in Table 3 were obtained with an immersion of about 1·5 foot; the modifications in the effective sections of the stream must be ascribed to the difference in the immersion. In both cases the co-efficient K_1 shows that the wheel worked as a reaction turbine at the best speeds, since had free deviation taken place the co-efficient must have been less than 1 and greater than 0·8484. At higher speeds than the best with an immersion of 0·375 foot and less, the value of K_1 is what might be expected with free deviation, and is very nearly the same as the calculated value.

In a turbine constructed like that of Haenel, with buckets of equal section throughout, the relative velocity of flow and pressure is nearly constant throughout the bucket, the pressure being that of the atmosphere; hence the water leaves the guide passages at approximately atmospheric pressure, and, apart from differences in the hydraulic resistances, the velocity of flow must be the same as if free deviation took place.

It will be seen on reference to the tables of results that as the speed of the wheel *increases*, the quantity of water consumed *decreases;* this, as explained in dealing with Rittinger's turbine trials, is due to the negative value of the angle a_1.

During one part of the experiments, the ventilating pipes communicating with the guide passages—referred to in the description of the turbine—were open, and during another part closed, but no appreciable difference was noticed in the performance of the wheel.

In Fig. 129 the efficiencies and quantities of water per second consumed by the wheel are plotted as ordinates with the number of revolutions per second as abscissæ, and for these values the mean curves have been drawn. The dotted portion of the curve of efficiency indicates only the *probable* form, as no experimental data were obtained below a certain speed.

In the following tables 1, 1A, 2, and 3, will be found the results of the trials.

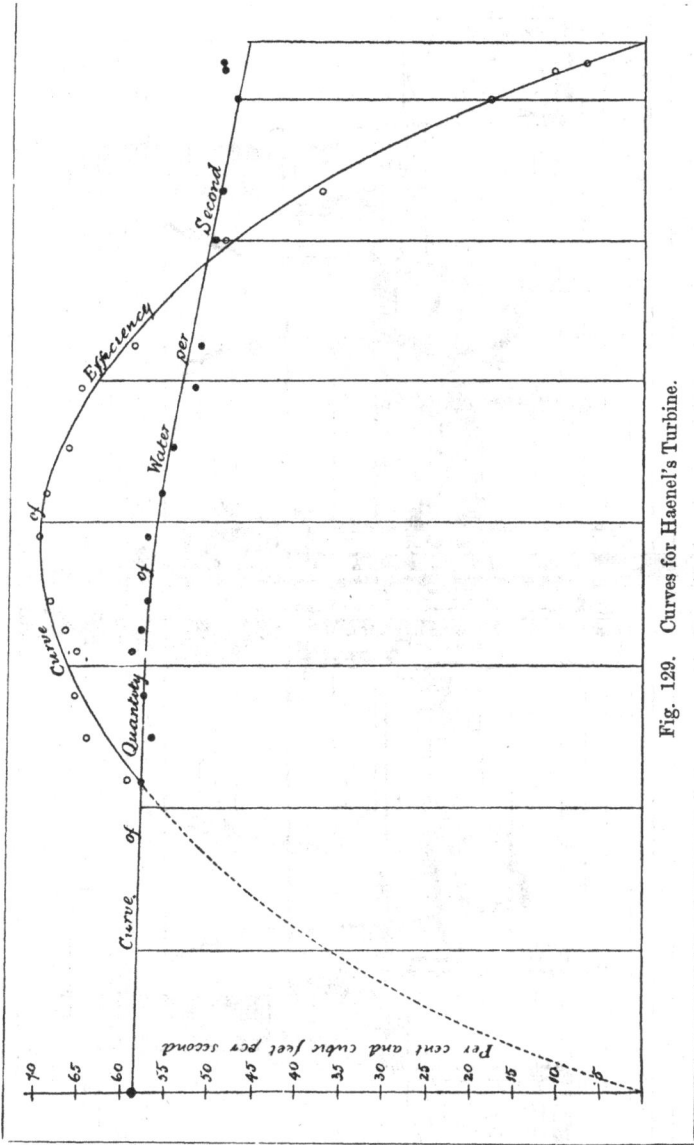

Fig. 129. Curves for Haenel's Turbine.

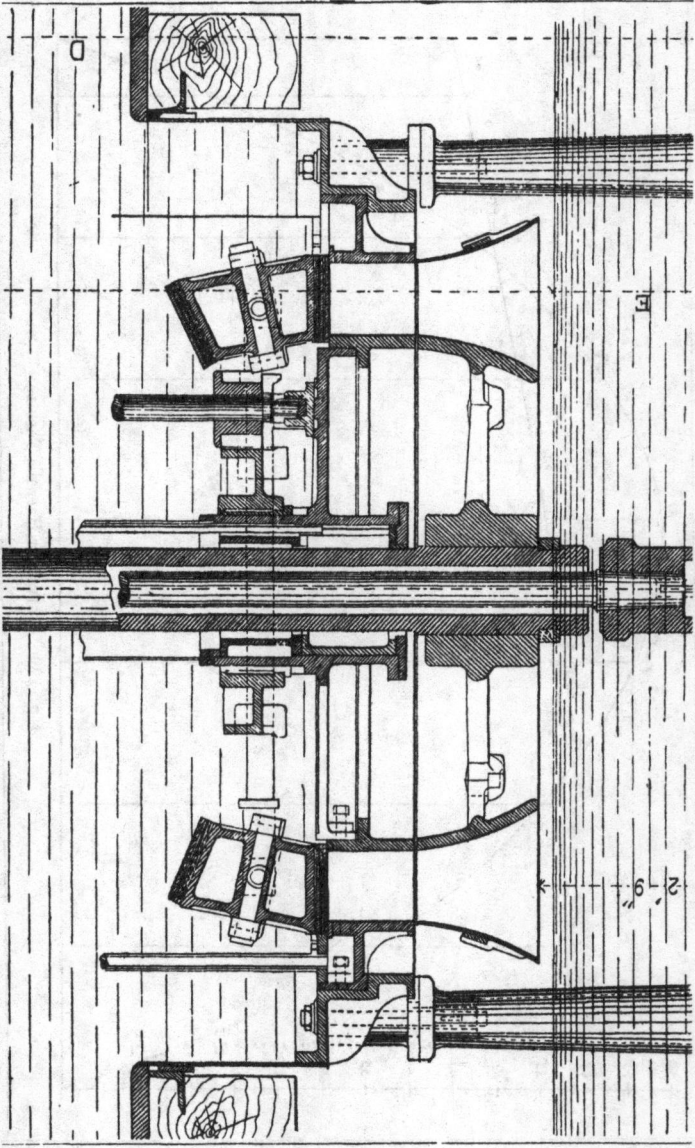

2'-9"

Results obtained with Haenel's Turbine.

TABLE 1.

Number of guide passages open.	Total head. h	(+) Clearance of wheel above tail-water or (−) depth of immersion.	Quantity of water used per second. Q	Number of revolutions per minute. n	Available energy of water per second. E_a	Power developed as measured by brake per second. W_e	Efficiency. ε_e
	Feet.	Feet.	Cubic ft.		Ft. lbs.	Ft. lbs.	
3	4·2292	− 1·9792	—	15·5	0	0	0
6	4·4063	− 1·6970	—	30	0	0	0
8	4·3333	− 1·8333	—	36	0	0	0
12	4·2292	− 2	—	42	0	0	0
16	3·8333	− 2·0417	—	44	0	0	0
16	4	− 2·1666	18·790	32·5	4960·6	2858·9	0·5763
16	3·9583	− 2·1666	18·914	32	4941·3	2814·8	0·5697
20	3·75	− 2·0833	—	47	0	0	0
22	3·9375	− 2·1666	28·308	32	7356·4	4691·4	0·6377
22	4	− 2·1666	27·728	31·5	7320·2	4618·2	0·6309
22	4·2083	− 1·9166	30·201	31	8388·3	5194·2	0·6192
22	3·6666	− 2·3333	34·676	0	8391·5	0	0
24	3·8125	− 2·1875	—	51	0	0	0
24	3·6666	− 2·4166	24·730	50·5	5984·5	0	0
24	3·6042	− 2·4166	31·580	27·417	7511·9	4593·6	0·6115
26	3·8333	− 2·1805	33·841	32	8561·8	4959·6	0·5793
26	3·5104	− 2·5	32·981	28·75	7641·3	4817·1	0·6304
26	3·625	− 2·4583	25·646	52	6135·7	0	0
28	3·8125	− 2·3333	—	54·5	0	0	0
28	4	− 2·0833	36·716	32·25	9693	6416·7	0·6620
28	3·9792	− 2·1042	36·716	30·5	9642·6	6387·9	0·6625
32	3·6666	− 2·3333	—	55	0	0	0
32	3·6355	− 2·3333	41·966	34·25	10069	6456	0·6412
32	3·9583	− 2·	46·162	33·5	12060	7719·0	0·6400
32	3·9583	− 2·	46·162	32	12060	8042·5	0·6669

TABLE 1A.

Number of guide passages open.	Total head.	(+) Clearance of wheel above tail-water or (−) depth of immersion.	Quantity of water used per second.	Number of revolutions per minute.	Available energy of water per second.	Power developed as measured by brake per second.	Efficiency.
	h		Q	n	E_a	W_e	ε_e
	Feet.	Feet.	Cub. ft.		Ft. lbs.	Ft. lbs.	
2	6·1042	0·1666	—	20·75	0	0	0
3	6·	0·1666		49·4	0	0	0
4	6·2292	0·1666	—	50·5	0	0	0
4	6·2083	0·0833	4·928	31·5	2019·1	1121·6	0·5555
4	6·0833	0·125	4·980	27	1999·5	1131	0·5656
6	6·2083	0·0833	7·868	34·5	3223·8	1806·4	0·5604
8	6·2083	0·0833	—	59	0	0	0
8	6·1666	0	10·520	36·5	4281·5	2675·6	0·6249
8	6·2083	0	10·177	32	4169·9	2680·8	0·6429
12	6·2083	0	—	61	0	0	0
12	6·0971	− 0·0833	17·023	34·5	6850·1	4335·5	0·6329
12	6	0	19·091	0	7559·7	0	0
14	5·875	− 0·1666	20·354	49·25	7892	3713·4	0·4705
14	5·9375	− 0·2086	20·508	35·25	8036·7	5315·6	0·6614
14	5·9375	− 0·2086	20·806	33·75	8153·2	5301·5	0·6502
14	5·9583	− 0·2086	21·441	32·166	8431·7	5052·6	0·5992
14	5·9166	− 0·2086	22·292	30·166	8705	5054·4	0·5806
16	6·1875	0	24·967	64	10196	0	0
16	6·	− 0·1666	24·007	34·25	9506·6	6169	0·6489
16	6·	− 0·0833	27·537	0	10905	0	0
20	6·125	− 0·0833	—	66	0	0	0
20	5·8333	− 0·25	29·723	33·5	11443	7717·9	0·6745
24	6·	− 0·0833	—	67	0	0	0
24	5·8333	− 0·25	31·076	34·5	11964	7587	0·6341
28	5·7917	− 0·2817	39·011	49	14912	8825·8	0·5919
28	5·6666	− 0·375	47·907	33	17917	11404	0·6365
32	5·875	− 0·2083	46·162	69·5	17899	0	0
32	5·5417	− 0·4583	56·889	33·5	20807	13331	0·6407

TABLE 2.

Number of guide passages open.	Total head. h (Feet.)	Clearance of wheel (+) above tail-race or (−) depth of immersion.	Quantity of water used per second. Q (Cub. ft.)	Number of revolutions per minute. n	Available energy of water per second. E (Ft. lbs.)	Power developed by brake as measured by brake per second. W_e (Ft. lbs.)	Efficiency. e_e	Gross power developed by turbine, including shaft-friction. W (Ft. lbs.)	Hydraulic efficiency. e	Theoretical velocity of flow due to head over guide orifices. V (Ft. per sec.)	Velocity of flow from guide orifices. c (Ft. per sec.)	Velocity of mean circumference of wheel. w (Ft. per sec.)	Ratio of flow to theoretical velocity. K_1	Ratio of velocity of mean circumference to theoretical velocity. K_2	Remarks.
4	6·4583	0	5·281	29	2250·8	1457·7	0·6476	1598·3	0·7101	18·328	14·679	7·592	0·8009	0·4142	In calculating the velocity of flow from the guide passages, the *effective value* has been taken by allowing for the obstruction of the flow by the ends of the wheel-vanes. In making this allowance the period during which the ends of guide- and wheel-vanes cover each other has been taken into account when the obstruction is not so great as at other times. $c = \dfrac{Q}{A'}$, for A', the area of out-flow from the guide passages.
8	6·3958	−0·1042	11·969	27·5	5052·5	3340·6	0·6612	3474	0·6876	18·400	16·636	7·199	0·9041	0·3913	
12	6·2916	−0·2086	18·160	36·5	7540·9	5351·2	0·7096	5528·2	0·7331	18·400	16·827	9·556	0·9145	0·5194	
16	6·2500	−0·2086	24·768	31·5	10217	6963·9	0·6816	7133·6	0·6982	18·338	17·212	9·163	0·9391	0·4999	
24	6·0000	−0·375	40·011	72·5	15844	10556	0·6662	10708·8	0·6759	18·186	18·536	8·247	1·0192	0·4535	
32	6·0833	−0·2817	49·227	70	19765	1214·8	0·0615	1566·4	0·0703	18·186	17·105	18·98	0·9406	1·0437	
”	6·0416	−0·2817	48·918	72	19506	1658·8	0·085	2008	0·1029	18·114	16·997	18·85	0·9383	1·0406	
”	6·0208	−0·3125	46·693	63·5	18555	2932·2	0·158	3271·7	0·1763	18·114	16·224	18·326	0·8957	1·0117	
”	5·9166	−0·3333	48·918	60	18861	6649·7	0·3596	6957·7	0·3689	17·97	16·782	16·624	0·9339	0·9251	
”	5·875	−0·375	49·918	53	18968	8796·5	0·4638	9087·5	0·4791	17·97	16·997	15·708	0·9459	0·8741	
”	5·8125	−0·4375	50·714	49·5	19455	11101	0·5706	11358	0·5838	17·97	17·621	13·875	0·9806	0·7722	
”	5·7916	−0·4583	51·363	45·25	19633	12441	0·6337	12681	0·6459	17·97	17·843	12·959	0·9932	0·7212	
”	5·75	−0·5	54·066	42	20518	13268	0·6467	13487·4	0·6573	17·97	18·786	11·846	1·0454	0·6592	
”	5·7291	−0·5308	55·177	39	20864	14074	0·6746	14277·7	0·6844	17·97	19·176	10·995	1·0672	0·6119	
”	5·7291	−0·5308	57·070	34·5	21579	14703	0·6813	14892·1	0·6901	17·97	19·830	10·21	1·1035	0·5682	
”	5·7291	−0·5308	56·709	32·5	21443	14451	0·6739	14618·3	0·6817	17·97	19·705	9·032	1·0966	0·5036	
”	5·7291	−0·5308	57·586	31	21774	14294	0·6565	14451·6	0·6637	17·97	20·009	8·509	1·1135	0·4735	
”	5·7291	−0·5308	58·701	28	22196	14284	0·6435	14434·3	0·6503	17·97	20·396	8·116	1·1350	0·4516	
”	5·75	−0·5	57·432	25	21796	14075	0·6458	14210·8	0·652	17·97	19·956	7·33	1·1105	0·4079	
”	5·75	−0·5	56·709	22	21521	13614	0·6326	13735·2	0·6382	17·97	19·705	6·545	1·0966	0·3642	
”	5·75	−0·5	57·586		21774	12902	0·5925	13008·7	0·5974	17·97	20·009	5·76	1·1135	0·3205	
”	5·75	−0·5	58·701	0	22196	0	0	0		17·97	19·206	0	1·0688	0	

TABLE 3.

No. of guide passages open	Total head, h (Feet)	Clearance of wheel above tail-race (+) or (−) depth of immersion	Quantity of water used per second, Q (Cub. ft.)	No. of revolutions per minute, n	Available energy of water per second, E (Ft. lbs.)	Power developed as measured by brake per second, W_1 (Ft. lbs.)	Efficiency, η	Gross power developed by turbine, including shaft friction, W (Ft. lbs.)	Hydraulic efficiency, e	Theoretical velocity of flow due to head over guide orifices, A (Ft. per sec.)	Velocity of flow from guide orifices, c (Ft. per sec.)	Velocity of mean circumference of wheel, a (Ft. per sec.)	Ratio of velocity of flow to theoretical velocity, v'	Ratio of velocity of mean circumference to theoretical velocity, x'
16	4·9166	−1·5416	20·806	22·25	6751·4	4230·7	0·6266	4353·2	0·6448	17·530	14·458	6·61	0·8248	0·3771
16	4·9583	−1·5416	20·255	25·5	6628·5	4241·2	0·6398	4350·3	0·6563	17·604	14·076	5·891	0·7996	0·3346
24	5·0208	−1·4791	25·381	55·25	8410·4	10414	0·1238	1309·3	0·1557	17·714	14·006	14·465	0·6638	0·8165
24	4·9687	−1·5	25·452	53	8346·7	2220·1	0·266	2477·1	0·2968	17·622	11·792	13·875	0·6691	0·7874
24	4·9583	−1·5	25·373	46·5	8303·1	3895·6	0·4692	4121·1	0·4963	17·678	12·841	12·174	0·6677	0·6915
24	5	−1·5	27·718	36·25	9147	5152·2	0·5633	5351	0·585	17·530	13·025	10·734	0·7264	0·6072
24	4·9166	−1·5416	28·115	41	9123·5	6073·7	0·6657	6249·5	0·685	17·604	13·770	9·49	0·7431	0·5414
24	4·9583	−1·5	29·722	32	9726·5	6702·1	0·6891	6857·3	0·705	17·604	13·770	8·378	0·7822	0·4759
24	4·9583	−1·5	29·722	26·5	9726·5	6660·2	0·6847	6789·7	0·698	17·567	15·300	6·938	0·7822	0·3941
24	4·9583	−1·5	33·024	23·75	10762	6963·9	0·6471	7079·1	0·6578	17·604	16·208	6·218	0·8709	0·3539
24	4·9375	−1·5308	37·152	0	12158							0	0·9207	0
32	4·9583	−1·5	35·871	60·5	11690	1117	0·0955	1410·4	0·1207	17·567	12·464	15·839	—	0·9016
32	4·8958	−1·5833	35·871	58·5	11690	24504	0·2096	2734·1	0·2339	17·567	12·464	15·315	0·7095	0·8718
32	4·9375	−1·5208	34·893	53·5	11659	4482	0·3844	4741·5	0·4067	17·788	12·124	14·006	0·6816	0·7874
32	4·9375	−1·5208	35·871	48·5	11936	6094·7	0·5106	6329·9	0·4687	17·751	12·464	12·697	0·7021	0·7153
32	5·0625	−1·4166	38·047	44	12503	7373·3	0·5896	7585·7	0·6087	17·641	13·220	11·519	0·7494	0·653
32	4·9792	−1·4375	39·011	41	12874	8587·1	0·667	8785·9	0·6825	17·678	13·555	10·734	0·7668	0·6072
32	5	−1·4375	41·611	36·25	13732	9110·6	0·6635	9286·4	0·6763	17·641	14·458	9·49	0·8179	0·5368
32	5	−1·4375	43·074	33	14166	9676·2	0·6836	9336·2	0·6949	17·604	14·967	8·639	0·8484	0·4897
32	4·9792	−1·4375	43·074	28·25	14096	9466·7	0·6716	9603·7	0·6813	17·604	14·967	7·396	0·8602	0·4201
32	4·9583	−1·4583	46·990	26·125	15506	9848·9	0·6351	9975·6	0·6433	17·678	16·327	6·839	0·9236	0·3869
32	5	−1·4166	49·227	23	16245	9436·6	0·5809	9548·1	0·5877	17·678	17·105	6·021	0·9676	0·3406
32	5	−1·4166	50·11	0	16536					17·678	16·396	0	0·9275	0

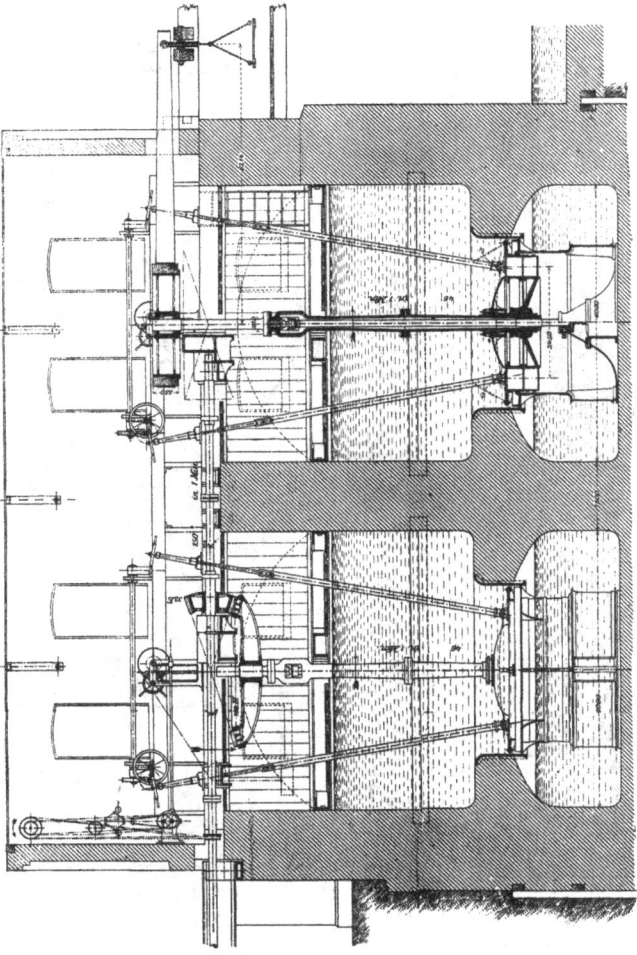

Fig. 131.

JONVAL TURBINES AT GÖGGINGEN.

Jonval Turbines at the Doubling and Sewing-Thread Mill, Goeggingen.[1]

The Goeggingen manufactory on increasing their works ordered two turbines from the Augsburg Maschinenfabrik to utilize the water-power of the Wertach.

The available fall is 13·45 feet, and the quantity of water per second 423·8 cubic feet, equivalent to 647 horse-power.

The makers of the turbines guaranteed an efficiency of 75 per cent. when all the passages in the wheel were open, and 70 per cent. when half were closed.

According to preliminary measurements, a minimum quantity of water of 176·58 cubic feet per second could be relied on, in which case the tail-water level was about 3·94 inches below the average, while, when floods occurred, the upper water-level rose by as much as 1 foot 7·69 inches; these circumstances had to be taken into account.

To meet these requirements two Jonval turbines of equal power were supplied, each of which could be regulated, in stages of $\frac{1}{12}$, down to half the greatest quantity of water.

Under the special conditions obtaining, automatic regulation is unnecessary, and the adjustment is effected through a series of sector-flaps hinged on the outside of the guide-vane casing, and operated by hand-wheels, as shown in the illustration (Fig. 131).

Both turbines have the same dimensions, namely, mean diameter 8 feet 0·46 inches, width of buckets 1·475 feet = 1 foot 5·72 inches (0·367 of mean radius). Under ordinary conditions they make forty-six revolutions per minute.

[1] *Zeitschrift des Vereins Deutscher Ingenieure*, vol. xxx., p. 781.

They have each thirty-six cast-iron guide-vanes and thirty-eight wheel-vanes of steel plate, and the depth of the guides and wheels is respectively 12·99 and 11·81 inches.

The wheel is surrounded by a welded wrought-iron ring, and keyed to the hollow cast-iron shaft by a complete truncated cone in halves.

The main bearing is of the overhead (Fontaine) type carried by a fixed wrought-iron spindle. This bearing is accessible through two openings in the hollow shaft, and the latter is in two lengths bolted together by flanges. The guide-casing has a dished top with an adjustable bearing of lignum-vitæ, through which the shaft passes.

Below the wheel is a short tube, from which the water flows out under a balanced circular sluice. This tube, in the event of the tail-water being very low, would act to a slight extent as a suction-tube. The position of the sluice can be regulated by hand. There are six flaps of wrought-iron arranged consecutively round one half of the circumference for closing the guide passages; each flap covers three passages. An arrangement is made by means of which air is allowed free access to the guide passages, when the latter are closed, through tubes in connection with the hollow hinges of the flaps.

The weight of the complete turbine with regulating apparatus and circular sluice is about twenty-six tons.

Water Measurements.

The quantity of water passing through the turbine was measured by means of current meters, as it was impossible to use a weir or tumbling bay. Two instruments were employed, one manufactured by Elliott Brothers and the other by Ertel and Son.

For the purpose of measurement the whole cross-section of the channel was divided into twenty-four rectangular parts, eight lengthways and three in depth, in the middle of each of which the velocity was observed. The total width of the channel was about 19 feet 8·22 inches, the depth 5 feet 4·72 inches.

Both the instruments used were carefully tested and their constants determined before use, at the hydrometric experimental station under the management of Dr. Bauernfeind at Munich.

As a further check on the accuracy of the observations, the curves of equal velocity for the section were constructed and from these the quantity of water flowing per second calculated. The agreement of the results obtained by the two methods was highly satisfactory.

Brake Tests.

	Proportion of guide passages open.	Head.	Quantity of water per second.	Available power.	Number of revolutions per minute.	Effective power measured on brake.	Efficiency.
		h	Q	E_a	n	W_e	ϵ_e
		Feet.	Cubic feet	Horse-power.		Horse-power.	
1	All	13·583	210·155	324·19	45·52	269·46	0·832
2	All	13·509	214·202	328·44	44·92	265·91	0·810
3	All	13·823	213·827	330·11	39·34	266·69	0·808
4	All	13·675	213·860	331·89	51·76	266·11	0·801
5	All	14·540	229·780	379·23	45·40	293·13	0·773
6	$\frac{9}{12}$	13·726	158·334	246·57	45·29	196·18	0·795
7	$\frac{7}{12}$	13·709	161·489	251·20	40·42	196·86	0·784
8	$\frac{6}{12}$	13·590	129·605	199·92	44·35	149·82	0·749
9	$\frac{5}{12}$	13·800	111·169	174·08	45·63	130·68	0·751
10	$\frac{4}{12}$	13·770	109·597	169·94	40·28	128·51	0·756
11	Turbine held fast	13·228	122·054		0	0	0

The construction and dimensions (in millimètres) of the brake used for determining the power given off are clearly shown in the illustrations.

In the preceding table the leading results of the trial are given.

Experiments with 480 horse-power Turbines at the Wood-fibre Manufactory of the Muenchen Dachauer Company, at Olching.[1]

This installation, illustrated by Fig. 132, consists of two turbines, one of the Jonval and the other of the Girard type, with vertical axes, enclosed in wooden chambers; the latter were originally made for a smaller quantity of water, but to avoid costly alterations it was decided to construct the new motors, with which the tests were carried out, to suit existing arrangements. Under ordinary circumstances, for about nine months of the year the available quantity of water is 16 to 17 cubic mètres (565 to 600 cubic feet) per second; in dry seasons this sinks to 10 to 11 cubic mètres (350 to 390 cubic feet), and in very extreme circumstances, which, however, only last a short time, to 7 cubic mètres (250 cubic feet). The fall is 2·9 mètres (9 feet 6 inches), and invariable.

To suit these conditions, the installation was so designed that with the normal quantity of water both turbines work together with *full* admission. When the amount of water decreases, the Girard wheel (which is the smaller) is regulated by closing some of the guide passages.

[1] *Zeitschrift des Vereins Deutscher Ingenieure*, vol. xxxii., p. 125.

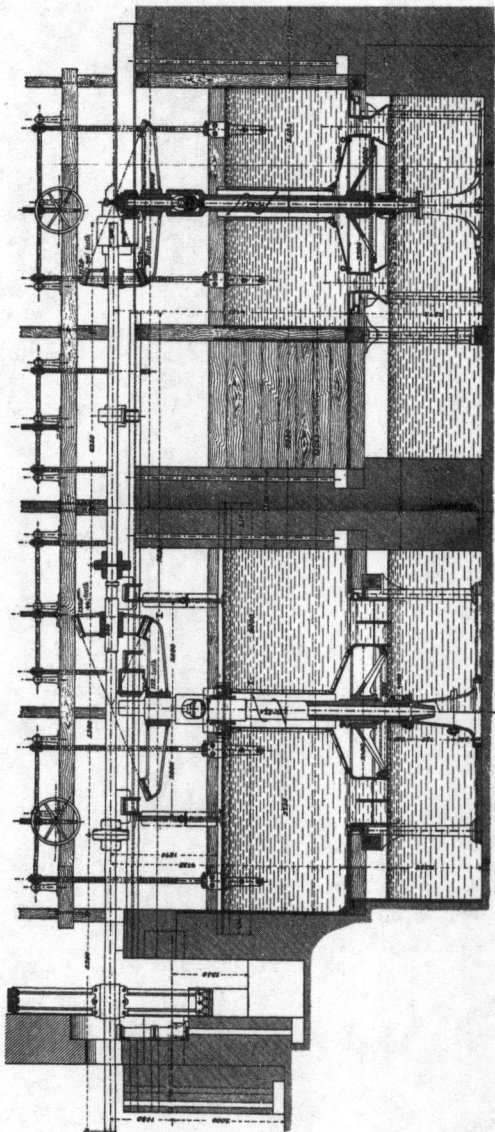

Fig. 132.

JONVAL AND GIRARD TURBINES AT OLCHING.

The manufacturer, Mr. J. G. Landes of Munich, guaranteed the following results :—

1. With both turbines working together at full power, with the normal quantity of water . . 75·0 per cent.

2. With the Jonval wheel full, and the water admitted to ⅔ of the guide passages of the Girard wheel . . . . . 74·6 per cent.

3. With the Jonval wheel full, and the Girard wheel with ⅓ admission . . . . . . 73·8 per cent.

4. With the Jonval wheel full, and the other closed . . . 75·0 per cent.

The Jonval turbine was designed to develop at least 300 horse-power under all circumstances, and the other motor to utilize the remainder of the available power.

Dimensions of Turbines.

	JONVAL WHEEL.			GIRARD.
	Inner Division.	Outer Division.	Totals or Means.	
Angle of outflow from guides	71° 10′	71° 10′	71° 10′	66° 0′
Angle of outflow from buckets	73° 25′	73° 20′	73° 23′	54° 0′
Angle of inflow into buckets .	9° 25′	9° 25′	9° 25′	11·81 in.
				at top
Clear width of buckets . .	16·34 in.	13·78 in.	30·12 in.	29·14 in.
				at bottom
Mean diameter . .	9 ft. 6·3 in.	12 ft. 3·6 in.	11ft. 1·86 in.	9 ft. 6·38 in.
Area of guide passages .	11·524 sq. ft.	13·853 sq. ft.	25·377 sq. ft.	
Outflow area of buckets .	9·253 sq. ft.	12·002 sq. ft.	21·255 sq. ft.	
Number of guide-vanes .	23	30		
Number of wheel-vanes .	29	38		
Thickness of guide-vanes .	0·59 in.	0·59 in.		
Thickness of wheel-vanes .	0·51 in.	0·51 in.		

The Jonval wheel is divided by a partition between the outer and inner casing into two concentric parts of unequal width, so proportioned that the power developed in each is nearly the same.

Brake Tests.

For measuring the effective power of the motors a brake of the usual "Prony" type was employed; the brake-wheel had a diameter of 2·2 mètres (7 ft. 2·61 in.) and a width of 0·55 mètres (1 ft. 9·66 in.). Very careful arrangements were made for keeping the wheel and brake-blocks cool during the experiments. The only available place for applying the brake was on the horizontal first-motion shaft, driven direct by bevel-gearing from the vertical turbine-shafts, and to arrive at the power actually given off by the latter, it was necessary to allow for the friction of the gearing and horizontal shaft. This was done by calculation from the actual weight of the machinery, allowing a co-efficient of friction of 0·07 for the bearings, and 0·13 for the bevel-gearing working with iron upon iron.

Water Measurements.

The measurement of the quantity of water used took place in the head-race by means of current meters. For this purpose the total section of the channel was divided into twenty-eight rectangles, fourteen in the width and two in the depth. Three instruments were used simultaneously at different points of the stream section, one by Elliott Brothers, another by Ertel and Son, and the third, with electric signal, by Amsler. In consequence of there being as yet no weir in the tail-race by which the level of the water in the latter could be regulated, this level varied according to the quantity of water used, and, in consequence, during experiment No. 1, the Girard wheel was *immersed* to a depth of 2 inches, while in trials 3 and 4 the Jonval wheel was 3·86 inches *above* the tail-water level. This of course affected to some extent the efficiency of both wheels, but no allowance has been made for this in the results given, which are as follows :—

Number of revolutions of *Jonval* wheel per minute.	Number of revolutions of *Girard* wheel per minute.	Horse-power measured on brake, after allowing for friction of horizontal shaft.	Available head.	Quantity of water used per second.	Available energy in horse-power.	Efficiency.	Remarks.
n	n	W_e	h	Q	E_a	e_a	
		Horse-power.	Feet.	Cubic feet	Horse-power.	Per Cent.	
27·10	25·55	472·24	9·308	585·86	629·35	75·0	Both tur-
27·69	26·10	481·70	9·433	583·57	624·72	77·1	bines to-
							gether.
28·87	—	365·62	9·991	404·12	458·33	79·8	Jonval tur-
28·28	—	364·43	9·991	404·62	458·92	79·4	bine alone.

If the height of the lower edge of the Jonval wheel above the tail-water surface were allowed for by deducting that distance from the available head, the efficiency would be **83** per cent. instead of 79·8.

The object of dividing the Jonval wheel into two concentric portions in this case is partly to strengthen the vanes and partly to allow the vane angles at different diameters to be independently fixed. Each part can then be treated as a separate wheel, and constructed, if necessary, with a degree of reaction differing from the other, although in the turbine in question this does not appear to have been done except to a slight extent.

The vane-angles are nearly the same at the mean diameter of each part; this could not be the case were there no division.

Francis's Experiments.

The well-known American engineer, Mr. James B. Francis, carried out in 1851 a series of very complete experiments with two large turbines, of which a full account is given in his work, *Lowell Hydraulic Experiments.* One of the turbines in question was an outward-flow radial (Fourneyron) wheel, the other an inward-flow radial wheel; the former of these was made for the Tremont Mills, the latter for the Boott Cotton Mills, both at Lowell, Massachusetts, where the experiments were carried out.

Experiments with Tremont Turbine.

Description of Turbine.

The turbine tested was, as above stated, of the Fourneyron type with vertical shaft.

The Wheel consisted of a central plate of cast-iron and of two crowns, to which the buckets were attached. The lower crown was fastened to the central plate by set screws; the upper and lower crowns were precisely alike, and each $9\frac{1}{2}$ inches wide.

The Wheel-vanes were of wrought-iron $\frac{9}{64}$ inch thick, let into grooves in the crowns, and secured by tenons in mortices cut in the crowns, and riveted over. The vanes terminated $\frac{1}{4}$ inch short of the edges of the crowns, for the purpose of allowing the wheel to be handled with less risk of injuring the ends of the vanes.

The Guide-Vanes were of plate-iron $\frac{1}{10}$ inch thick, secured by tenons to the bottom of a cast-iron cylindrical casing, to the flanged upper end of which was bolted the cast-iron mouth-piece of the curved supply-pipe. Up through the centre of the casing passed a vertical cast-iron tube protecting the shaft from contact with the water. The upper corners of the guides near the wheel were connected by a ring or garniture composed of thirty-three pieces of cast-iron, one to fill each space between the guide-vanes, riveted to the latter and to each other.

The Regulating-Sluice or gate was cylindrical, of cast-iron, passing between the orifices of the guide passages and the inlets to the buckets. The upper part of the cylinder was stiffened by a rib, to which were attached three brackets, each connected with a vertical wrought-iron rod; by these rods the gate was raised or lowered. The rods were actuated by an arrangement of levers, racks, and pinions. The weight of the gate was counterbalanced by weights attached to the levers.

The cylindrical portion of the gate had a thickness of one inch.

The Supply-Pipe, conducting the water from the head-race to the wheel, was in its upper part of wrought-iron plates $\frac{3}{8}$ inch thick riveted together, and had a diameter of 9 feet where the water entered horizontally, diminishing in the curved portion to its junction with the cast-iron mouth-piece before referred to.

The Shaft was of wrought-iron, 8 inches diameter for the main part of its length; it was suspended from a collar bearing and steadied laterally by bearings of cast-iron lined with babbitt-metal, one in the cast-iron shaft-tube, the other in the floor of the turbine-house above the supply-pipe. The collar-bearing was carried upon a gimbal or hollow frame resting on adjusting screws supported by a cast-iron girder connected with the walls of the turbine-house. At its lower end the shaft had a neck running in a step made in three parts, and lined with case-hardened wrought-iron; this step was furnished with adjusting screws, by means of which the shaft could be moved slightly in any direction horizontally. This step did not bear any portion of the weight of the motor, but only served to retain the lower end of the shaft in place.

Dimensions of Turbine.

The following were the leading dimensions, and proportions of the turbine :—

Angle of outflow from guides	.	$a_1 = 62°$.
Angle of outflow from buckets	.	$a_2 = 68°$.
Angle of inflow into buckets	.	$a_1 = 0°$.
Diameter of wheel where water enters	. . . . .	$2 r_1 = 6.750$ ft.
Diameter of wheel where water leaves	. . . .	$2 r_2 = 8.292$ ft.
Width of guide passages where water leaves	. . .	$c = 0.9707$ ft.
Width of buckets where water enters	. . . .	$e_1 = 0.9368$ ft.
Width of buckets where water leaves	. . . .	$e_2 = 0.9314$ ft.
Number of guide-vanes	. .	$z = 33$.
Number of wheel-vanes	. .	$z_1 = 44$.

Thickness of guide-vanes $t = \frac{1}{10}$ in. $= 0\cdot0083$ ft.
Thickness of wheel-vanes
$$t_1 = t_2 = \frac{9}{64}\text{ in.} = 0\cdot0117\text{ ft.}$$
Measured outflow area of guide
   passages .   .   .   . $A^1 = 6\cdot5371$ sq. ft.
Measured   outflow   area   of
   buckets .   .   .   . $A_2{}^1 = 7\cdot6869$ sq. ft.

With respect to the outflow areas A^1 and $A_2{}^1$ of the guide passages and buckets, some explanation is necessary of the manner in which these have been calculated.

Francis determined these areas by ascertaining the mean values of the shortest distances $a\ b$ and $c\ d$, Fig. 133, between two adjacent vanes, and multiplying them with the width of the passages or buckets and their corresponding number. It is open to question whether or not this method gives the true areas. The ends of the guide-vanes are comparatively slightly curved, and the actual direction of the stream leaving each passage is assumed to be parallel with the tangent at a, the extreme end of the vane-curve on the *concave* face of the vane. This stream would consequently be bounded by the parallel (dotted) lines through a and b, and its sectional area on leaving the orifices would be *less* than if all the filaments of water followed similar courses relative to the wheel. As has been previously pointed out, it would be reasonable to suppose that the fluid occupies the largest section possible, and, in the case of the turbine in question, this must happen when it follows similar curves through all

Fig. 133.

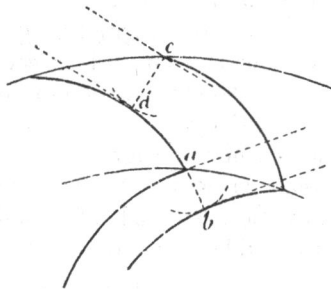

points of the circumference. Under such conditions the area A^1 would be found by the well-known formula:

$$A^1 = e \ (2 \ \Pi \ r \ cos \ a \ - \ z \ t).$$

Experiment, as will be subsequently shown, proves that calculated results do not agree so well with practice on the latter hypothesis as when Francis's method of computing the areas is adopted, and therefore the dimensions given above and in Table A have been calculated by Francis's method. (*Vide* p. 122 *et seq.*)

In the Tremont turbine the guide-vanes did not extend to the outer circumference of the casing, as room had to be left for the regulating-gate between the guide-vanes and buckets. On this account no allowance has been made for the obstruction of the wheel-vanes in estimating the effective area of outflow A from the guides, as the distance of the edges of the wheel-vanes from the guide passage orifices is very considerable. The co-efficient of contraction assumed is 0·9, and with this the effective area of the guide passages

$$A = 5 \cdot 8834 \text{ square feet.}$$

The same co-efficient has been used in calculating the outflow area from the buckets

$$A_2 = 6 \cdot 9182 \text{ square feet.}$$

From the dimensions given the following ratios result:

$$\frac{A}{A_2} = 0 \cdot 850 \ ; \ \frac{r_1}{r_2} = 0 \cdot 814 \ ; \ \frac{e_1}{e_2} = 1 \cdot 006.$$

Method of conducting Experiments.

The chief quantities to be ascertained were as usual the actual power developed, the quantity of water passing through the wheel, and the head, but in addition to these an arrangement was made for observing the absolute direction in which the water left the wheel, and the temperature was also noted.

For each class of data a separate observer was appointed, and the time of each observation recorded.

Measurement of Head.

The level of the water in the head-race was observed immediately before entering the supply-pipe, and the level of the tail-race in the wheel-pit. For these observations hook-gauges were employed, enclosed in boxes to which the water was admitted only through a hole in the bottom, in order to prevent the surface where the measurements were taken from being affected by oscillations and other disturbances. Both gauges had the same zero point, so that the difference in the readings gave the desired head at once.

Measurement of Power developed.

For ascertaining the effective power given off by the wheel a Prony brake was employed, applied to a pulley attached to the vertical shaft of the turbine. The friction-pulley was of cast-iron, 5 feet in diameter, 2 feet wide on the face, and 3 inches thick. The brake was of maple wood, made in two blocks drawn together by bolts, one of the blocks being prolonged to form the lever. The end of the lever was supported in a scale at the end of one arm of a vertical bell-crank lever, to the other arm of which the weights were hung. To regulate the motion of the brake and prevent sudden shocks, a hydraulic regulator or dashpot was employed in connection with the bell-crank lever in a manner now well known.

To ascertain the velocity of the wheel a counter was attached to the top of the vertical shaft, so arranged that a bell was struck at the end of every fifty revolutions of the wheel. The friction-pulley was lubricated and cooled by four jets of water, the mean quantity used during two trials being 0·0288 cubic feet per second.

In some of the experiments with heavy weights and low velocities, oil was used instead of water, as the friction was found to be greater with the former.

Observation of Absolute Direction of Outflow.

For ascertaining the absolute direction in which the water left the wheel, the following apparatus was employed : a rectangular vane was placed near the circumference of the wheel, keyed to a vertical shaft which turned freely on a step resting on the wheel-pit floor. The upper end of the shaft carried a pointer which indicated the position of the vane on a graduated semicircular scale divided into 180 degrees. When the vane was parallel with a tangent to the circumference of the wheel drawn through the point nearest to the axis of the vane, and the latter pointed in the direction of the motion of the wheel, the pointer indicated 0°; when the vane was in a radial position relative to the wheel the pointer indicated 90°.

Measurement of Quantity of Water.

The quantity of water passing through the turbine was gauged by a weir erected for the purpose at the mouth of the wheel-pit. To quiet the surface of the water before reaching the point at which the measurements in connection with the weir were made, a grating was placed across the wheel-pit. The measuring-weir had two bays of nearly equal length, one being 8·489 feet, the other 8·491 feet.

The crest of the weir was of cast-iron, planed on the upper edge and on the up-stream face for a width of $1\frac{1}{8}$ inch below the top; the upper edge was $\frac{3}{8}$ inch thick and practically horizontal. The up-stream corner was sharp. The ends of each bay were of wood, and of the same form as the crest near the edges. The ends of the bays projected a mean distance of 1·235 feet from the walls of the wheel-pit, and from the central pier dividing the weir into two parts.

The crest of the weir was about 6·5 feet above the floor of the wheel-pit which formed the canal of approach.

The depth of water or head over the weir was observed by means of a hook-gauge enclosed in a box, the distance of the gauge from the weir being about 5·5 feet.

The formula used for calculating the quantity of water from the head h over the weir was that of Francis:

$$Q = 3\cdot33\ (L - 0\cdot1\ n\ h)\ h^{\frac{3}{2}}.$$

1 — Number of the experiment.	4 — Height of the regulating gate in inches.	8 — Number of revolutions of the wheel per second. $\frac{n}{60}$	10 — Useful effect, or the friction of the brake in pounds avoirdupois raised one foot per second. We	13 — Total fall acting upon the wheel in feet. h	14 — Depth of water on the weir in feet.	15 — Quantity of water which passed the weir, in cubic feet per second. Q	16 — Total power of the water in pounds avoirdupois raised one foot per second. E	17 — Ratio of the useful effect to the power expended. e_e	18 — Velocity due to the fall acting on the wheel in feet per second. V	19 — Velocity of the interior circumference of the wheel in feet per second. v_3	20 — Ratio of the velocity of the wheel to the velocity due to the fall acting on the wheel. K_2	23 — Direction of the water leaving the wheel as indicated by the vane. deg.	m.
4	11·49	1·59651	33348·3	12·554	2·0383	156·6470	122663·3	0·27187	28·4169	33·8553	1·19138	18	3
6	,,	1·46149	51680·6	12·653	1·9989	152·2682	120174·8	0·43005	28·5287	30·9921	1·08635	22	56
8	,,	1·30933	66845·5	12·720	1·9536	147·2942	116864·8	0·57199	28·6041	27·7653	0·97067		
10	,,	1·18460	77154·6	12·800	1·9815	144·8734	115667·0	0·66704	28·3670	25·1203	0·87546	29	49
13	,,	1·78404	0·	12·510	2·0989	163·4313	127527·3	0·	28·8319	37·8319	1·33366	35	37
14	,,	1·06744	83969·8	12·856	1·9098	142·5180	142842·2	0·73475	28·7566	22·6359	0·78716	41	26
16	,,	0·99945	86314·4	12·890	1·9048	141·9762	114150·8	0·75614	28·7946	21·1940	0·73604	46	18
18	,,	0·94507	87392·7	12·880	1·8908	140·4657	114818·8	0·77442	28·7835	20·0409	0·69626	48	26
20	,,	0·91116	87819·8	12·886	1·8866	140·0066	112532·3	0·78040	28·7902	19·3219	0·67113		
22	,,	0·89713	88076·2	12·898	1·8834	139·6678	112364·5	0·78394	28·8036	19·0243	0·66048		
23	,,	1·78094	0·	12·429	2·0834	161·6944	125355·0	0·	28·2750	37·7662	1·33567	50	37
25	,,	0·88496	88189·9	12·899	1·8775	139·0291	111859·4	0·78840	28·8047	18·7662	0·65150	58	10
30	,,	0·85106	88278·9	12·903	1·8697	138·1892	111218·1	0·79375	28·8092	18·0474	0·62645		
31	11·48	1·78971	0·	12·539	2·0891	162·3283	126960·2	0·	28·3999	37·9521	1·33650	61	54
32	,,	0·83624	88320·8	12·915	1·8704	138·2668	111384·0	0·79294	28·8225	17·7330	0·61525	86	12
34	,,	0·78401	87947·8	12·941	1·8687	138·0869	114630·0	0·78903	28·8515	16·6254	0·57624		
35	,,	0·74211	87066·5	12·939	1·8652	137·7076	111387	0·78340	28·8493	15·7371	0·54549	99	25

No.	x												
37	,,	0·64568	83965·5	12·940	1·8412	135·1415	109077·1	0·76978	28·8504	13·6922	0·47459	131	18
38	,,	1·78571	0·	12·500	2·0834	161·6944	126071·1	0·	28·3557	37·8674	1·33544	139	45
40	,,	0·60000	82109·8	12·973	1·8380	134·7976	109077·0	0·75277	28·8872	12·7235	0·44045	147	25
41	,,	0·53232	78492·2	12·977	1·8282	133·7538	108265·8	0·72499	28·8916	11·2882	0·39071		
43	,,	0·	0·	12·797	1·8460	135·6536	108280·4	0·	28·6906	0·	0·		
45	,,	1·78333	0·	12·471	2·0864	162·0237	126034·8	0·79104	28·3228	37·8168	1·33521		
46	,,	0·83207	88603·9	12·954	1·8737	138·6244	112009·3	0·78904	28·8660	17·6447	0·61126	60	52
48	,,	0·80321	88240·1	12·948	1·8723	138·4690	111832·0	0·78723	28·8593	17·0327	0·59020	66	27
50	,,	0·76693	87865·8	12·952	1·8694	138·1559	111613·5	0·34589	28·8638	16·3058	0·56492	89	44
51	8·55	1·48331	39452·4	12·758	1·9173	143·3319	140607·7	0·67462	28·6468	31·4548	1·09802	12	0
53	,,	1·14177	74827·2	12·909	1·8656	137·7518	110917·6	0·72485	28·8159	24·2121	0·84023	22	19
55	,,	1·02459	80215·4	12·950	1·8586	137·0026	110664·7	0·74975	28·8616	21·7272	0·75281	28	56
57	,,	0·93071	81911·6	12·965	1·8408	135·0974	109252·2	0·76049	28·8783	19·7365	0·68344	37	47
59	,,	0·86538	82195·5	12·999	1·8240	133·3014	108082·5	0·76260	28·9161	18·3511	0·63463	47	30
61	,,	0·80429	81786·2	13·026	1·8117	131·9960	107246·3	0·75012	28·9461	17·0556	0·58922	59	39
63	,,	0·70779	79788·1	13·028	1·8013	130·9932	106366·6	0·	28·9484	15·0091	0·51848	94	4
64	,,	1·71808	0·	12·720	1·9742	149·5470	118652·1	0·	28·6041	36·4332	1·27370		
65	5·65	1·54799	0·	13·170	1·7160	121·9685	100194·6	0·	29·1057	32·8262	0·87363		
66	,,	1·35135	29054·7	13·077	1·6829	118·5511	96699·5	0·30046	29·0028	28·6564	0·85216	6	30
68	,,	1·08460	53142·4	13·176	1·6409	114·2599	93904·9	0·56592	29·1123	22·9997	0·81823	12	32
70	,,	0·92348	58727·1	13·253	1·6139	111·5197	92188·4	0·63703	29·1973	19·5831	0·79628	21	17
72	,,	0·77482	58776·2	13·311	1·5793	108·0452	89707·1	0·65520	29·2611	16·4306	0·76979	39	2
74	,,	0·68627	56253·1	13·326	1·5541	105·5341	87720·9	0·64127	29·2776	14·5530	0·49707	69	27
76	,,	0·45662	46056·1	13·412	1·5034	100·5410	84110·0	0·54757	29·3719	9·6830	0·32966	144	56
77	9·96	0·85470	86207·6	12·883	1·8620	187·3618	110380·8	0·78100	28·7868	18·1246	0·62961	55	52
79	,,	0·78918	86164·4	12·912	1·8544	136·5469	109973·0	0·78350	28·8192	16·7351	0·58069	74	28
80	2·875	1·24902	0·	13·347	1·2914	80·4534	66979·0	0·	29·3006	26·4865	0·90396	0	30
82	,,	0·96022	21256·6	13·395	1·2492	76·6213	64018·1	0·33204	29·3533	20·3622	0·69369	4	32
84	,,	0·67189	27985·1	13·478	1·1960	71·8750	60424·6	0·46314	29·4441	14·2480	0·48390	20	9
86	,,	0·46243	24462·9	13·556	1·1497	67·8158	57342·0	0·42661	29·5292	9·8061	0·33208	81	40
88	,,	0·	0·	13·516	1·0623	60·3593	50886·6	0·	29·4856	0·	0·		
90	1·00	0·61958	4998·8	13·985	0·7798	38·2210	33340·8	0·14993	29·9928	13·1386	0·43806		
91	,,	0·68886	3427·7	14·001	0·7846	38·5699	33683·5	0·10176	30·0099	14·6079	0·48677		
92	,,	0·38760	7815·6	14·020	0·7653	37·1733	32508·0	0·24042	30·0303	8·2193	0·27370		

Results of Experiments.

The more important results of the trials of the Tremont turbine are given in the preceding table, compiled from the original.

For the *best speed* the experimental results will be compared with those obtained by calculation on three assumptions as regards the method of computing the areas of outflow A and A_2.

(1) When determined on the assumption that all filaments of water follow the same relative course:

$$A = 7\cdot6424 \text{ square feet,}$$
$$A_2 = 4\cdot4809 \quad \text{,,} \quad \text{,,}$$

and

$$\frac{A}{A_2} = 1\cdot705.$$

(2) With A as above and A_2 as determined by Francis:

$$A = 7\cdot6424,$$
$$A_2 = 6\cdot9182 \text{ square feet (allowing}$$

for contraction), and

$$\frac{A}{A_2} = 1\cdot105.$$

(3) When both areas are estimated in the manner adopted by Francis:

$$A = 5\cdot8834,$$
$$A_2 = 6\cdot9182,$$
$$\frac{A}{A_2} = 0\cdot850.$$

The following table shows side by side the experimental data and the figures obtained by calculation under all three hypotheses.

The angles, it should be remembered, vary according to

the way in which the outflow areas are computed, and the values for them employed in calculation must be those corresponding to the *mean direction of flow*, not necessarily nor generally the vane-angle.

| | A = 7·6424. | | | | A = 5·8834 | |
| | (1) $\frac{A}{A_2} = 1\cdot705$. | | (2) $\frac{A}{A_2} = 1\cdot105$. | | (3) $\frac{A}{A_2} = 0\cdot850$. | |
	Experi-mental.	Calcu-lated.	Experi-mental.	Calcu-lated.	Experi-mental.	Calcu-lated.
Co-efficient of flow K_1 =	0·628	0·586	0·628	0·748	0·815	0·856
Co-efficient of velocity K_2 =	0·6264		0·6264		0·6264	
[1] For inflow without shock K_2 =		0·564		0·564		0·734
For radial absolute outflow K_2 =		0·848		0·523		0·523
Mean value K_2 =	0·6264	0·706	0·6264	0·543	0·6264	0·628
Purely Theoretical values.						
For inflow without shock K_2 =		0·526		0·672		0·756
For radial absolute outflow K_2 =		0·792		0·623		0·549
Mean value K_2 =		0·659		0·647		0·652

From the above figures it will be seen that the *third* method of determining the outflow areas gives by far the best agreement with experiment both as regards the velocity of flow and the velocity of rotation. The *mean* of the calculated speeds for inflow without shock and radial outflow agrees very closely with experiment; the difference between these speeds shows that the wheel has not been designed for simultaneous fulfilment of both conditions.

A further confirmation of the correctness of the third method of calculation adopted is afforded by the observed directions of outflow of the water from the wheel at certain speeds.

On reference to the table of results obtained with the

[1] Calculated from *observed* quantity of water.

Tremont turbine, it will be seen that at the *best speed* (line 30) the angle at which the water left the wheel, as indicated by the vane used, was 58° 10′; as determined by graphical construction from what has been termed the experimental velocity of flow and the actual speed, this angle is about 64°. This is a very fair agreement under the circumstances, and proves that the data from which the velocity of flow has been ascertained are nearly correct, as well as the corresponding relative angle of outflow a_2.

At a velocity of rotation of **0·545** of the velocity due to the head, the angle of absolute outflow observed was 99° 25′; while according to calculation (from observed data) radial outflow (corresponding to an angle of 90°) would have occurred for $K_2 = 0·523$.

It is remarkable that the purely theoretical mean values of K_2 for all the three methods of calculation are very nearly the same.

The quantity of water passing through the turbine as calculated theoretically by the third method is greater than by the first.

Experiments with Turbine at Boott Cotton Mills.

This turbine, like that at Tremont, was made to the designs of Mr. Francis in 1849, being one of a pair intended to develop about 230 horse-power with a fall of 19 feet. It was of the radial inward-flow type.

The Wheel was of cast-iron with plate-iron vanes $\frac{1}{4}$ inch thick, the upper of the two crowns enclosing the buckets being attached to a central plate with boss keyed to the shaft and curved downwards towards the centre so as to form a kind of inverted cone, in order to guide the water after leaving the buckets from the horizontal to the vertical direction.

The Guide Apparatus was formed by two flat rings or crowns, containing the guide-vanes between them, surrounding the wheel. The guide-vanes were of plate-iron $\frac{3}{16}$ inch thick.

The Regulating-Gate was a cast-iron cylinder, as in the case of the Tremont turbine, passing between the guide-orifices and buckets. This gate was made to slide on the circumference of a disc supported from four brackets resting on columns, the disc having the object of relieving the wheel from the pressure of the column of water above it. The gate was kept water-tight by means of leather packing against the circumference of the disc; the latter also contained the lower bearing for the turbine-shaft.

The water was conducted to the wheel-pit (or fore-bay) by a wrought-iron riveted pipe about 130 feet long and 8 feet in diameter, constructed of plates $\frac{3}{8}$ of an inch thick.

The Shaft was of wrought-iron suspended from a collar-bearing, and steadied laterally by two bearings adjustable by means of set screws. The top of the fore-bay was closed to prevent the water rising to its natural level by about 6 or 7 feet, so that the shaft had to pass through a stuffing-box.

Dimensions of Turbine.

The following were the leading dimensions and proportions of the turbine:—

Angle of outflow from guides . . $a = 78°$.
Angle of outflow from buckets . $a_2 = 75°$.

Angle of inflow into buckets .   . $a_1 = 30°.$
Diameter of wheel where water
    enters .   .   .   . $2\,r_1 = 9\text{·}338$ ft.
Diameter of wheel where water
    leaves .   .   .   . $2\,r_2 = 7\text{·}987$ ft.
Width of guide passages where
    water leaves .   .   . $e = 1\text{·}0066$ ft.
Width of buckets where water
    enters .   .   .   . $e_1 = 1\text{·}0066$ ft.
Width of buckets where water
    leaves .   .   .   . $e_2 = 1\text{·}2300$ ft.
Number of guide-vanes.   .   . $z = 40.$
Number of wheel-vanes   .   . $z_1 = 40.$
Thickness of guide-vanes $t = \frac{3}{16}$ in. $= 0\text{·}0156$ ft.
Thickness of wheel-vanes $t_1 = t_2 = \frac{1}{4}$ in. $= 0\text{·}0208$ ft.
Measured area of guide passages   . $A_1 = 5\text{·}904$ sq. ft.
Measured outflow area of buckets   . $A_2{}^1 = 6\text{·}8092$ sq. ft.

The areas A^1 and $A_2{}^1$ are those obtained by taking the
shortest distance between two vanes, measured from the
extremity of one of them, as the depth of the outflow
section of each passage or bucket. This gives a *larger*
area than the assumption that every filament of water
follows the same relative course. The angle of outflow
is given by the direction of a perpendicular to the plane
of the outflow section in the centre of the latter, as elsewhere explained; the angles of flow are therefore not
the same as the vane-angles.

In the present case, owing to the considerable space
between the extremities of the guide-vanes and the entrance to the buckets to allow room for the regulating-
gate, some uncertainty exists as to the true value of the
angle a. If the water continued to follow the direction
in which it issues from the guide passages, a portion of

it, instead of coming in contact with the buckets, would
simply pass round the out-
side of the wheel, as indicated in the sketch, Fig.
134. This of course is
impossible, as fresh water
keeps flowing in, so that
the fluid must be deflected
to some extent before en-
tering the wheel; what
this deflection, and there-

Fig. 134.

fore the *absolute* direction of inflow is, it is impossible to
say with certainty. It has, in the present instance, been
assumed that the angle *a*, which the direction of flow
of the water immediately before entering the buckets
forms with a radial line to any point of entry, is *the same*
as the angle at which it leaves the guides; if the course
remained unchanged, the former would be *greater* than
the latter—in the case of some portions of the fluid, 90°.

The effective area of outflow from the guide passage,
after allowing for contraction, is

$$A = 5\cdot3136 \text{ square feet,}$$
while $$A_2 = 6\cdot1283 \text{ ,, ,,}$$

From the dimensions given, the following ratios
result :

$$\frac{A}{A_2} = 0\cdot867,$$

$$\frac{r_1^2}{r_2} = 1\cdot169,$$

$$\frac{e_1}{e_2} = 0\cdot82,$$

EXPERIMENTS ON THE BOOTT CENTRE-VENT WATER-WHEEL.

Number of the experiment. (1)	Height of the regulating-gate in inches. (4)	Number of revolutions of the wheel per second. $\frac{n}{60}$	Useful effect, or the friction of the brake in pounds avoirdupois raised one foot per second. We (10)	Total fall acting upon the wheel in feet. h (1)	Depth of water on the weir in feet. (14)	Quantity of water passing the wheel in cubic feet per second. Q (15)	Total power of the water in pounds avoirdupois raised one foot per second. E (16)	Ratio of the useful effect to the power expended. e (7)	Velocity due to the fall acting on the wheel in feet per second. V (18)	Velocity of the exterior circumference of the wheel in feet per second. (19)	Ratio of the velocity of the exterior circumference of the wheel to the velocity due to the fall acting on the wheel. (20)
1	3	0·59289	23211·8	14·603	1·2964	67·532	61493·4	0·37747	30·648	17·393	0·56750
2	,,	0·87829	12073·5	14·672	1·2619	64·887	59364·4	0·20338	30·721	25·766	0·83871
3	,,	0·72314	20032·3	14·568	1·2821	66·432	60347·0	0·33195	30·612	21·214	0·69301
4	,,	0·54455	22447·2	14·158	1·2845	66·614	58821·6	0·38161	30·178	15·975	0·52937
5	,,	0·49929	22630·5	14·197	1·2899	67·029	59351·7	0·38129	30·219	14·647	0·48470
6	,,	0·44944	22026·8	14·143	1·2881	66·889	59002·2	0·37332	30·162	13·185	0·43714
7	,,	0·25445	16129·1	14·244	1·2943	67·368	59858·1	0·26946	30·269	7·465	0·24661
8	,,	1·01420	0·	14·300	1·2115	61·083	54486·4	0·	30·329	29·753	0·98101
9	6	1·01010	22952·9	14·288	1·5145	84·998	75732·8	0·30308	30·316	29·633	0·97746
10	,,	0·95986	28807·9	14·226	1·5308	86·355	76008·0	0·37604	30·250	28·159	0·93086
11	,,	0·93220	31814·1	14·197	1·5395	87·080	77093·2	0·41267	30·219	27·347	0·90496
12	,,	0·90090	34476·0	14·193	1·5467	87·685	77007·2	0·44424	30·215	26·429	0·87470
13	,,	0·85627	38243·0	14·194	1·5539	88·285	78143·8	0·48939	30·216	25·120	0·83184

14 ,,		0·69444	45135·3	13·778	1·5762	90·166	77480·4	0·58254	29·770	20·372	0·68433
15 ,,		0·59791	46402·8	13·606	1·5943	91·697	77812·1	0·59634	29·584	17·540	0·59291
16 ,,		1·17647	0·	13·945	1·4180	77·112	67076·7	0·	29·950	34·513	1·15237
17 ,,	9	1·10565	19779·8	13·518	1·6418	95·762	80736·3	0·24499	29·488	32·436	1·09997
18 ,,		0·99338	35931·0	13·373	1·6734	98·490	82145·2	0·43741	29·329	29·142	0·99362
19 ,,		0·90090	48212·7	13·369	1·6955	100·418	83728·2	0·57582	29·325	26·429	0·90125
20 ,,		0·82493	56195·7	13·396	1·7184	102·421	85571·4	0·65671	29·354	24·200	0·82442
21 ,,		0·78628	59226·4	13·379	1·7230	102·825	85800·0	0·69029	29·336	23·066	0·78629
22 ,,		0·74627	61168·8	13·342	1·7308	103·517	86138·0	0·71013	29·295	21·893	0·74731
23 ,,		0·72464	62065·4	13·322	1·7337	103·769	86218·8	0·71986	29·273	21·258	0·72620
24 ,,		0·71111	62752·2	13·334	1·7328	103·689	86229·3	0·72774	29·286	20·861	0·71232
25 ,,		0·69869	63199·3	13·304	1·7389	104·229	86483·7	0·73077	29·253	20·497	0·70067
26 ,,		1·25392	0·	13·702	1·5981	92·018	78648·0	0·	29·688	36·785	1·23907
27 ,,	12	0·70922	74979·3	13·400	1·8316	112·525	94057·5	0·79716	29·359	20·806	0·70868
28 ,,		0·69930	75347·2	13·431	1·8367	112·987	94662·2	0·79596	29·393	20·515	0·69796
29 ,,		0·67935	74580·9	13·331	1·8320	112·562	93603·9	0·79677	29·283	19·929	0·68058
30 ,,		0·67189	75153·2	13·378	1·8368	112·996	94296·4	0·79699	29·335	19·711	0·67193
31 ,,		0·66007	75208·3	13·386	1·8377	113·071	94415·2	0·79657	29·343	19·364	0·65990
32 ,,		0·64865	75249·2	13·383	1·8379	113·164	94471·1	0·79653	29·340	19·029	0·64856
33 ,,		0·63636	75142·9	13·356	1·8443	113·090	94219·0	0·79753	29·311	18·668	0·63692
34 ,,		0·62435	75103·3	13·381	1·8511	113·673	94881·9	0·79154	29·338	18·316	0·62431
35 ,,		0·61350	75202·0	13·405	1·8476	114·293	95571·2	0·78687	29·364	17·998	0·61291
36 ,,		0·59242	73993·4	13·321	1·8087	113·969	94703·1	0·78132	29·272	17·379	0·59371
37 ,,		0·	0·	13·537	1·8072	110·454	93270·1	0·	29·508	0·	0·
38 ,,		0·	0·	13·575		110·325	93422·4	0·	29·550	0·	0·
39 ,,		1·28505	0·	13·596	1·6884	99·795	84635·0	0·	29·573	37·698	1·27477

The method of conducting the experiments and the apparatus employed were practically the same in the case of the Boott turbine as for the Tremont turbine, and require no special description.

Results of Experiments.

The more important results of the trial of the turbine at Boott are given in the preceding table.

Comparing experimental with calculated results, for the best speed it will be seen that while the observed value of $K_1 = 0.726$, the value as calculated from the observed efficiency, after allowing for shaft friction, is $K_1 = 0.659$.

The experimental co-efficient of speed $K_2 = 0.637$; that calculated from the observed velocity of flow, for the assumption that the inflow is without shock, is $K_2 = 0.797$; the agreement in this case is not good, the actual velocity of rotation being much *less* than that corresponding to inflow without shock.

For radial outflow $K_2 = 0.711$, a much closer correspondence.

The purely theoretical value of K_2, calculated with the theoretical value of K_1, is for inflow without shock,

$$K_2 = 0.724,$$

for radial outflow $K_2 = 0.645$.

The latter of these figures is very near that obtained by experiment, the purely theoretical values thus showing a closer agreement with the facts than those calculated from the observed velocity of flow.

The imperfect agreement of the calculated with the experimental results is probably due to the actual absolute direction of flow of the water on entering the wheel being different to that assumed; the true angle

a less than 78°. This would make the calculated speed less than the velocity above stated, and at the same time increase the calculated value of K_1, bringing it into closer agreement with actual results.

The difference between the values of the co-efficient K_2 as determined for inflow without shock and radial outflow shows that the turbine was not exactly designed in accordance with the principles adopted by the author; the difference is not sufficient to cause any serious discrepancy between theory and practice. In such cases, as elsewhere pointed out, it is safest to take the mean of the values of K_2 for inflow without shock and radial outflow.

Holyoke Testing Station.

At Holyoke in Massachusetts, in the United States, there is a large permanent installation for testing turbines, the only one of the kind in the world, of which the following account is taken from the pamphlet published by the Holyoke Water Power Company, to whom the establishment belongs:—

"The Holyoke Water Power Company controls the flow of the Connecticut river at Holyoke, Mass., on a fall of nearly 60 feet. Above its dam is a drainage area of 8144 square miles, as measured on the best maps in existence at the present time. At this date (1883) there are nearly 15,000 horse-power in use by day, and over 8000 by night, of which part are 'permanent powers,' held by the parties using them under indentures, and subject to annual rental, and the balance are 'surplus powers,' held by the parties using them subject to withdrawal at short notice. Surplus powers are paid for from day to day, and according to the amount used. In times

of drought, if used after prohibition, the parties so using
are liable to a heavy penalty.   There are at date 139
water-wheels in use in Holyoke, of which fifty-nine run
about ten hours per day, and eighty run from Sunday
midnight to Saturday midnight, or 144 hours per week.
The number of distinct establishments operated by these
wheels is about seventy.   Observations are taken, giving
the opening of the speed-gates of each wheel and the head
acting upon it, once a day, and once during each night.

" These are carefully preserved, and from them are computed the amounts discharged by each establishment during
the quarter year, and the ' surplus power ' thus shown to
have been used is paid for at the end of the quarter.

" This system results in economy in the use of water,
where otherwise there would be great wastefulness.   In
times of low water it restricts the use of water, and, when
need be, confines all parties to their indentured quantities.

" For the purpose of making the necessary experiments
on the wheels of their tenants before they are set in the
mills, the Holyoke Water Power Company has built at a
large outlay a permanent testing flume. . . .   Wheels
are tested here both for power and for amount of water
discharged.   They are usually tested at five or six different openings of the speed-gate, ranging from wide open
to the opening at which the discharge is one-half that at
full opening, and at five or six different velocities of
revolution at each gate opening, making some thirty odd
experiments in all on each wheel.   The final result is,
that for all practical purposes the water-wheel has been
converted into a water-meter, and its discharge may be
known under any of the conditions under which it is
found in the mill.   Besides this, its efficiency, or the value
of the wheel as a water-motor, is also known.   Owing to
the permanent and special arrangements made for testing

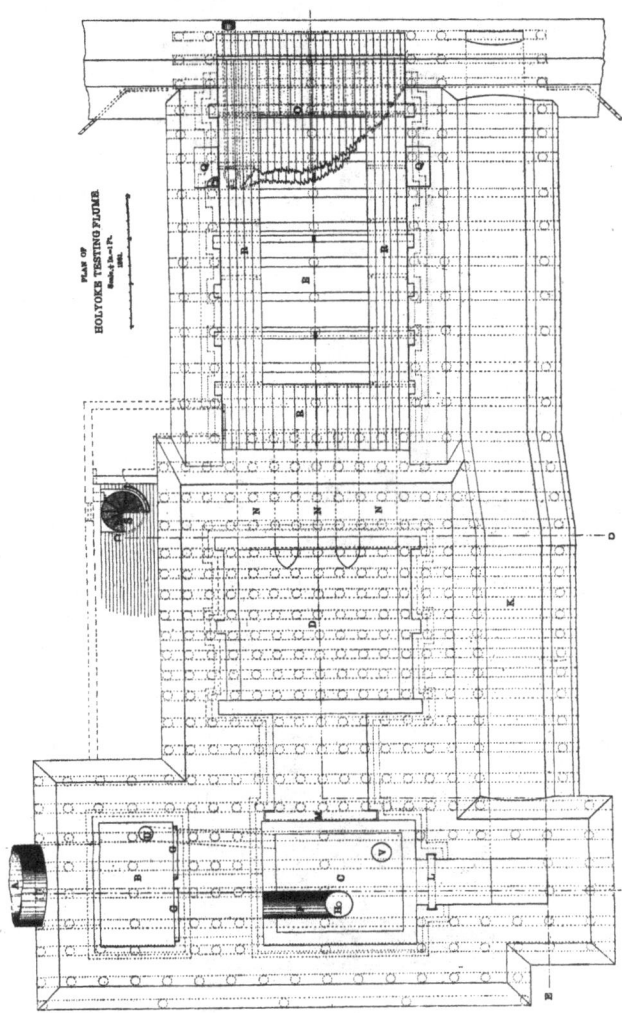

PLAN OF
HOLYOKE TESTING FLUME.
Scale, 8 to 1 Ft.
1881.

HOLYOKE TESTING FLUME.

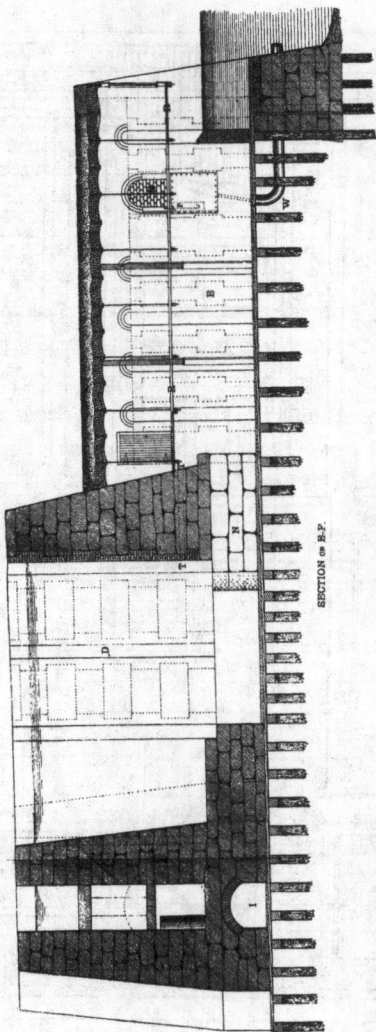

Fig. 135.

SECTION ON B-P.

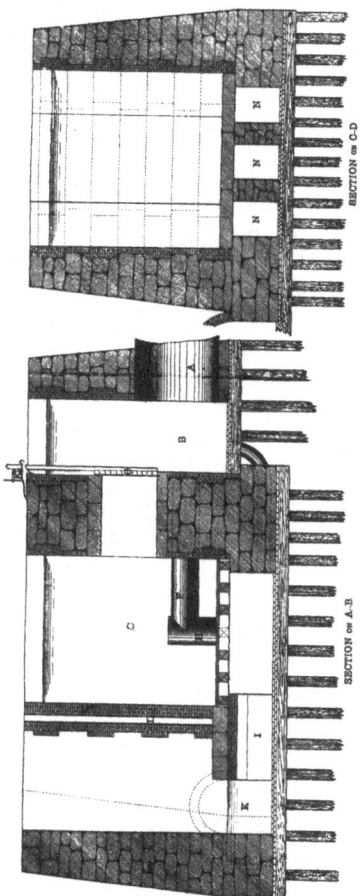

SECTION on C-D

SECTION on A-B

wheels and the large number of wheels tried, the company are able to carry out the tests at much less expense than would be possible if the experiments were made in the mill or manufactory where the wheel is used, and the result is that large numbers of wheels are sent from all parts of the United States to be tested at Holyoke.

"The underground portion of the testing flume (*vide* Fig. 135) consists in the main of a trunk or penstock *A*, bringing the water; a sort of vestibule *B*; an ante-chamber *C*; the wheel-pit *D*; and the tail-race *E*. The trunk *A* is of boiler-iron, about 9 feet in diameter, laid low down in the ground so as to pass longitudinally under the centre of a street. . . . The object of the vestibule *B* is simply to afford opportunity for the application of two head-gates (sluices) *G G*. Besides these, there is a head-gate at the point of entry of the trunk into the canal whence it takes the water. A small trunk *F*, about 3 feet in diameter, takes water from this vestibule independently of the gates *G*, and leads it to a turbine-wheel *H*, set in an iron casing, in the chamber or pit *C*; so that this wheel can run even when *C* is empty. This wheel discharges through the floor at the bottom of *C*, thence through the arch *L* and the supplementary tail-race *K* into the second level canal. It is used to operate the repair shops, also to lift and lower the gates *G*. The chamber *C* is bounded on one side by a tier of stop-plank *L*, and on another side by a tier of stop-plank *M*. The object of the stop-plank *L* is to afford a waste-way out of the chamber *C*. This is of special use in regulating the height of the water when testing under low heads; also to skim off the oil floating on the water, which has dropped down from the dynamometer. The water thus washed over the plank *L* falls directly into the tail-race *K*, and passes into the second level. The stop-planks *M* come into use when testing scroll or cased wheels. In that event, D is empty of water, and the wheel-case in question is attached to the planks M, over a proper opening cut through the same. Large case-wheels could be set in the centre of D and the water be led to them by a short trunk or penstock, leading from an opening cut in the planks M. Flume wheels are set in the centre of the floor of D, and D is filled with water. They discharge through the floor of D (not shown on the drawing) and out of the three culverts N N N into the tail-race E. At the down-stream end of this tail-race is the measuring-weir O, the crest being formed of a piece of planed wrought-iron. It can be used with or without end contractions. The depth of the water on the weir is measured in a quarter-cylinder P, set in a recess Q, fashioned into the sides of the tail-race. These recesses are water-tight, and the observer is thus enabled to stand with the water-level about breast-high, or at a convenient height for accurate observation. The methods of measuring water over this weir are those described in *Lowell Hydraulic Experiments,* by James B. Francis. . . . In this country (America) these experiments are well known, and the formula which was derived from them has met with general acceptance. . . . A platform R surrounds the tail-race, and is suspended from the iron beams that roof it in. Above the tail-race is a street. The wheels to be tested arrive on this street, are lifted from the wagon by a travelling windlass that runs out on a framework over the street, are run into the building and lowered into the wheel-pit D. Winding stairs lead into a passage-way that leads in turn to the platform R. In the well-hole of these stairs is set up the glass tube which measures the head of water upon the wheel. It is connected with the pit D or the chamber C by means of pipes running through a cast-iron pipe T, built into the masonry dam which forms the down-stream end of the wheel-pit D.

" The power is weighed by a Prony brake, consisting of a cast-iron pulley surrounded by a hollow brass band, through which a stream of water is allowed to circulate, and a bent lever and weights to hold the band and pulley in its place. Five sizes of pulley and band are used, to weigh different sizes of wheels. The limits of power weighed have been from 11·4 to 230 horse-power. Heads have varied from 4 to 19 feet. To enable the observer at the brake-wheel, the one at the head-gauge, and the one at the measuring-weir to take simultaneous observations at intervals of one minute, an electric clock is set up in the testing-flume which rings three bells simultaneously at intervals of one minute, or of half-a-minute if desired.

" The whole structure is built in a durable and efficient manner. The pits and tail-race are all lined with brick laid in cement.

" The stone masonry was intended by careful work and grouting to be water-tight without the brick lining, and the brick lining was then carefully laid up with joints full of mortar as an extra precaution. As a consequence the front of the wall forming the down-stream side of the pit D is barely damp with 20 feet head of water upon it. The floor of the pit D is built so tight that an exact measurement of the leakage of the wheel-gate could be made if desired."

The following trials of " Collins," " Boyden," and " Hercules " turbines were carried out at Holyoke, and the author is indebted to the engineer of the Holyoke Water Power Company for most of the dimensions and other particulars regarding the construction of these turbines.

The greatest length of the measuring-weir at Holyoke is about 20 feet.

In the following tables the dimensions are given in feet, and the available and effective power in horse-power unless otherwise stated.

Tests of a "Collins" Turbine.

In 1883 experiments were made at Norwich, Connecticut, on a turbine of the Jonval type constructed by Messrs. J. P. Collins and Co., to determine the efficiency of the wheel with and without a suction-tube (or "draft-tube," as it is called in America).

Fig. 136.

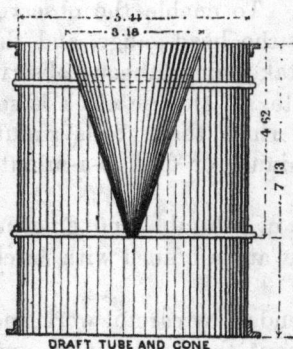

DRAFT TUBE AND CONE

The suction-tube was tried first with an interior inverted cone, as shown in Fig. 136, to give a gradually increasing area, and, secondly, without this cone.

Weir 8 feet long, with end contractions.

Dimensions of Turbine.

$a = 72\frac{1}{4}°$.　　$A_1 = 2·912$ sq. ft.　　$2\,r_m = 4·170$ ft.

$a_1 = 12\frac{1}{4}°$.　　$A_2 = 2·822$　„

$a_2 = 71°$.　　$z = 24$.

$e = 0·836$ ft.　　$z_1 = 30$ (brass vanes).

Effective area A, after allowing for obstruction by wheel-vanes, $= 2·889$.

$$\frac{A}{A_2} = 1·023.$$

60″ Collins Turbine (Jonval type).

Proportional gate opening.	Total head. h	Quantity of water per second. Q	Number of revolutions per minute. n	Horse-power measured on brake. W_e	Efficiency. e	Ratio of velocity of rotation to $\sqrt{2gh}$. K_2	Depth on weir.	
1·000	16·58	64·31	82·00	96·24	79·71	0·551	1·867	
,,	·57	64·35	85·25	96·41	79·85	0·573	·868	
,,	·55	64·40	89·17	97·03	**80·40**	0·600	·869	
,,	·56	64·40	93·00	96·65	80·03	0·625	·869	
,,	·56	64·49	99·00	95·02	78·57	0·665	·871	
0·748	·79	58·50	81·50	86·69	77·94	0·544	·750	
,,	·78	58·50	85·67	87·46	78·68	0·572	·750	
,,	·77	58·50	90·00	87·48	78·75	0·601	·750	
,,	·77	58·54	95·67	87·73	78·92	0·639	·751	
,,	·77	58·44	105·00	86·65	78·08	0·701	·749	
0·600	·86	53·61	77·90	69·53	67·93	0·519	·650	
,,	·89	53·61	83·50	70·44	68·70	0·556	·650	
,,	·89	53·61	86·00	68·87	67·17	0·572	·650	Leakage
,,	·90	53·61	89·75	69·68	67·92	0·597	·650	0·48 cubic
,,	·95	53·71	94·25	70·87	68·75	0·626	·652	foot per
0·503	17·18	45·14	69·67	56·22	64·02	0·460	·468	second.
,,	·18	45·10	75·67	57·82	65·91	0·499	·467	
,,	·19	45·06	80·17	57·83	65·93	0·529	·466	
,,	·19	45·06	84·75	57·51	65·57	0·559	·466	
,,	·18	45·06	88·67	56·37	64·31	0·585	·466	
0·303	·41	34·25	71·00	34·72	51·42	0·465	·218	
,,	·42	34·21	76·00	36·24	53·70	0·498	·217	
,,	·41	34·17	79·00	36·22	53·77	0·518	·216	
,,	·40	34·12	83·50	35·73	53·15	0·548	·215	
,,	·43	34·21	88·67	35·23	52·18	0·581	·217	
0·161	·84	24·80	53·00	19·12	38·16	0·343	0·984	
,,	·84	24·72	59·17	19·53	39·11	0·383	·982	
,,	·85	24·70	65·67	19·27	38·60	0·425	·981	
,,	·85	24·67	71·17	18·71	37·52	0·461	·980	
,,	·85	24·70	75·67	18·04	36·14	0·490	·981	

60″ *Collins Turbine*, with suction-tube.

Proportional gate opening.	Total head.	Quantity of water per second.	Number of revolutions per minute.	Horse-power measured on brake.	Efficiency.	Ratio of velocity of rotation to $\sqrt{2gh}$.	Depth on weir.	
	h	Q	n	W_e	e_e	K_2		
		With inverted cone.						
1·000	16·58	65·13	77·50	97·12	79·46	0·521	1·880	
,,	·56	·18	86·00	101·46	83·05	·578	·881	
,,	·57	·30	90·25	102·62	**83·78**	·606	·884	
,,	·58	·37	95·87	102·56	83·60	·644	·885	
,,	·58	·37	106·00	101·73	82·92	·712	·885	Leakage 0·30 cubic foot per second.
0·548	·96	52·09	74·00	69·66	69·66	·492	·613	
,,	·98	·01	79·25	70·73	70·76	·526	·612	
,,	17·19	·36	87·33	73·14	71·79	·576	·619	
,,	·01	51·99	92·25	71·62	71·54	·612	·611	
,,	·01	22·01	99·50	69·95	69·85	·660	·612	
0·297	·40	36·11	67·00	38·50	54·13	·439	·257	
,,	·63	·26	75·67	39·78	54·98	·493	·261	
,,	·61	·22	82·12	39·16	54·23	·535	·260	
,,	·61	·19	87·12	38·34	53·15	·568	·259	
,,	·63	·22	93·25	37·05	51·26	·608	·260	
		Without inverted cone.						
1·000	16·56	64·57	78·00	97·75	80·75	0·524	1·869	
,,	·57	·70	85·50	100·87	83·12	·575	·871	
,,	·56	·88	89·87	102·18	84·01	·604	·875	
,,	·55	·99	96·00	102·70	**84·34**	·646	·877	
,,	·56	·88	105·50	101·25	83·25	·709	·875	Leakage 0·30 cubic foot per second.
0·548	17·01	51·04	68·87	64·83	65·97	·457	·591	
,,	·00	·04	76·00	67·83	69·05	·504	·591	
,,	16·99	50·99	82·50	69·09	70·45	·548	·590	
,,	17·00	·92	89·75	69·68	71·10	·595	·589	
,,	·02	·89	98·50	69·25	70·62	·653	·588	
0·297	·53	34·58	62·37	35·84	52·23	·407	·221	
,,	·53	·53	69·92	36·76	53·64	·457	·220	
,,	·52	·53	77·00	36·72	53·61	·503	·220	
,,	·51	·53	82·00	36·09	52·73	·536	·220	
,,	·51	·66	89·33	35·50	51·66	·584	·223	

It will be seen that without a suction-tube the maximum efficiency was 80·40 per cent.; with a suction-tube and inverted cone 83·78 ,, ,, ; and with a suction-tube without cone 84·34 ,, ,, ; the suction-tube, therefore, increased the efficiency, but no advantage was obtained by using the inverted cone, although it might have been anticipated that this would act as a diffuser. The reason for this failure was probably due to the enlargement of the area not being sufficiently gradual, so that practically there was the same sudden change of velocity of the water as when no inverted cone was used.

Test of a 96-inch Collins Turbine.

The following are stated to be the best results ever recorded at Holyoke with a turbine of the Jonval type; no details of the dimensions of the wheel are given, but the results are interesting as affording the material for a fair comparison between the parallel-flow and the mixed-flow turbine so much in vogue in America. Both types of wheel are tested under much the same conditions at Holyoke, and the same method of observation and calculation for determining the quantity of water used is employed, so that the effect of any important errors arising from differences in the method pursued is thus eliminated.

96″ *Collins Turbine (Jonval type).*

Proportional gate opening.	Total head.	Quantity of water per second.	Number of revolutions per minute.	Horse-power measured on brake.	Efficiency.
	h	Q	n	W_e	ϵ_e
1·00	16·59	113·46	63·38	131·49	85·06
0·778	16·75	101·67	63·00	160·83	83·27
0·631	16·95	87·66	58·93	133·56	79·26
0·547	17·12	77·19	54·32	111·36	74·64
0·422	17·26	65·36	49·65	84·40	65·96

Experiments with "Boyden" Turbine.

The following are the particulars of dimensions and results of tests made with a Boyden turbine, a wheel of the Fourneyron type, for a mill of the Merrick Thread Company, on April 26th, 1882.

Dimensions of Turbine.

[1] $a = 61°$. $A = 6·814$ sq. ft.   [3] $d_g = 73·45$ in.

$a_1 = 0°$. $A_2 = 5·660$ sq. ft.   [2] $r_1 = 73·60$ in.

[2] $a_2 = 61°$.   [2] $r_2 = 90·00$ in.

$e = 9·875$ in.

$e_1 = 8·64$ in.   Height of buckets at lowest point.

$e_2 = 9·125$ in.   Height of buckets at outer end.

$z = 34$ (brass vanes).

$z_1 = 54$ (brass vanes).

The weir was 11 feet long and practically tight, the leakage being only 0·05 cubic foot per second.

[1] Mean angle of flow ; vane-angle $= 66°$.
[2] Mean angle of flow ; vane-angle $= 64°$.
[3] Diameter at guide-orifices.

Boyden Turbine (*Fourneyron type*).

Proportional gate opening.	Total head.	Quantity of water per second.	Number of revolutions per minute.	Horse-power measured on brake.	Efficiency.	Ratio of velocity of rotation to $\sqrt{2gh}$.	Depth on weir.	
	h	Q	n	W_e	ϵ_e	K_2	*North.*	*South.*
1·000	16·61	145·35	57·00	216·79	79·17	0·560	1·789	1·421
,,	·57	146·18	59·33	216·55	78·82	·584	·790	·433
,,	·60	147·10	63·50	222·04	80·17	·624	·798	·439
,,	·62	148·32	66·50	220·28	78·79	·653	·807	·448
,,	·64	149·87	71·12	222·50	78·66	·697	·824	·454
0·883	·66	142·04	55·67	203·19	75·71	·546	·761	·399
,,	·63	142·82	58·75	207·23	76·93	·577	·765	·407
,,	·65	143·27	62·50	210·87	77·94	·614	·769	·410
,,	·64	144·93	65·17	209·88	76·73	·640	·783	·421
,,	·63	145·74	68·37	209·71	76·29	·671	·790	·426
0·773	·74	136·12	56·00	188·94	73·11	·548	·716	·353
,,	·72	137·40	58·75	189·21	72·61	·576	·722	·367
,,	·69	138·18	62·33	191·17	73·09	·611	·729	·372
,,	·73	139·65	66·17	192·81	72·76	·648	·736	·388
,,	·73	140·66	69·25	193·29	72·42	·678	·749	·390
0·662	·80	128·86	55·75	169·28	68·95	·545	·654	·303
,,	·81	130·69	59·67	172·04	69·04	·583	·658	·312
,,	·81	130·89	62·00	171·15	68·58	·606	·665	·324
,,	·80	132·22	66·17	172·51	68·47	·647	·673	·337
,,	·76	133·22	69·67	170·95	67·51	·682	·685	·340
0·442	17·10	105·65	50·63	113·36	55·32	·490	·456	·127
,,	·08	106·55	54·17	114·64	55·54	·525	·461	·137
,,	·04	107·51	58·50	114·84	55·27	·568	·464	·151
,,	·10	108·03	59·17	116·15	55·44	·573	·470	·153
,,	·09	108·50	60·33	111·03	52·79	·585	·474	·157
,,	·08	110·67	66·50	114·23	53·28	·645	·490	·177

Tests of a 36-inch " Hercules " Turbine.

Inward and Downward Flow.

These experiments were made in 1883 at Holyoke for the manufacturers of the wheel. The performance at the first test was so much better than the makers had anticipated, that it was decided to repeat the test. As the result of the second trial was equally good, the con-

D D

structors feared there might be some fault in the arrangements or the method of calculation. Professor Thurston
was therefore requested to investigate the matter, and if
possible ascertain whether the reported values of the
efficiency might be safely accepted. The third trial, made
under Professor Thurston's personal superintendence,
practically confirmed the results previously obtained. In
the following table only the particulars of the final test
are given.

The weir was 20 feet long without end contractions.

Dimensions of Turbine.

a $= 75\frac{1}{4}°$, measured with centre line of guides.
$a_1 = - 8°$
$A^1 = 4\cdot752$ sq. ft. d_1 mean $= 36$ in. $z = 24$.
$A_2^1 = 7\cdot925$ sq. ft. $z_1 = 17$.

Description of Turbine.

The "Hercules" turbine is of the inward and downward-flow type; the water on entering the wheel has a
downward inclination, but the radial component prepon-

Fig. 137.

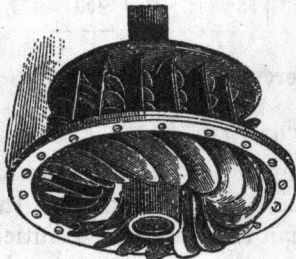

derates. During its passage
through the wheel the course
of the fluid becomes more vertical, but only a portion of it
near the outer circumference
assumes entirely the axial direction. Fig. 137 shows a
perspective view of the wheel
itself, and Fig. 138 an outside
view of the guide apparatus
and casing. In Chapter XII., in connection with the

Fig. 138.

"Hercules" Turbine.

subject of "American Turbines," the shape of the buckets is shown in axial section through the wheel.

The outer edges of the bucket-vanes are inclined, so that the outer surface of the wheel where the water enters forms a truncated cone, the mean diameter of which is 3 feet. At the inflow the vanes are tapered almost to a knife-edge.

The regulation is effected by a cylindrical gate between the guide passages and wheel. Both guide passages and buckets are divided by transverse partitions into several parts, so as to form practically as many turbines superimposed, which can be shut off in succession by the gate. This reduces the loss incidental to working at part gate. The outflow areas as given are calculated on the assumption that all the elementary filaments do *not* follow similar relative paths, but that the stream from each bucket or guide passage has the outflow area corresponding to the shortest distance between two adjacent vanes.

As with all radial turbines, to allow for contraction, the *effective* outflow area from the guides has been taken as only 90 per cent. of the measured area. Since there is a clear space between the guide orifices and the inflow orifices of the buckets (to allow room for the gate), the area of outflow, where measured, is not diminished by the obstruction of the wheel-vanes, and therefore no allowance has been made for this in computing the effective area.

A comparison of calculated with experimental results for the 36-inch "Hercules" turbine will be found under the heading of "American Turbines."

36″ "*Hercules*" *Turbine.*

Proportional gate opening.	Total head.	Quantity of water per second.	Number of revolutions per minute.	Horse-power measured on brake.	Efficiency.	Ratio of velocity of rotation to $\sqrt{2gh}$.	Depth on weir.
	h	Q	n	We	ee	K	
1·000	16·82	100·20	Still	0	0	0	1·305
,,	·90	89·87	123·50	145·30	84·39	0·588	·214
,,	·94	89·66	129·10	145·59	84·55	·614	·212
,,	·95	89·00	135·33	146·02	85·38	·644	·206
,,	·96	88·33	140·62	145·72	**85·80**	·669	·200
,,	·98	87·79	144·80	143·88	85·14	·688	·197
0·806	16·84	80·56	112·33	128·73	83·70	·536	·129
,,	·85	79·88	117·00	129·09	84·60	·558	·123
,,	·88	79·38	123·00	129·71	85·39	·586	·118
,,	·88	78·63	130·25	131·01	87·07	·621	·111
,,	·92	78·00	137·00	130·28	**87·08**	·652	·105
,,	·94	77·38	143·00	129·01	86·82	·680	·099
0·647	17·05	68·53	111·50	107·39	81·07	·529	·014
,,	·05	67·83	120·00	110·46	84·25	·569	·007
,,	·09	67·25	126·50	111·04	85·22	·599	·001
,,	·15	66·52	134·67	111·65	86·33	·637	0·994
,,	·16	65·72	140·40	109·55	85·69	·664	·986
0·489	·24	56·11	106·60	83·18	75·85	·503	·888
,,	·25	55·54	115·60	85·27	78·50	·545	·882
,,	·22	54·80	123·00	85·48	79·90	·581	·874
,,	·28	54·07	130·00	84·79	80·05	·613	·866
,,	·27	53·31	136·00	82·90	79·43	·641	·858
0·379	·62	45·64	104·00	64·66	70·93	·485	·774
,,	·61	45·20	112·75	65·29	72·36	·526	·769
,,	·65	44·50	122·67	65·06	73·06	·572	·761
,,	·66	43·76	129·40	63·89	72·93	·603	·753
,,	·65	43·24	135·00	61·72	71·34	·629	·747
1·000	9·59	68·01	88·67	60·00	81·14	·561	1·009
,,	·62	67·41	93·33	60·31	82·03	·589	·003
,,	·61	67·02	96·80	60·19	82·43	·612	0·999
,,	·65	66·31	103·33	60·47	83·36	·651	0·992
,,	·65	65·62	109·00	59·80	83·30	·687	0·985

Trial of a " Humphrey " Turbine.

In December 1880, the "Humphrey" Machine Company, of Keene, New Hampshire, agreed to supply the Tremont and Suffolk Mills with a new turbine of their own design to replace a Fourneyron turbine of about 160 horse-power, the new wheel being intended to develop about 270 horse-power with an average fall of 13 feet. The payment for this was to depend upon the result of a trial carried out by the well-known hydraulic engineer, Mr. James B. Francis.

Description of Turbine.

The "Humphrey" turbine in question is of the inward and downward (mixed) flow type shown in Figs. 139, 140, and 141.

The wheel *A* has thirteen vanes *a a*, &c., which project below the casing. *B* is a stationary cylindrical casing enclosing the wheel; it has twelve openings *q*, through which the water enters. *C* is the guide apparatus, which contains twelve guide-vanes, and at the same time serves as a regulator; it is enclosed by the curved sides *v v*, and runs on spherical rollers *b*. It can be turned by pinions *D D* gearing with a toothed ring on the circumference, which receive their motion from a hand-wheel *G* through worm and bevel gearing *E* and *F*.

The vanes *a* are of gun-metal, and are secured to the inner central casing by flanges and screws.

The turbine-shaft runs in a collar-bearing at the upper end, and at the lower end in a footstep-bearing, which is also capable of supporting the weight, if necessary, in the absence of the collar-bearing.

Fig. 140.

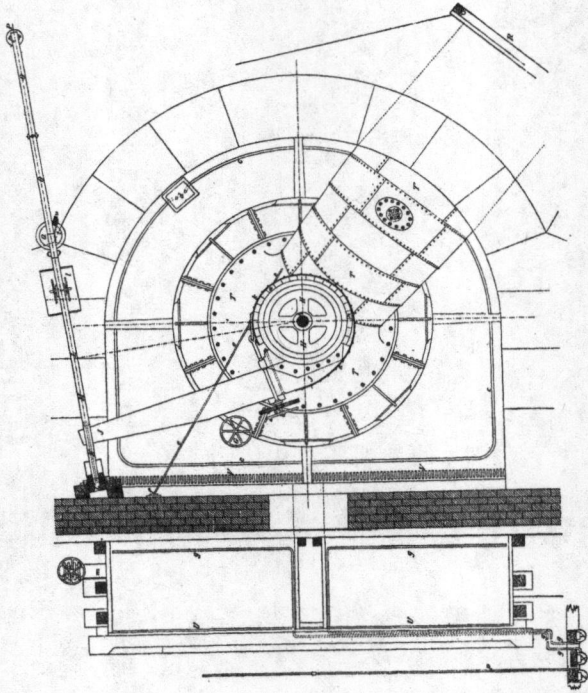

Plan.

"HUMPHREY" TURBINE.

Fig. 139.

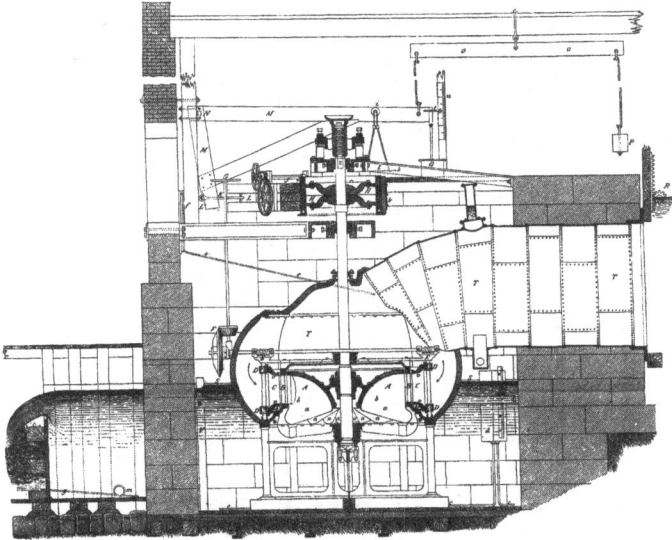

Sectional Elevation.

Fig. 141.

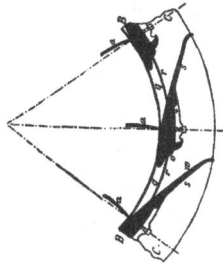

Part Horizontal Section through
Guides and Wheel showing Regulator.

The water from the head-race *R* enters the turbine through the wrought-iron tube *T*, which was employed for the old Fourneyron wheel, and is consequently rather too small for the new one. The estimated loss of head from this cause amounts to about four inches ; but during the experiments the head was measured immediately over the wheel, so that the source of loss referred to was eliminated.

Brake Measurements.

During the trial the main driving-wheel was removed from the turbine-shaft, and in its place the brake-pulley *H* substituted.

The latter is of cast-iron, containing a number of channels for the cooling water supply. *J* represents the brake-lever, which acts on the bell-crank lever *M* through the rod *K* suspended on knife-edges *L* and *L'*. The lever *M* rests on the knife-edge *N* and carries the pan *l* ; the weight of the latter and of the bell-crank lever are balanced by the weight *P* hanging from the lever *O*.

The diameter of the brake-pulley was 5·44 feet, and the breadth 2·7 feet. The effective length of the brake-lever *J* was 15·9 feet, and the lengths of the arms of the bell-crank lever, *L' N* and *N i* respectively, 6·015 and 12·013 feet, so that the lever-arm to be used in the calculation was 31·755 feet, equivalent to a circumference of 199·523 feet.

Water Measurements.

In order to observe the level of the water in the turbine - chamber immediately below the wheel, where the fluid is in a state of violent agitation, a pipe $c\,c$ was laid on the floor of the chamber, both ends of which were connected with a water-tight vessel d containing a scale. The pipe $c\,c$, of $1\frac{1}{2}$ inches internal diameter, was perforated on its upper side by twenty-two holes of about $\frac{3}{16}$ inch diameter.

It was agreed that as upper water-level was to be taken the level shown in the pipe e connected with the chamber immediately above the wheel.

The pipe e terminated in a glass-tube, parallel with which was placed a scale f.

The head introduced into the calculations is the difference between the levels in e and d.

The quantity of water used was measured by two weirs $U\,U$, both of which extended over the whole width of the channels conducting the water to them, so that there was no end contraction; the widths were respectively 11·22 and 10·98 feet. To prevent any disturbance of the flow over the weir by the water in the wheel-chamber, gratings were introduced into the tail-race between the weirs and the chamber, which caused an appreciable depression of the water-level and a nearly uniform flow. The height of the water below the weirs was so regulated that the level was sufficiently deep below the sill not to influence the outflow.

The quantity of water passing over the weirs was calculated by Francis's well-known formula,

$$Q = 3\text{·}33\,L\,H^{\frac{3}{2}},$$

in which

L denotes the length of the weir,

H the head of water above the sill, measured some distance from the weir, both dimensions in feet.

The head of water over the sills of the weirs was

measured by means of hook-gauges in water-tight vessels, connected by pipes with the water above the weirs. The hook-gauges were adjustable by micrometer screws, and allowed of very accurate observations.

The leakage during the tests was found to be only 0·01 cubic foot per second at most.

Dimensions of Turbine.

Outer diameter 8·1. $\qquad r_1 = 4·05$.
Inner diameter 2·0.

The *mean* diameter of outflow is uncertain, owing to the peculiar construction of the wheel, which allows the water on the outer side to follow its own course.

$$a = 75°. \qquad e = 2·086 \text{ feet.}$$
$$a_1 = 10°. \qquad e_2 = \text{indefinite.}$$
$$a_2 = \text{not given.}$$

Humphrey Turbine.

Opening of regulator.	Head of water.	Quantity of water per second.	Available power.	Number of revolutions per minute.	Velocity of circumference of wheel.	Work performed measured on brake.	Efficiency.	Ratio of velocity at circumference to $\sqrt{2gh}$.
	h	Q	E	n	w_1	We	ee	K_2
Per cent.	Feet.	Cubic feet.	Ft. lbs.			Ft. lbs.	Per cent.	
100	12·478	207·82	161744	47·424	20·113	132475	81·9	0·7301
82·5	12·839	181·59	145421	51·600	21·884	111527	16·69	0·7831
77·24	12·983	180·27	145985	48·000	20·357	111733	76·54	0·7245
68·21	13·103	166·80	136326	45·840	19·441	99084	72·68	0·687
52·90	13·456	131·52	110387	50·64	21·477	67363	61·02	0·7508
40·66	13·688	110·70	94520	45·588	19·339	53065	56·14	0·6702
27·94	14·069	79·26	69558	55·680	23·614	18517	26·62	0·8073
19·15	13·998	59·02	51534	9·704	21·070	8264	16·04	0·7225
11·38	14·088	37·91	34013	37·206	15·777	3093	9·09	0·5336

The preceding table has been compiled from the original
table of results, which has been much condensed, only the
best out of several sets of observations at each opening of
the regulator being given.

By mutual agreement between the users and makers of
the turbine tested, it was agreed to measure the head
available immediately over the wheel, in order to elimi-
nate the loss due to friction in the conducting-pipe, which
belonged to a former turbine of less power, and was consequently too small for the quantity of water supplied.

In doing this, however, no allowance has been made for
the velocity of the water ; at the point where the pressure
was measured this velocity was about 3·458 feet per second
in the first experiment, in which the maximum efficiency
was obtained, equivalent to a head of about 0·2 foot, which
should be added to the head given in the table to arrive
at the effective available head over the turbine.   The
corresponding efficiency in that case is 80·59 instead of
81·9 per cent., but as no allowance has been made for
friction of bearings the *hydraulic efficiency* will probably
be somewhat in excess of the latter value.

Experiments with "Little Giant" Turbine.

This series of tests is the only one made in Europe with
an American turbine of which the author has been able to
obtain any record; it was conducted by Professors Veith
and Kronauer of Zürich.

The turbine is of the mixed-flow type, the water entering from the outside radially, and leaving axially ; it is
divided in the middle by a partition at right angles to
the axis, which separates the water entering into two portions, which flow out on opposite sides of the wheel.   That
part of each bucket through which the water passes

after it has assumed an axial direction of flow stands out from the side of the wheel, in shape like a semicircular cowl; the construction is plainly shown in Fig. 142.

The water in its passage through the buckets appears to undergo a rather abrupt deflection of its course from the radial to the axial direction. The turbine is furnished with a guide apparatus.

Fig. 142.

(1) *Turbine half open.*

Fall.	Quantity of water per second.	Head above the weir.	Number of revolutions per minute.	Effective work.	Available energy.	Efficiency.	Proportion of velocity of rotation to $\sqrt{2gh}$.
h	Q	H	n	W_e	E	c_e	K_2
Feet.	Cubic feet.	Feet.		Horse-power.	Horse-power.		
1) 5·479	0·404	0·151	340	0·1568	0·2505	0·62	0·44
2) 5·479	0·444	0·161	323	0·1655	0·2742	0·60	0·42
3) 5·479	0·404	0·151	370	0·1517	0·2505	0·60	0·48
4) 5·560	0·388	0·147	386	0·1183	0·2436	0·48	0·50

With a load of 1·213 pounds on the brake-lever the turbine came to a standstill, and the quantity of water flowing through was 0·459 cubic foot, with a head of 5·446 feet.

(2) *Turbine fully open.*

	Fall. h	Quantity of water per second. Q	Head above the weir. H	Number of revolutions per minute. n	Effective work. W_e	Available energy. E	Efficiency. ϵ_e	Proportion of velocity of rotation to $\sqrt{2gh}$. K_2
	Feet.	Cubic feet.	Feet.		Horse-power.	Horse-power.		
1)	5·381	0·954	0·269	424	0·2446	0·5824	0·42	0·55
2)	5·623	1·024	0·282	390	0·4206	0·6528	0·65	0·50
3)	5·610	1·095	0·295	310	0·3972	0·6904	0·57	0·40
4)	5·610	1·095	0·295	294	0·4063	0·6904	0·58	0·38
5)	5·610	1·130	0·302	264	0·3925	0·7200	0·55	0·34

With a load of 3·307 pounds on the brake-lever the turbine came to a standstill, and the quantity of water flowing through was 1·095 cubic foot, with a head of 5·446 feet.

The above results are in each case the mean of four tests.

Water Measurements.

The water used was measured by means of a weir, the fall taking place over the whole width. The quantity of water Q was determined by the formula

$$Q = 0·431 \, L H \sqrt{2 \, gH},$$

where L is the width of the weir and H the head above its edge ; $L = 1·978$ foot.

The depth of water behind the weir during the experiments was 0·8525 foot.

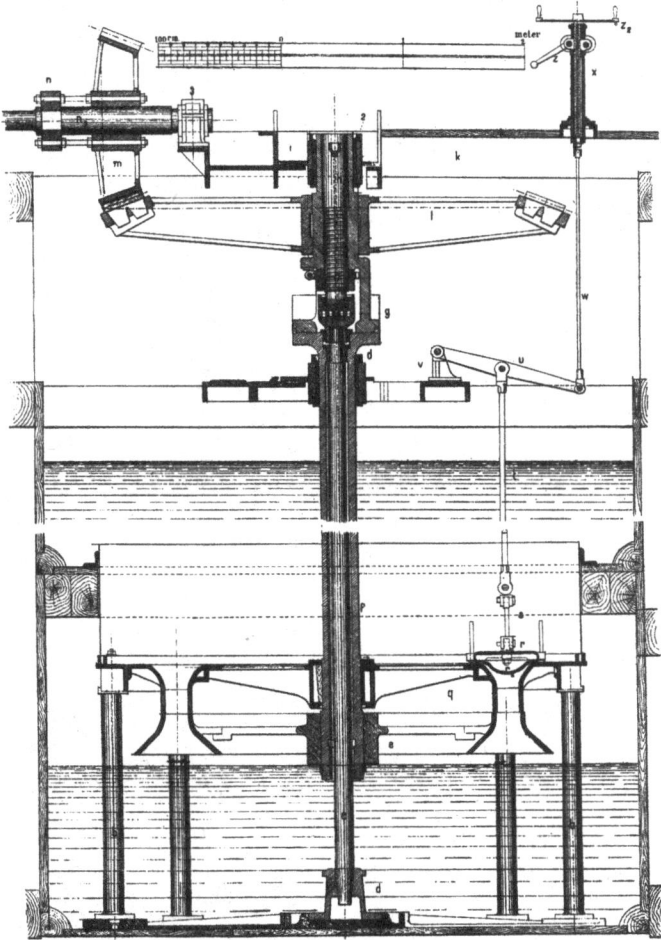

Fig. 143.

GIRARD TURBINE AT VARZIN.

Trial of Girard Turbines at Varzin.

At the Campmühle at Varzin (the property of Prince Bismarck) two Girard turbines, each of 200 horse-power, are employed to drive a wood-pulp manufactory. The turbines were constructed by the Gräflich Stolberg'schen Machine Works at Magdeburg.

The fall available is 13 feet 1½ inches, and the quantity of water for each turbine 176·6 cubic feet per second. The makers guaranteed an efficiency of 75 per cent. with full admission, and 72 per cent. with half admission. To test the performance of these motors experiments were made by Professor Dr. Zeuner, which from the care and completeness of method with which they were carried out may be regarded as in all respects a pattern of what such trials should be.

Description of the Turbines.

Both turbines are exactly alike in their construction (shown in Fig. 143), and enclosed in a wooden chamber divided by a partition into two independent compartments. Each motor can be shut off separately by a sluice from the head-race.

The axes of the wheels are vertical, and power is transmitted to the horizontal first-motion shaft by bevel gearing.

The turbines are of the axial or parallel-flow type, with ventilated buckets.

The guide apparatus rests on a circular cast-iron frame supported by eight cast-iron columns standing on a bed-plate on the floor of the tail-race; the water enters the guide apparatus through a wrought-iron cylindrical casing connecting it with the head-race and also supported by the columns.

The guide-vanes are of steel-plate cast into the lower part of the walls of the guide-casing, which are here cylindrical; towards the top this casing is enlarged— the section being bell-mouthed—and divided into eight segments by cast-iron radial partitions, each of which can be closed by a valve. These valves serve to regulate the turbine. The inner casing of the guide apparatus supports a frame with arms, the spaces between which are filled in by wrought-iron plates; the frame carries the lower bearing for the turbine-shaft. This bearing consists of a cast-iron cylinder with four compartments containing blocks of lignum-vitæ, which can be adjusted by metal wedges with screws; a leather collar held by the cover of the bearing prevents leakage, and the water necessary for lubrication enters through a hole in the cover.

The turbine-wheel itself consists of the cast-iron casing with steel vanes, and a boss and arms, cast separately, bolted to flanges on the casing. This is keyed to the shaft, and to prevent the possibility of falling if it should become loose, a ring made in halves is let into the shaft below the boss.

The shaft is hollow, of cast-iron, with the main bearing, supporting the weight, above water. Into the upper part of the shaft is fixed the wrought-iron prolongation, the end of which rests upon the step, while to it is keyed the bevel driving-wheel. The step itself is carried upon the top of a wrought-iron spindle fixed into a taper socket on the bed-plate.

The wrought-iron prolongation of the turbine-shaft is screwed, and can be adjusted to take up wear by means of a nut which is turned by a small worm and worm-wheel, the latter forming part of the nut.

The turbine is regulated by eight valves of cast-iron,

already referred to; these valves are made water-tight by india-rubber packing held in place by dished cast-iron plates; they can be raised and lowered by rods actuated from screwed spindles in connection with gearing.

The following are the leading dimensions, &c. of the turbine and gearing:—

Mean diameter of guides and wheel . .	8 ft.	11·1 in.
Radial width of guide passages at outflow.	0 ft.	8·11 in.
Radial width of wheel-buckets at inflow .	0 ft.	8·5 in.
Radial width of wheel-buckets at outflow .	2 ft.	1·59 in.
Depth of wheel	1 ft.	1·78 in.
Number of guide-vanes		64.
Number of wheel-vanes		70.
Thickness of guide-vanes		0·20 in.
Thickness of wheel-vanes		0·31 in.
Calculated number of revolutions per minute		30.

Bevel Gearing.

Number of teeth in bevel mortice-wheel on the turbine-shaft	136.
Number of teeth in iron pinion on first-motion shaft	54.
Breadth of teeth	13·78 in.
Pitch of teeth	3·47 in.
Number of revolutions per minute of first-motion shaft	75.

Brake Measurements of Power.

The brake employed to measure the horse-power of the turbine was of the usual "Prony" type. The brake-pulley was keyed to the horizontal first-motion shaft, it being impracticable to attach it direct to the vertical turbine shaft; the diameter of this pulley was 8 feet 2·4 inches, and the breadth 1 foot 3·75 inches.

The end of the brake-lever pressed against the platform of a weigh-bridge, by which the load acting on the former was registered.

Water was employed for the lubrication of the brake-blocks. During the trial of the turbine with full admission a large quantity of water was required to keep the blocks cool.

The weight of the brake-pulley was 6788 lbs., that of the brake-band, blocks and lever complete 2595 lbs. Part of this latter weight was carried by the shaft and part by the weighing-platform, and before commencing the trials the portion of the weight carried by the platform was carefully ascertained by supporting the brake on a knife-edge laid on the highest point of the pulley, the brake-band being slackened so that none of the blocks were in contact with the pulley.

The weighing-machine was then adjusted so as to balance the free end of the lever, and the weight required for this noted, and in calculating the horse-power deducted from the total pressure on the weigh-bridge.

To arrive at the power actually given off by the vertical turbine-shaft, it was necessary to make allowance for the friction of the horizontal first-motion shaft and of the gearing.

This was done by calculation. For determining the power absorbed by the friction of the bearings the weight carried by these and its approximate distribution was taken into consideration, as well as the pressure exerted by the bevel-wheels. For the shaft-friction the co-efficient $f_1 = 0·07$ was employed in the formula for the work lost,

$$L_1 = f_1 \frac{n\Pi}{60} P d,$$

where P denotes the total pressure on a bearing, n the number of revolutions per minute of the shaft, and d the diameter of the neck.

For friction of the gearing, a co-efficient $f_2 = 0.1$ was used in the formula,

$$L_2 = f_2 \frac{l}{a} \, W,$$

where W is the power transmitted, l the length of a tooth, and a half the length of the arc of contact of the teeth.

Water Measurements.

The river Wipper, which supplies the power for the turbines, is dammed by a weir, from which the water enters each turbine-chamber by two sluices. During the experiments only one of these was open, and for the purpose of measuring the water behind this sluice a conducting channel of timber was specially constructed, having a width of 7 feet 4 inches, a depth of sides of 8 feet $2\frac{1}{2}$ inches, and a length of 23 feet, nearly. A section of this channel, at a distance of $19\frac{1}{2}$ inches behind the sluice, was selected in which to measure the velocity of the current. This was done by means of "Woltmann" current meters which had been carefully tested. The section was divided into sixteen equal rectangular parts, four in the width and four in the depth; in the centre of each of these the velocity of flow was observed, and from the results the mean velocity calculated.

When the turbine was working with full admission the mean velocity was 3.151 feet per second, and 1.481 feet with half admission.

The depth of the water in the channel varied slightly

E E

during the trial, and in order to take these variations into account, the duration of the trial, 1 hour 20 minutes, was divided into four periods, during each of which the average depth was ascertained and used in calculating the section of the stream.

As the variation in the depth was relatively slight, Professor Zeuner considered it justifiable to assume that it had no appreciable influence on the velocity of flow.

According to previous agreement, the *effective head*, measured to the lower edge of the wheel-vanes, was to be taken into account in determining the efficiency, and not, as should from a scientific point of view be the case, the total head from the upper to the lower water-level.

As only *one* of the sluices admitting water to the turbine-chamber was open, the inflow was one-sided, and in consequence of this a whirling motion was set up, causing a depression in the surface of the fluid above the motor. To avoid any uncertainties due to this, the head was measured to the level of the water in the conducting channel immediately before entering the chamber, and not, as often happens, in the latter.

Results.

In the following tables the results of the trials with full and half admission are given. To the particulars given in the original tables the author has added the available *total* head, measured from the head-race to the tail-race level, and the efficiency ϵ referred to this head.

Some further supplementary experiments were also carried out by Professor Zeuner to ascertain (1) the effect of allowing the turbine to *wade*, that is, to work partly immersed, and (2) the reduction in the efficiency caused by running the wheel at a higher and lower speed than the best.

The result of these experiments are also stated in a tabular form.

It will be seen that the power of the turbine is reduced by increasing the immersion; this is what would be anticipated, since free deviation turbines should work clear of the tail-water.

The experiments on the variation of the power with the velocity show that 30 revolutions per minute, the speed for which the turbine was designed, really is the best speed, as the work done decreases above and below this.

Results of Experiment on the Effects of Immersion.

Depth of immersion or clearance. h	Number of revolutions per minute. n	Effective horse-power. W_e	Remarks.
—0·446	29·5	105·571	
—0·446	29	101·617	
—0·478	29	99·448	
—0·217	33	108·123	With half admission.
+0·118	30	109·604	
+0·373	30	111·842	
+0·520	30	113·646	

Results of Experiment on Varying the Speed.

Depth of immersion or clearance. h_2	Number of revolutions per minute. n	Effective horse-power. W_e	Remarks.
+0·380	25·5	216·545	
+0·380	30	221·120	With half admission.
+0·377	34·5	215·589	
+0·380	36	213·489	

Results of Trial with Full Admission.

Total available head. h	Clearance over tail-water. h_2	Effective head. h	Quantity of water used per second. Q	Available horse-power from effective head. E_a	Number of re-volutions per minute. n	Effective horse-power deve-loped. W_e	Efficiency re-ferred to effec-tive head. e_e	Efficiency re-ferred to total head. e
Feet.	Feet.	Feet.	Cubic feet.					
13·03	0·220	12·81	174·713	253·964	29	202·81	0·798	0·784
13·19	0·276	12·92	177·136	259·594	30·37	209·35	0·806	0·789
13·40	0·354	13·05	180·163	266·713	30	209·13	0·784	0·763
13·55	0·387	13·16	182·811	273·023	30·62	216·36	0·792	0·769
					Mean :	209·41	0·795	0·776

Remarks.

According to contract the fall was only to be measured to the lower edge of the wheel, that is, not to include the clearance above the tail-water ; the efficiency e_e calculated at the trials was referred to the head thus measured, which is called the effective head h_1. For purposes of comparison the total head $h = h_1 + h_2$ has been added to the table as well as the efficiency e referred to it.

Results of Trial with Half Admission.

Total available head. h	Clearance over tail-water. h_2	Effective head. h_1	Quantity of water used per second. Q	Available horse-power from effective head. E_a	Number of re-volutions per minute. n	Effective horse-power deve-loped. W_e	Efficiency re-ferred to effec-tive head. e_e	Efficiency re-ferred to total head. e
Feet.	Feet.	Feet.	Cubic feet.					
13·59	0·076	13·51	89·666	137·488	29·62	109·032	0·793	0·786
13·61	0·066	13·54	89·949	138·188	30·37	111·359	0·805	0·801
13·63	0·062	13·57	90·305	139·065	30	111·871	0·804	0·800
13·51	0·072	13·58	90·376	139·243	29·83	111·842	0·803	0·807
					Mean :	111·024	0·801	0·798

Remarks.

A negative (−) sign before the figure giving the clearance denotes that the turbine-wheel was *immersed* to the depth stated.

Radial Outward-flow Impulse Turbines at Terni.

At Terni in Italy all the machinery of the large steel-works there is driven by a number of turbines acting under a head of 595½ feet. All these motors, which were made by Messrs. J. J. Rieter and Co. of Winterthur (Switzerland), are of the radial impulse outward-flow type with horizontal axes, and vary in capacity from 20 to 1000 horse-power. In the following list the leading particulars are given:—

Horse-power.	Effective head.	Quantity of water per second.	Revolutions per minute.	Inner diameter.
	Feet.	Cub. ft.		Ft. in.
1000	595·5	19·77	180 to 240	7 10½
800	,,	15·89	200	8 2·4
500	,,	9·89	240	6 5·9
350	,,	7·06	200	7 10½
150	,,	3·00	250	6 4·7
50	,,	0·99	850	1 10¼
50	,,	0·99	850	1 10¼
40	,,	0·85	450	3 6·1
40	,,	0·85	450	3 6·1
30	,,	0·60	600	2 7·5
20	,,	0·42	450	3 6·1

In Fig. 144 a perspective view of one of the smaller motors is shown, while Figs. 145 and 146 represent a section and plan of the 800 horse-power turbine employed for driving the rail-mill.[1] The following are the leading dimensions and particulars of this latter motor:—

Angle of outflow from guide-
passages . . . . $a = 70°$.

[1] For these illustrations the author is indebted to the courtesy of the editors of *Industries.*

Fig. 144.

Angle of inflow into wheel-buckets	$a_1 = 54°$.
Angle of outflow from wheel-buckets . . .	$a_2 = 70°$.
Internal diameter of wheel .	$2\,r_1 = 8$ ft. 2·4 in
External diameter of wheel .	$2\,r_2 = 9$ ft. 5 in.
Width of guide-passages . .	$c = 4·91$ in.
Width of buckets at outflow .	$c_2 = 16·14$ in.
Number of guide-passages .	$z = 2.$
Number of wheel-buckets .	$z_1 = 100.$
Maximum proportion of circumference over which water is admitted . . .	$\varpi = \frac{1}{36}.$
Diameter of supply-pipe . .	$= 24$ in.
Diameter of auxiliary pipe .	$= 8$ in.

Fig. 146.

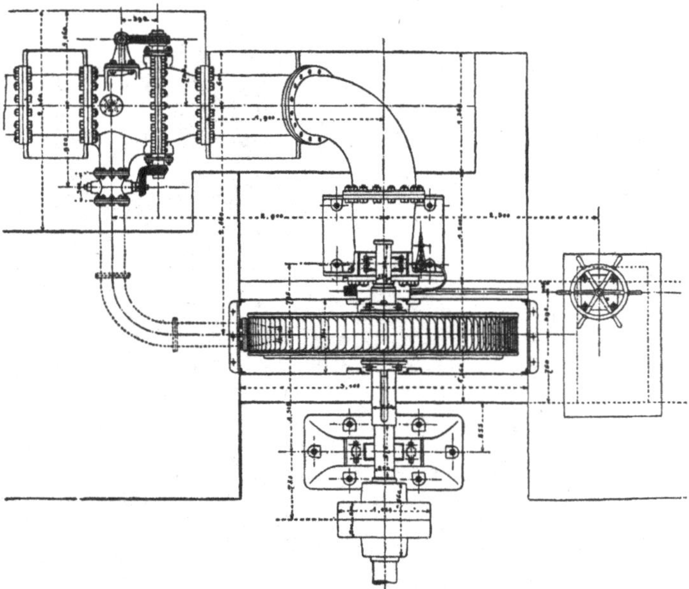

PLAN OF 800 H.-P. TURBINE AT TERNI.

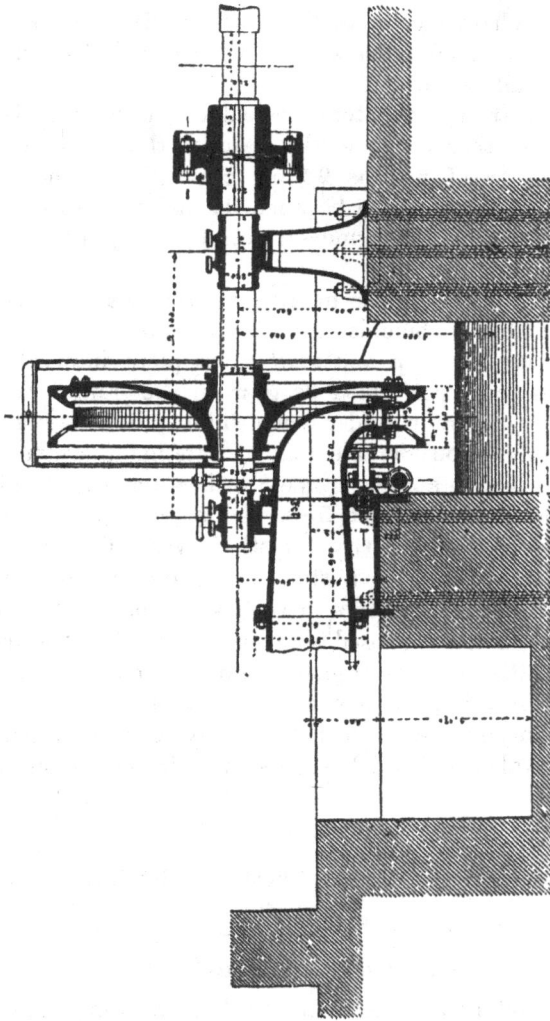

Fig. 145.

Section of 800 Horse-power Turbine at Terni.

The radial velocity of the water leaving the wheel is 34 feet per second, representing a loss of only 3·6 per cent. of the available head.

The quantity of water is calculated as 85 per cent. of that theoretically due to the head at the guide-orifices; the velocity of flow as 92 per cent. of the theoretical velocity; the difference between 92 and 85 being due to the obstruction of the guide-orifices by the passing wheel-vanes.

The dimensions given in the illustrations are in mètres.

Since this turbine is used for a rolling mill train, it is necessary to stop and start at frequent intervals, the reversals being performed by gear on the train. To stop the flow of water under such high pressure suddenly would entail considerable risk of bursting the pipes, and to avoid this an auxiliary pipe 8 inches diameter has been provided, branching off from the main supply pipe on the pressure side of the throttle-valve. This auxiliary pipe is provided with a stop-cock geared to the spindle of the main throttle-valve in such a manner that it opens when the latter is closed, and *vice versâ*. The closing of the throttle-valve is effected through worm-gearing by hand. By this arrangement the motion of the water in the main supply-pipe is not entirely arrested when the throttle-valve is closed, but finds an outlet by the auxiliary pipe.

Trial of Radial Outward-flow Impulse Turbine at Immenstadt.

Description of Turbine.

This motor was constructed to drive a string manufactory at Immenstadt, and is of the same type as those·

used at Terni previously described. It is intended to develop a maximum of 400 horse-power with a head of 570·8 feet, and a supply of water of 8·5 cubic feet per second.

The majority of the vanes are of steel-plate 0·2 inch thick, cast in with the wheel-casing; but there are besides thirty-six strong vanes of cast-iron to support one side of the casing from the other, which is connected with the arms and boss. The regulation of the water-supply is effected by a segmental slide with a rack on the back worked by a pinion through a worm-wheel and worm. The face of the slide fits against the inflow edges of the guide-vanes, and closes or opens the guide-passages as required; it can be controlled either from a governor or by hand. The principal dimensions of the turbine are as follows :—

Inner diameter	. . . $2\,r_1$ =	7 ft. 10$\frac{1}{2}$ in.
Outer diameter . . .	$2\,r_2$ =	8 ft. 11·1 in.
Width of guide-passages	. e =	4·32 in.
Width of buckets at inflow	. e_1 =	4·72 in.
Width of buckets at outflow	. e_2 =	15·75 in.
Number of wheel-vanes .	. z_1 =	110.

The water is conducted to the motor through supply-pipes having a total length of 4642 feet 4 inches, of which 3481 feet are of cast-iron and the remainder wrought-iron. The cast-iron pipes have a diameter of 21·66 inches, and are in lengths of about 13 feet, the metal being 0·71 inch thick. The wrought-iron pipes are 18·9 inches in diameter, in lengths of 19 feet 8 inches, and 0·2 to 0·47 inch thick.

The velocity of the water in the cast-iron pipes, with the largest possible supply, does not exceed about 4·125 feet per second, while in the wrought-iron pipes it is 5·44 feet per second.

Water Measurements.

The quantity of water used by the wheel during the trial was measured over a weir or bay, the formula employed to determine the co-efficient being that of Braschmann, according to which

$$Q = \left(0\cdot3838 + 0\cdot0386 \div \frac{L}{B} + \frac{0\cdot00174}{h}\right) L\,H\sqrt{2gH}$$

(*vide* "Measurement of the Quantity of Flowing Water"). As a check the results were compared with those obtained by Castel and with Weisbach's formula.

The head immediately before the guide-passages was measured by means of a pressure-gauge, and found (as above stated) to be 570·8 feet.

Measurement of Power developed.

The power exerted by the turbine was measured with a brake in the ordinary way, and in addition to this the work expended on shaft-friction was also determined.

Results of Trial.

The results of the experiments are given in the accompanying table. At the time they were carried out the maximum supply of water for which the wheel was designed (8·5 cubic feet) was not available, so that the turbine could not be tested to its full power.

In the account of the trial from which the preceding particulars have been taken, the vane-angles are not given, but as the effective head is very nearly the same as for the 800 horse-power turbine of the same type at Terni, it is safe to assume that the angles are also nearly identical with those of the latter:

a = 70°, angle of outflow from guides.
a_1 = 54°, angle of inflow into buckets.
a_2 = 70°, angle of outflow from buckets.

Radial Outward-flow Girard Turbine with Horizontal Axis at Immenstadt.

Head.	Opening of regulator.	Quantity of water per second.	Available power.	Number of revolutions per minute.	Effective power measured on brake.	Efficiency.	Power absorbed by friction of shaft.
h		Q	E_{hp}	n	W_e	e_e	W_f
Feet.		Cubic Feet	Horse-power.		Horse-power.		Horse-power.
570·84	10	2·111	136·8	211	81·5	0·595	5·0
,,	15	3·350	217·1	211	144·3	0·665	4·8
,,	20	4·375	283·4	214	196·8	0·694	4·7
,,	25	5·187	336·0	216	253·9	0·755	4·5
,,	15	3·156	204·5	240	129·2	0·632	5·6
,,	15	,,	,,	213	137·5	0·672	4·9
,,	15	,,	,,	194	140·2	0·686	4·4
,,	31	6·804	440·9	210	336·8	0·764	

Head measured with pressure-gauge immediately before turbine.

From observed quantities and measurements the velocity of flow from the guide-orifices was found to be

$$c = 0.948\sqrt{2gh} = \sqrt{\frac{2gh}{1 + 0.113}}.$$

For calculating the effective area of the guide passages, the above co-efficient must of course be reduced by the usual amount to allow for obstruction by the wheel-vanes, the value given having been determined when the guide orifices were free.

Turbines at Schaffhausen for Electric Power Transmission.[1]

Five new turbines have recently been added to those already in existence for about thirty years, for utilizing the fall of the Rhine at Schaffhausen. The older turbines have always been and still are used for driving wire ropes which transmit the power developed to the opposite bank of the river. The new turbines drive dynamos from which the power is transmitted by electric cables to various points; each develops at least 300 H.P.

Two of the turbines were supplied by Messrs. J. J. Rieter and Co. of Winterthur. These work dynamos from which the current is transmitted through copper wires to Schaffhausen, on the opposite bank of the Rhine, for the purpose of driving a spinning manufactory. Of the remaining three turbines, supplied by Messrs. Escher, Wyss and Co. of Zürich, two are employed in providing electricity for aluminium works adjoining the turbine house.

In every case the dynamos are driven by ropes from counter-shafts connected by mortice bevel-gearing with the vertical turbine-shafts. Figs. 147, 148 illustrate the arrangement and design of Messrs. Rieter and Co.'s turbines. These are of the Jonval type, each with two concentric sets of buckets, an arrangement adopted on account of the variations occurring in the available head and quantity of water.

In time of flood, in summer, the head is only about 9 ft. 10 in., while the flow of water amounts to 388 cubic feet per second. When the river is low the maximum

[1] *Zeitschrift des Vereins Deutscher Ingenieure*, vol. xxxvii., No. 46, November 18, 1893, p. 1416.

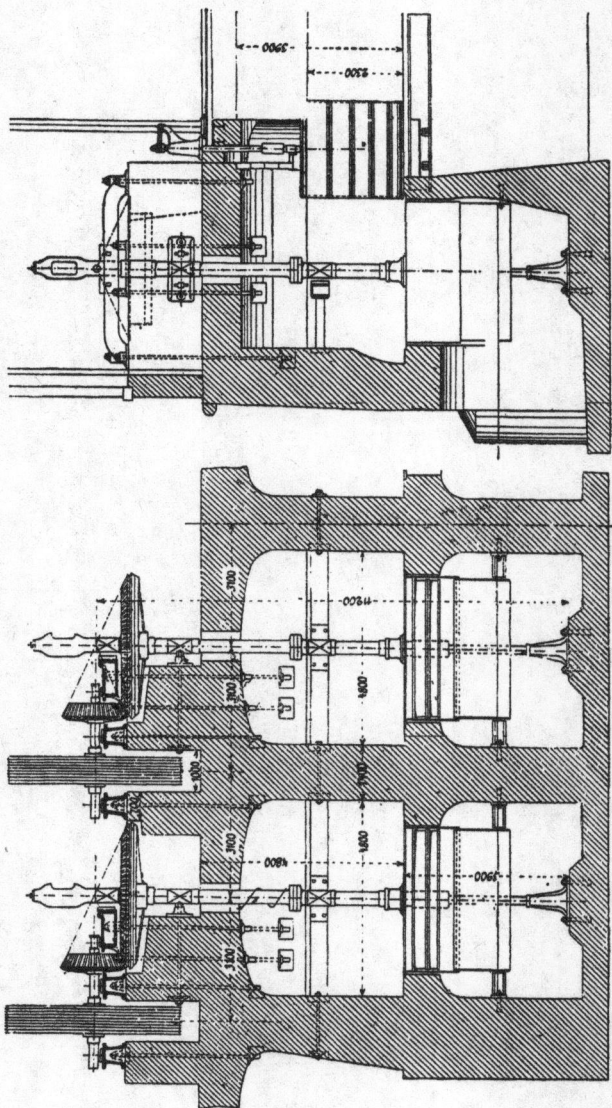

Fig. 147.

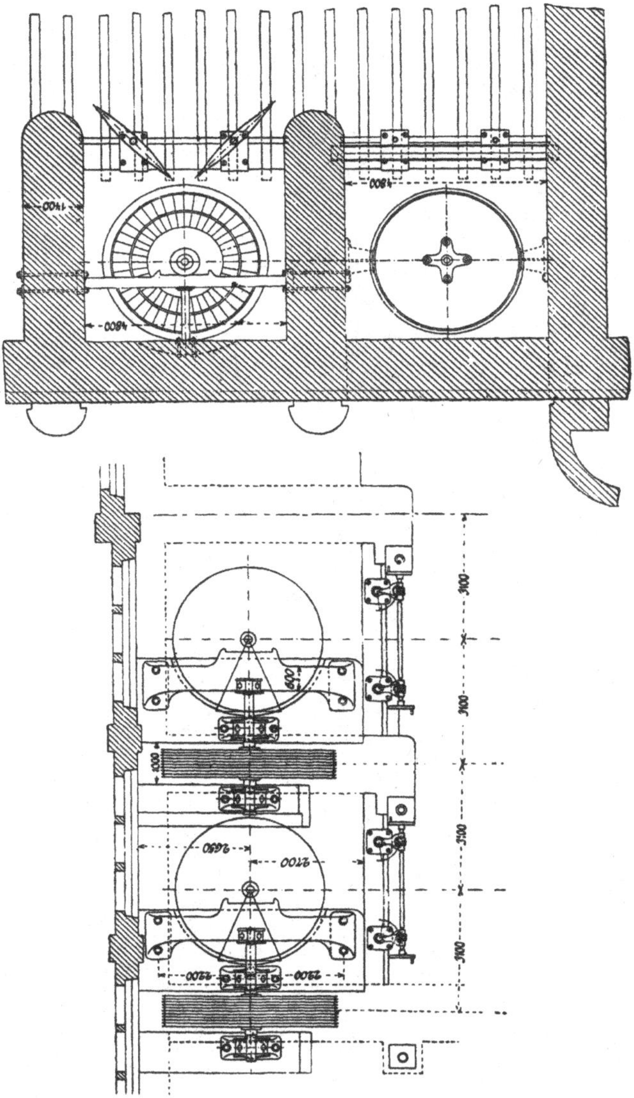

1400
4800
4800
3100
3100
3100
3100
600
1000
2650
2700
2200
2200

head is 14 ft. 9 in., and the corresponding quantity of water

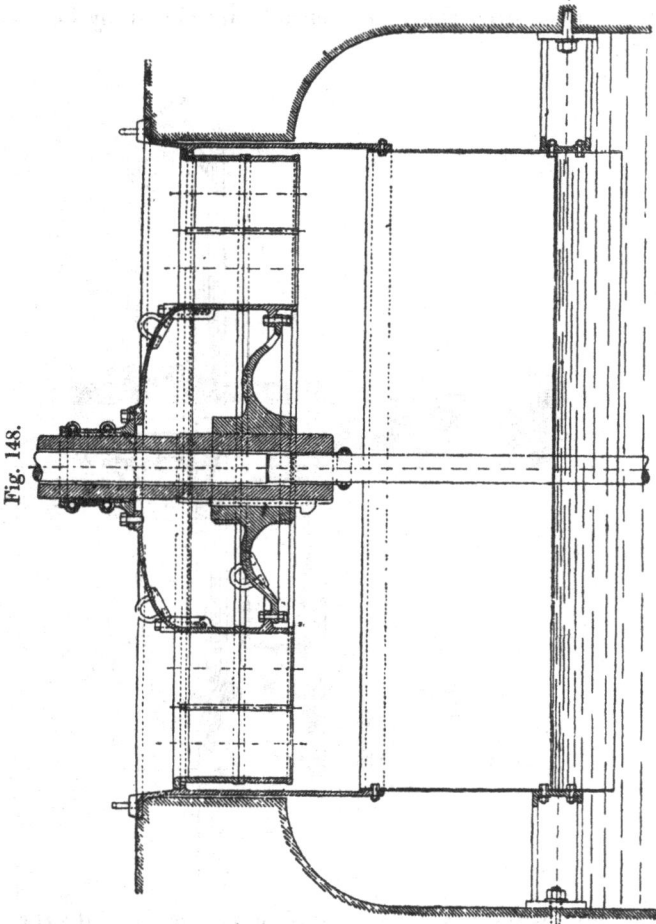

Fig. 148.

is 240 cubic feet per second; the change in the head is due to the rise of the tail-water level. When the water

is low—with the maximum head—the outer set of buckets
only is used, while in floods both sets come into action.
When not required the inner buckets are closed by hand-
covers.

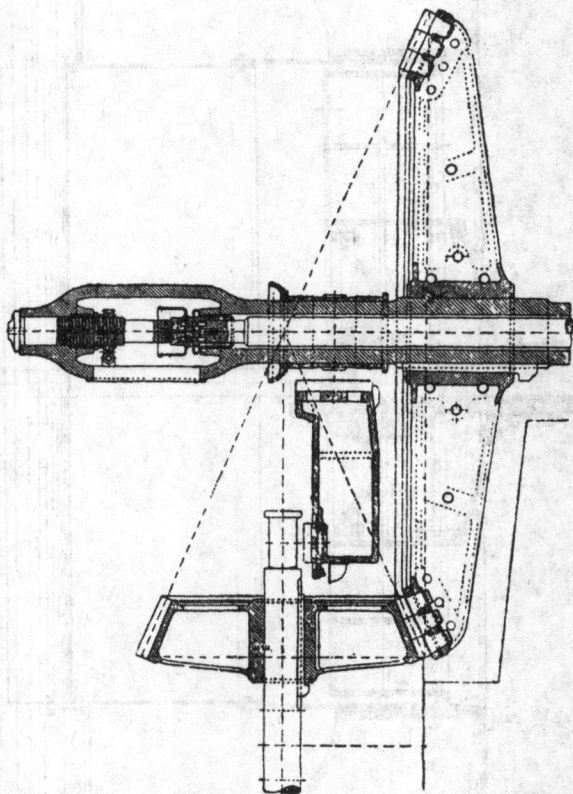

Fig. 149.

The guide-ring fits into a cast-iron casing, let into the
masonry, and carrying at its lower end a wrought-iron
draft-tube. The lower edge of the wheel is 4 ft. 7 in.

above the tail-water level when the latter is at its lowest.

Each turbine wheel is keyed to a hollow cast-iron shaft supported on an overhead bearing. The shaft is made in two parts, with a total length of 32 ft. 1·84 in.; the outer diameter is 13 in., the inner diameter 7·47 in.

The wrought-iron column, about which the shaft revolves, fits into a cast-iron socket secured by four lewis-bolts to the floor of the tail-race.

The column carries at its upper end an oil-bath in which the shaft-pivot revolves, as clearly shown in Fig. 149. The lower end of the pivot is secured to a cast-iron washer, which runs on a similar washer attached to the top of the column; both washers have a diameter of 7·08 in., and the total load on their bearing surface, including water pressure, is 50,706 lbs.

To steady the shaft there are four bearings, of which two are below water and have lignum vitæ bushes.

The bevel mortice-wheel keyed to the upper end of the turbine-shaft has 132 teeth, with a pitch of 3·34 in., a pitch diameter of 11 ft. 8·32 in., and a breadth of 13·78 in. The pinion has 56 cast-iron teeth, with a pitch diameter of 4 ft. 11·53 in.

Each rope-pulley driving a dynamo has a diameter of 13 ft. 1·48 in., and 10 grooves for ropes each 1·97 in. diameter.

The water is admitted to the turbine-chambers by balanced gates turning about vertical spindles, as shown in Fig. 147.

The three turbines of Messrs. Escher, Wyss and Co. are designed for similar conditions, and are of the same type as those of Messrs. J. J. Rieter. They run, however, at a speed of 48 revolutions per minute, are smaller in diameter, and have a greater width of buckets.

Instead of working with a draft-tube, each turbine is placed at the lower end of a cylindrical casting forming part of the turbine-chamber.

With low water the wheel is just covered, but in times of flood the tail-water rises round the outside of the casing.

A section through the wheel and guide-casing is shown in Fig. 150.

Fig. 150.

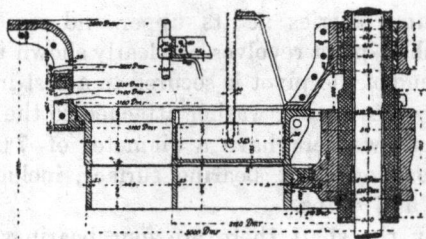

The inner set of guide-passages can be closed by vertical slides.

The dynamos driven from the Rieter turbines are six-pole compound dynamos, running at a speed of 200 revolutions per minute, and each developing 330 amp. with a pressure of 630 volts, equivalent in round numbers to 350 H.P.; of this, 565 H.P. is transmitted to the spinning manufactory.

The accompanying table, which has been compiled from data furnished to the author by the makers, contains the more important dimensions, ratios, and quantities for one of Messrs. Rieter's turbines.

The outer ring is so designed as to work at the best speed with the maximum head of 14·75 feet; its area is calculated for a quantity of water somewhat in excess of the minimum actually occurring, namely, 270 cubic feet

Details of Jonval Turbine at Schaffhausen (by Messrs. J. J. Rieter and Co.).

	Outer Ring.			Inner Ring.			Outer and Inner Ring combined.
	Outside.	Mean.	Inside.	Outside.	Mean.	Inside.	
E_{hp} = Effective Horse-power	...	296	...	...	...	...	296
h = Heads, feet ...	...	14·75	...	...	9·833	...	9·833
Q = Quantity of water per second, cubic feet	...	270	...	...	198	...	420·4 calculated with mean co-efficient K_1 and lower head.
a = Guide outflow angle	66°	63°	60°	64°	59½°	54°	
a_1 = Wheel inflow ,,	40°	36½°	33°	20°	17°	14°	
a_2 = ,, outflow ,,	71½°	69°	66½°	67½°	63½°	59½°	
d = Diameters, feet	11·155	9·9083	8·6617	8·4975	7·185	5·8725	
$e = e_1 = e_2$ = Widths, feet...	...	1·247	...	...	1·3125	...	
z = Number of guide-vanes	...	40	...	...	40	...	
z_1 = Number of wheel-vanes ...	...	42	...	...	42	...	
t = Thickness of guide-vanes, inches	...	0·59	...	...	0·59	...	
t_1 = Thickness of wheel-vanes, inches	...	0·55	...	...	0·55	...	
h_o = Depth of guides, inches	...	10·63	...	...	10·63	...	
h_o^1 = Depth of wheel, inches	...	10·63	...	...	10·63	...	
A^1 ... ... sq. ft.	...	15·073	...	...	12·400	...	
A^1_2 ... ... ,, ,,	...	11·338	...	...	10·443	...	
A ... ... ,, ,,	...	13·908	...	...	11·292	...	25·200
A_2 ... ... ,, ,,	...	11·338	...	...	10·443	...	21·781
$\dfrac{A}{A_2}$ = Area-ratios ...	...	1·227	...	...	1·081	...	1·15
n = Revolutions per minute	...	46	...	...	46	...	46
V = Velocity due to head, feet per second	...	30·816 (1)	...	...	25·164 (2)	...	
K_1 = Co-efficient of velocity	...	0·630	...	...	0·697	...	0·663 Mean of (1) and (2).
K_2 = Co-efficient of speed	...	0 774	...	...	0·688	...	0·814
w = Velocity of rotation, feet per second	26·867	23·865	20·862	20·467	17·305	14·143	20·504
c = Velocity of flow, feet per second ...	...	19·414	...	...	17·540	...	16·684 for lower head.
c_2 = Relative velocity of outflow, feet per second	...	23·812	...	...	18·966	...	

per second (the minimum being 240). The inflow takes place without shock, and the outflow is approximately vertical, w and c_2 being nearly equal.

The inner ring is designed to give the maximum efficiency with the minimum head of 9·833 feet, and under these conditions is calculated to consume 198 cubic feet of water per second.

It is clear that in a double turbine of this class, when the diameter and width of the inner ring have been fixed, that of the outer ring is also determined within comparatively narrow limits, and this of course settles the velocity of rotation. In calculating the dimensions of the outer ring, a method somewhat different to that employed for a simple turbine has to be followed.

In the first place (as usual) a may be chosen, but the other angles, the area ratio, and the diameter can only be determined by a process of trial and error.

If the inner ring is of normal design with $\dfrac{A}{A_2} = 1$ and a_1 a small angle, then for the outer ring $\dfrac{A}{A_2}$ must have a larger value, and a_1 must be greater in order that the conditions of inflow without shock and vertical outflow may be observed. This is the case in the present example. For the inner ring—as will be seen from the table—the area-ratio is 1·081, the angles a and a_2 are rather less than is usual, a_1 slightly greater, while the diameter is a little under the normal value for continental practice. For the outer ring the ratio $\dfrac{A}{A_2}$ and all the angles are greater, the angle a_1 being quite abnormal.

When the two rings are both at work with the maximum quantity of water and the minimum head, the inner ring only is working at the best speed. Under these conditions the available power is greater than with the maximum head and minimum quantity, and the guaranteed effective power (296 H.P.) can be obtained with an efficiency of 68 per cent. When only the outer ring is working (with the maximum head), an efficiency of 73·5 per cent. is necessary in order to develop the required power. The preceding efficiencies are calculated for the actual maximum and minimum quantities of flow 388 and 240 cubic feet per second respectively, and not for the quantities for which the turbine is designed (which are rather greater in order to allow some margin).

The Pelton Wheel.

The Pelton wheel, or, as it is popularly called, the "hurdy-gurdy," is a form of motor which has recently found great favour in the mining districts of California and other parts of America. Its construction is shown in Fig. 151.[1] It consists essentially of a series of double buckets attached to the circumference of a wheel; the water issues through a nozzle from the supply-pipe, and, striking the buckets in the middle, is deflected to both sides equally. The direction of the jet of fluid is tangential to the mean circumference, measured to the middle of the radial width of the buckets. A little consideration will make it evident that the Pelton wheel is an impulse turbine of the *axial* class, in which the angle a is 90°, and a_2 also near 90°. Figs. 152 and 153 show respectively a perspective view of and a section through one of the buckets. The relative velocity of inflow is $c - w = c_1$, and the relative velocity of outflow—apart from friction—the same $c_2 = c_1$, since there is no fall in traversing the

[1] For this illustration the author is indebted to the courtesy of the editors of *Engineering*.

Fig. 151. The Pelton Wheel.

buckets, and the velocity of rotation is the same where the water enters as where it leaves the wheel. To obtain the minimum absolute residual velocity, in this case O : $c_2 = w = c_1$, and hence $c = 2\ w$, or $w = \dfrac{c}{2}$ the best speed of the wheel.

As actually made, the relative angle of inflow a_1 is not quite 90°, which for absence of shock should on theoretical grounds be the case, but to attain this it would be necessary to bring the middle of the bucket to a knife-edge to avoid loss by impact against a blunt surface. To do this would be impracticable, and the loss caused by making

Fig. 152.   Fig. 153.

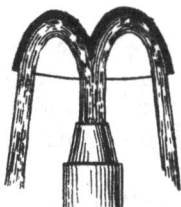
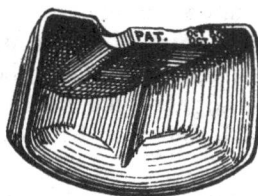

the edge flat would probably more than counterbalance any advantage obtained from the correct construction of the angle a_1.

Very high results are claimed for these turbines, one of which is said to have shown an efficiency of over 87 per cent.; without feeling bound to accept this result as absolutely correct, the author sees no reason why a Pelton wheel should not have a very fair efficiency, although there are causes of loss peculiar to its construction which must be pointed out.

In the earlier wheels, owing to the angle a_2 being 90°, the water on leaving a bucket has no velocity in an axial direction to enable it to clear the following bucket, and must be carried round some distance by the wheel before acquiring from gravity the motion necessary for the purpose. This was pointed out in a paper read by Mr. Hamilton Smith, junr., before the American Society of Civil Engineers, in which the author stated that "an examination of a hurdy-gurdy with either flat, recessed, or curved buckets, while at work, shows that the wheel 'carries' over a large amount of water, the force of which is consequently lost, and in fact becomes an additional load to lift."

In more recently constructed Pelton wheels this defect has been remedied by making the angle a_2 rather less than 90°. Obviously, as in the case of other impulse turbines, the thickness (or depth) of the jet must be taken into account in determining a_2.

Another cause of loss consists in the obstruction of the jet of water driving the wheel, not only by the central dividing edges of the buckets in a plane at right angles to the axis, but also by the outer edges parallel with the latter.

The usual method of regulating Pelton wheels is by throttling the jet, but recently nozzles have been introduced in which the jet is regulated by an accurately fitted internal cone, as illustrated by Figs. 154 and 155. For high powers several jets are sometimes employed, placed so far apart that a bucket supplied by the first jet is emptied before being acted on by the next jet.

In cases in which the water-supply varies at different times, the supply-pipes are so arranged that nozzles of different sizes can easily be fitted on to them.

The body of the wheel is generally of cast-iron, and the buckets of hard bronze. For small wheels the buckets are polished internally in order to reduce friction.

It is stated that Pelton wheels can be made ranging in

power from $\frac{1}{40}$ to 2000 h.p., and that the manufacturers guarantee an efficiency of from 80 to 85 %.

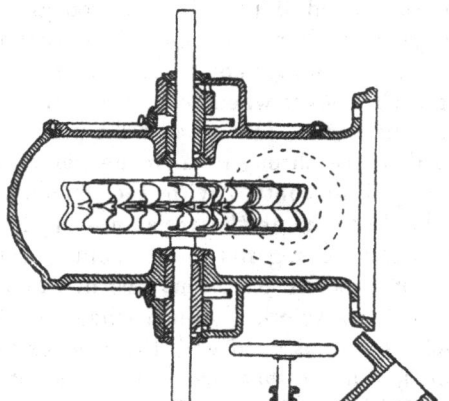

Fig. 155.

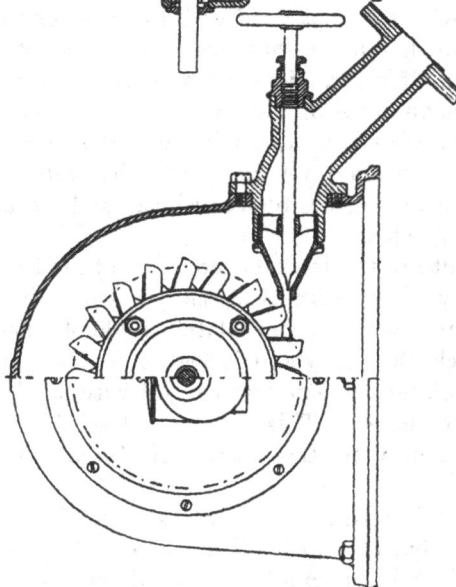

Fig. 154.

In designing a Pelton wheel it is obviously an advantage to pitch the buckets as far apart as practicable, in order to reduce the loss caused by impact of the jet with the edges of the buckets to a minimum.

On the other hand, if the pitch be too great, it is not possible to give the buckets such a form that the energy of the jet is advantageously utilized in the extreme position of each bucket when the jet is acting on it from the greatest distance. The greater the pitch, the greater also is the distance through which the water has to pass before coming into contact with the buckets, and this increases the loss due to the resistance of the air. The best proportions, as regards these points, can only be determined by experience. It may be mentioned, that as at present made in America, a 6-foot wheel has 24 buckets.

The buckets are not fixed with their longer edges radial to the wheel, but so that, when any bucket is in the position in which the jet first acts upon it, the central edge forms an angle of *not over* 90° with the direction of the jet, measured on the side next the nozzle. If the angle in question were over 90°, the water would be deflected inwards (towards the wheel), and its free escape possibly impeded.

The width of the buckets should be at least seven times as great as the thickness of the jet (where it leaves the nozzle) ; and, with a given relative angle of outflow a_2 and given pitch, the jet where it leaves the bucket must be thin enough not to come into contact with the back of the succeeding bucket. This point has been fully dealt with in connection with the design of impulse turbines in general.

It is worth noting that with a Pelton wheel the water enters the buckets without shock at *all* speeds of the wheel, although the outflow is axial only at one speed. Among the various installations of Pelton wheels now at work the following may be mentioned—

At the Idaho Mines in Nevada, where these motors were first employed on a large scale, there are now eighteen Pelton wheels of various sizes, driving winding-drums, stamps, air-compressors, pumps, &c.

At the Treadwell Mine in Alaska, a 7-foot wheel working with a fall of about 490 feet and a water-supply of 10·5 cubic feet per second, develops about 500 horse-power, and drives 240 stamps, 96 ore-mills, and 13 ore-crushers. The nozzle usually employed has a diameter of $3\frac{5}{16}$ inches, but when the water-supply is plentiful a nozzle of 4 inches diameter is used, and the wheel gives off about 735 horsepower.

The largest Pelton wheel yet made is said to be one driving air-compressors at a mine in Costa Rica. This has a diameter of about 14 feet 6 inches, and at 95 revolutions per minute with a fall of 390·5 feet is stated to develop 120 horse-power.

In a recent installation of the "Pelton Water Wheel Company," the wheel works with a fall of 2106 feet, and has a circumferential velocity of 181 feet per second. In this case the body of the wheel is made of steel, instead of cast-iron, on account of the great stress to which it is subject.

In 1891 a test was made by the "Chester Hydraulic Engineering Company, Ltd.," of a very small Pelton wheel, driven by water from the mains of the London Hydraulic Power Company, supplied under a pressure of 740 lbs. per square inch. The wheel had a diameter of about 18 inches over the tips of the buckets, and worked at 1600 revolutions per minute. The efficiency as ascertained by means of a brake was 70 per cent. The pressure measured by a gauge in connection with the main close to the motor was 660 lbs. per square inch, equivalent to a head of 1522 feet. The difference between this pressure and that of

740 lbs. per square inch observed near the accumulator was due to loss in the mains.

In California some trials with a Pelton wheel were carried out by Mr. E. Browne. The diameter of the wheel was about 15 inches, the width 1⅜ inch. The fall was 56 feet. With a nozzle of 10 millimetres internal diameter, the efficiency—according to the *Engineering and Mining Journal*—was 82·6 per cent., while with a nozzle of 9 millimetres diameter, the efficiency reached 82·5 per cent.

The Niagara Falls Installation.

Various projects for utilizing on a large scale some of the water-power available at the Falls of Niagara have long been under consideration, and at length one of these is actually being carried out and is approaching its completion.

The project in question contemplates as its leading feature the installation of a Central Power Station, from which the water-power utilized through turbines will be transmitted and distributed by electricity to neighbouring manufactories, and employed for lighting purposes.

The chief characteristic of the scheme, as carried out, is a large tunnel tail-race, into which all the turbines erected will discharge through branch tail-races.

Users of power on a large scale, instead of being supplied electrically from the central station, may construct turbines of their own discharging into the tunnel.

This installation is designed for utilizing 10,200 cubic feet of water per second, with an available head of 140 feet, equivalent—with an assumed efficiency of 75 per cent. for the turbines—to about 120,000 horse-power.

The head-race (or supply canal) starts at a point $1\frac{1}{2}$ miles above the American Fall; it has a width of 500 feet, a depth of 12 feet, and is 1500 feet long. Along the edge of this canal wheel pits, 160 feet deep, have been sunk for accommodating the turbines. From the bottom of these pits the branch tail-races drain into the tunnel. This tunnel is cut out of limestone rock, and it was expected that this would prove hard enough to render lining unnecessary. Experience has shown that such is not the case, and it has been found necessary to protect the rock with four rings of brickwork in cement. The tunnel has a horseshoe section, 21 feet high by 19 feet wide, with an area of about 385 square feet. The length is 6,700 feet, and the gradient 0·7 feet per 1000 feet. This unusually steep gradient is estimated to give a velocity of flow of 25 feet per second, which is extremely high.

The first two wheels for use in connection with the Central Power Station—for the Niagara Falls Power Company—are now approaching completion. They were designed by Messrs. Faesch and Piccard of Geneva (Switzerland), and are being made by the I. P. Morris Co. of Philadelphia.

Each of these wheels is to develop 5000 horse-power with a mean head of 136 feet.

The arrangement and construction of the turbines is shown in Figs. 156 to 162.[1]

Each turbine consists of two wheels of the Fourneyron type, keyed to the same shaft; into one of these the inflow is downwards, into the other upwards. The water is supplied through a steel pipe $7\frac{1}{2}$ feet diameter, Figs. 158 and 159, and flows through a cast-iron bend into the guide-vane chamber.

[1] The Author is indebted for these illustrations to the courtesy of the Editors of *Engineering.*

The lower wheel is so constructed that the pressure due to the head is not borne by the wheel itself, but by the guide-casing. On the other hand, the full pressure acts on the upper wheel and balances the superincumbent weight.

Each wheel with its guides is divided into three stages, and the regulation is effected by means of a balanced circular sluice surrounding the wheel. This sluice is actuated through levers and connecting rods from a governor—of peculiar construction patented by Messrs. Faesch and Piccard—placed at the top of the wheel-pit. The leading dimensions of the turbines are given in the illustrations. The water supply for each twin-turbine is 430 cubic feet per second, which, with a head of 136 feet measured to the centre between the pair of wheels and an efficiency of 75⅓ per cent., will give 5000 horse-power.

The wheels will make 250 revolutions per minute; this gives a co-efficient of speed for the inflow-circumferences of the wheels, $K_2 = 0.734$.

The main shaft of each turbine is a steel tube 38 inches in diameter, except at the guide-bearings, where it is solid, and has a diameter of only 11 inches. A thrust bearing carries the weight of the shaft and wheels when at rest; but when running it is relieved of its load by the pressure of the water on the upper wheel as already described.

To insure greater regularity in the speed, a fly-wheel 14½ feet diameter and weighing 10 tons will be mounted on the shaft. As its peripheral velocity will be about 11,400 feet per minute, it is being made of wrought iron.

Before arriving at a decision as to the type of turbine to be adopted for the Niagara installation, the promoters referred the matter to an International Commission. This Commission invited competitive designs from some of the leading turbine-makers of the world, with the result, that the design considered to be first in order of merit was

Fig. 156.   Fig. 157.

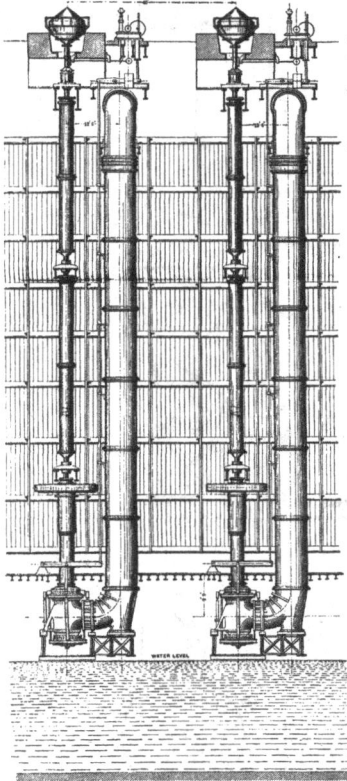
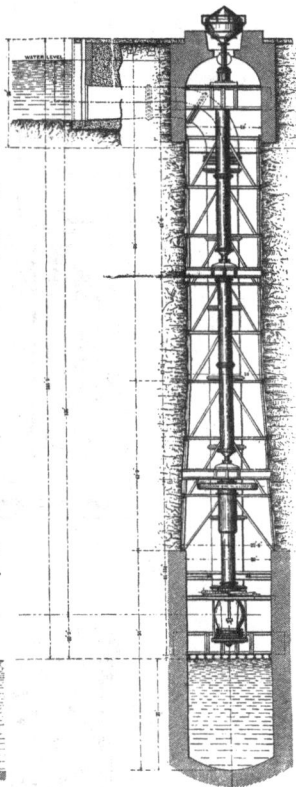

Fig. 158.

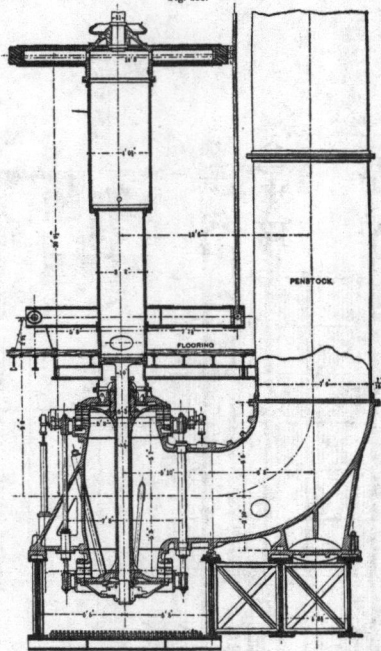

PENSTOCK.

FLOORING

Fig. 159.

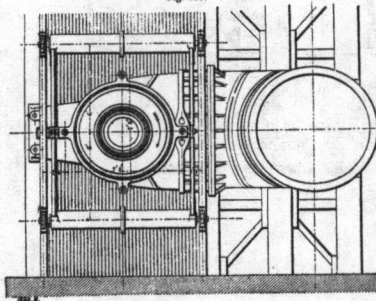

Fig. 160.

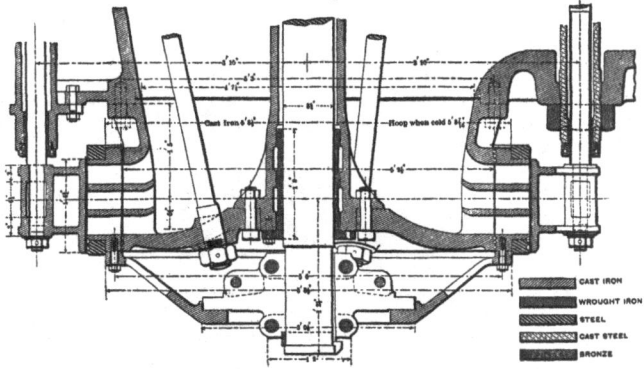

	CAST IRON
	WROUGHT IRON
	STEEL
	CAST STEEL
	BRONZE

Fig. 161.   Fig. 162.

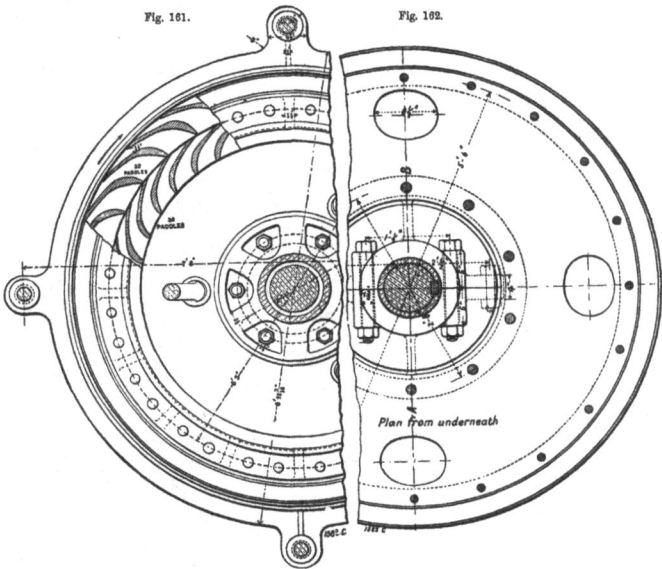

Plan from underneath

that of Messrs. Escher, Wyss and Co. of Zürich. Owing, it is understood, to the refusal of this firm to allow their designs to be carried out by American manufacturers, the scheme of Messrs. Escher, Wyss and Co. has not been practically adopted; as being admittedly the best of those submitted to the Commission, it however deserves notice.

Messrs. Escher, Wyss and Co. proposed the employment of twin Jonval turbines, keyed to the same shaft, the lower one discharging upwards and the other downwards. The water, however, after leaving the wheel-vanes was to be so directed, by a curved continuation of the wheel-casing, that the final discharge into the tail-race should be *horizontal*. With this arrangement the pressure of the water on the lower wheel would neutralize that on the upper wheel, and wholly or partially balance the weight of the revolving parts. The bearing was to be of the over-head Fontaine type. For regulating the turbine a balanced circular sluice throttling the discharge and controlled by an hydraulic governor was to be provided. In this case also the speed would have been 250 revolutions per minute, and the power developed 5000 horse-power.

The first hydraulic motors actually at work in connection with the Niagara installation are three turbines supplied to the Niagara Falls Paper Company by Messrs. R. D. Wood and Co. of Philadelphia. Each of these is designed to develop 1100 horse-power under a fall of 140 feet. The type of wheel adopted is the inverted Jonval, in this case also with the object of balancing the weight of the shaft and gearing. The regulation is effected by a circular sluice throttling the supply. Each wheel has a diameter of 4 feet 8 inches, and is designed to run at a speed of 260 revolutions per minute.[1]

[1] A description with illustrations of these turbines is given in *Engineering*, vol. lvii., 1894, no. 1476, p. 480.

Trial of a "Swain" Turbine at Boott Cotton Mills.[1]

In 1874, Mr. James B. Francis carried out a series of experiments on a "Swain" turbine, one of a pair, both of the same pattern, made to replace so-called centre-vent (inward-flow) wheels, which had been in use since 1849.

The apparatus used for measuring the power of the wheel was substantially the same as that employed in testing the above-mentioned centre-vent, and also the Tremont turbine previously described. For gauging the discharge of the water consumed, a weir without end contractions was adopted, with a length of 16·311 feet; in other respects the method of observation was similar to that followed in the earlier trials. The crest of the weir was 12·08 feet above the bottom of the canal of approach. The water had a free fall from the weir, the height of the water on the down-stream side of the weir being in no experiment less than 2·5 feet below the crest. For computing the quantity of water flowing over the weir, Francis's well-known formula

$$Q = 3 \cdot 33 \, L \, H^{\frac{3}{2}}$$

was used.

The experiments, of which the results are given in the adjoining table, were very carefully carried out in every respect, and perfectly independent of the makers of the turbines.

---

[1] *The Journal of the Franklin Institute*, 1875, vol. xcix., p. 249.

Description of the Turbine.

The "Swain" turbine tested was of the inward mixed-flow type, the outflow from the upper portion of the wheel being radial, while that from the lower portion had a varying vertical component, the outflow at the outer diameter being axial, but becoming more radial towards the centre.

The discharging edge of each bucket lay in a vertical plane passing through the axis of the wheel, being parallel to this axis from the under side of the crown to a point about 8½ inches below it, and from this point was continued in the form of a quadrant, with a radius equal to one-fifth of the outer diameter of the wheel, and having its centre in the cylinder forming the outer circumference of the wheel.

The regulating gate was formed by the lower side (or disc) of the guide-casing to which the guide-vanes were attached, the construction being such that the whole of this lower disc could be moved vertically, so as to increase or diminish the width of the guide-passages parallel with the axis of the wheel. In the upper crown of the guide-apparatus were slots to allow the passage of the vanes when the gate was raised.

Twenty-one of the guide-vanes were of bronze, 0·23 inch thick and 18·94 inches long, tapered at each end to 0·04 inch in thickness, with a bevel on each side one inch long; the remaining three of the guide-vanes were of cast-iron of sufficient thickness at the outer ends to allow the passage of the rods for adjusting the gate.

The inner edges of the guide-vanes were distant radially 1⅝ inch from the outer edges of the buckets. The water after leaving the wheel passed through a short

RESULTS OF EXPERIMENTS WITH "SWAIN" TURBINE.

Number of the experiment.	Height of the regulating gate.	Number of revolutions per second.	Useful effect or friction of the brake.	Total fall.	Depth of water on the weir.	Quantity of water passing the weir per second.	Available power of the water.	Efficiency.	Ratio of the velocity of the exterior circumference to velocity due to the head.
		$\frac{n}{60}$	W_e	h	H	Q	E_a	ε_e	K_2
	Inches.		Foot lbs.	Feet.	Feet.	Cubic feet.	Foot lbs.		
3	3·25	1·0438	38038·2	13·912	1·2025	71·623	62022·1	0·6133	0·6583
14	6·50	1·0503	74992·3	13·316	1·6596	116·126	96257·8	0·7791	0·6770
19	13·00	1·1111	107810·0	12·666	2·0983	165·092	130166·2	0·8282	0·7344
29	2·00	1·0080	21572·5	14·283	0·9611	51·177	45518·7	0·4739	0·6274
34	3·00	1·0526	34806·5	13·977	1·1617	68·009	59189·5	0·5881	0·6623
49	4·00	1·0019	47672·5	13·699	1·3362	83·894	71562·5	0·6662	0·6478
51	5·00	1·0204	59179·7	13·482	1·4794	97·736	82048·8	0·7213	0·6537
69	6·00	1·0362	69263·5	13·281	1·6016	110·092	91044·1	0·7608	0·6689
78	7·00	1·1055	76932·3	13·102	1·6894	119·268	97309·2	0·7906	0·7184
90	8·00	1·0683	84794·3	12·968	1·7916	130·253	105184·6	0·8061	0·6979
103	9·00	1·0810	91961·2	12·836	1·8695	138·839	110956·3	0·8288	0·7098
113	10·00	1·0721	95583·9	12·680	1·9224	144·774	114292·8	0·8363	0·7082
118	11·00	1·0943	99266·1	12·640	1·9815	151·501	119226·3	0·8326	0·7227
123	12·00	1·1267	102435·2	12·521	2·0266	156·703	122158·8	0·8385	0·7490
128	13·08	1·8060	20237·7	13·099	1·7000	120·392	98203·9	0·2061	1·1738
129	,,	1·6299	55189·4	12·880	1·8608	137·871	110581·6	0·4991	1·0683
130	,,	1·4614	80804·4	12·603	1·9608	149·133	117041·9	0·6904	0·9683
131	,,	1·3006	99109·8	12·480	2·0453	158·877	123471·8	0·8027	0·8660
132	,,	1·2115	103522·0	12·432	2·0654	161·225	124814·5	0·8294	0·8083
133	,,	1·1516	104652·9	12·372	2·0766	162·538	125223·8	**0·8357**	0·7701
134	,,	1·1364	105401·9	12·408	2·0854	163·572	126387·3	0·8340	0·7588
135	,,	1·1080	104857·4	12·366	2·0857	163·607	125986·6	0·8323	0·7412
136	,,	1·0874	104952·0	12·368	2·0911	164·243	126496·7	0·8297	0·7273
137	,,	1·0558	103885·0	12·385	2·0911	164·243	126670·6	0·8201	0·7057
138	,,	1·0618	104472·3	12·380	2·0957	164·785	127037·5	0·8224	0·7098
139	,,	1·0188	102160·1	12·372	2·0923	164·384	126646·6	0·8067	0·6813
140	,,	0·9928	101416·6	12·386	2·0935	164·526	126899·0	0·7992	0·6635
141	,,	0·9370	99238·3	12·381	2·0940	164·585	126893·2	0·7821	0·6264
142	,,	0·8532	95034·8	12·399	2·0978	165·033	127423·8	0·7458	0·5700
143	,,	0·6550	78892·7	12·399	2·0793	162·855	125741·9	0·6274	0·4376
144	,,	1·8405		13·173	1·6759	117·841	96666·1		1·1928
145	,,	0·		12·580	1·9840	151·788	118907·9	0·	0·
146	,,	0·		12·556	1·9832	151·696	118609·3	0·	0·

Remarks.

At *part gate* only the results at the *best speed* are given; in the original, the results of trials at various speeds for each opening of the gate are recorded, but have been here omitted for the sake of brevity.

divergent suction-tube, about 5 feet 9 inches in diameter and $21\frac{1}{2}$ inches long.

The step upon which the wheel revolved was a cylinder of white oak with conical ends, free either to revolve with the wheel or remain stationary, and resting on a cast-iron socket attached by three arms to the sides of the suction-tube. This step was enclosed in the hollow continuation of the wheel-boss, and supplied with water for lubrication by a pipe from below. The diameter of the step appears to have been very considerable—about 12 inches—in proportion to that of the wheel, and makes it probable that the loss by friction was greater than the average for such turbines.

The wheel-vanes were of bronze cast into the cast-iron wheel-casing.

Dimensions of Turbine.

Outside diameter . .	$2\,r_1 = 6$ ft.
Least inner diameter .	$2\,r_2 = 2$ ft. 8 in.
Width of guide passages .	$e = 13\cdot08$ in.
Width of buckets at inflow	$e_1 = 13\cdot285$ in.
Number of guide-vanes .	$z = 24.$
Number of wheel-vanes .	$z_1 = 25.$
Thickness of guide-vanes .	$t = 0\cdot23$ in. generally.
Thickness of guide-vanes .	$t = 0\cdot04$ in. at ends.
Measured outflow area from guide passages . .	$A_1 = 9\cdot880$ sq. ft.
Measured outflow area from buckets . . .	$A_2{}^1 = 9\cdot558$ sq. ft.
Mean angle of outflow from guides . . .	$a = 65°.$
Mean angle of inflow into buckets . . .	$a_1 = 0°.$

Mean angle of outflow from
buckets . . . $a_2 = 65°$.

With respect to the mean angle of outflow, it must be
remarked that at the outer diameter, where the outflow
is axial, a_2 is about 64°, while for that portion of the
wheel where the outflow is radial, a_2 is about 68°. From
the *outer* to the *inner* diameter, the depth of the passages
between the buckets, or the shortest distance between
them at the outlet, *decreases*, so that for a given width
a much larger proportion of the water leaves the wheel
through the *outer* part of the bucket orifices than through
the *inner*. Taking this unequal distribution into account,
the mean *effective* value of the angle a_2 must be nearer 64°
than 68°, and has been estimated at 65°.

To arrive at some conclusion as to the *mean effective
radius* at which the water may be assumed to leave the
turbine, it is also necessary to have regard to the distribution of the water, on the assumption that the relative
velocity of outflow c_2 is the same at all points. From a
study, as careful as the available information and illustrations permitted, of the proportions of the wheel, the
author has estimated the effective ratio of the outer to
the mean inner radius, $\frac{r_1}{r_{2m}}$, as 1·3, whence the mean
inner diameter is 4·615 feet.

The co-efficient of contraction for the guide-passage
orifices, from which all the water issues radially, may be
taken as somewhat less than that applicable to the bucket
orifices, from which a portion of the water flows in a more
or less axial direction; the values of the two co-efficients
may be assumed as respectively 0·9 and 0·95, in which
case the ratio of the effective outflow areas A and A_2 is
practically 1, since

$$A = 0\cdot9 \times 9\cdot88 = 8\cdot992 \text{ sq. ft.}$$
$$A_2 = 0\cdot95 \times 9\cdot558 = 9\cdot080 \text{ sq. ft.}$$

Calculated from the preceding value of A and the quantity of water, $Q = 162\cdot538$ cubic feet, consumed at the best speed—as given in the table of results—the velocity of flow

$$c = 18\cdot075 \text{ feet per second,}$$

and $\quad K_1 = 0\cdot644.$

The theoretical value for K_1 determined from the angles and proportions previously stated with an allowance of $3\cdot4\%$ for friction is

$$K_1 = 0\cdot638.$$

The agreement in this instance is excellent, especially considering the uncertainty as to the actual distribution of the outflowing water and the co-efficients of contraction.

In the "Swain" wheel the outer casing is continued downwards as far as the lowest part of the buckets, instead of being, as in the "Hercules" wheel, merely a narrow ring; on this account the effective outflow area A_2 has been taken as nearly equal to the measured area, since the water after entering the buckets in much the same way as in the "Hercules" turbine, and making an abrupt turn at the outer edge, has room to spread itself out again before leaving the buckets, and thus fill the latter.

The *experimental* best velocity was

$$w_1 = 21\cdot708 \text{ ft. per second,}$$

that for inflow without shock

$$w_1 = 16\cdot376 \text{ ft.,}$$

that for radial outflow

$$w_1 = 21\cdot289 \text{ ft.}$$

It will be seen that the last value agrees very closely with practice. It is obvious that the wheel was not designed for simultaneous inflow without shock and vertical outflow. The mean of the velocities on these two assumptions is 18·832 feet per second.

THE PURDON-WALTERS STREAM MOTOR.

For utilizing the kinetic energy of a flowing stream, where no fall is available, floating water-wheels of the old-fashioned type are still used on many rivers with rapid currents. Stream turbines are also occasionally employed. The latter are in all essential points similar to ordinary turbines, but must of course be arranged on horizontal shafts parallel with the general direction of the current.

Owing to the comparatively small amount of energy available in the form of velocity in a given stream-area, a turbine of very large diameter is necessary in order to develop a small amount of power, and obviously the diameter is limited by the depth of the stream.

In order to get over this difficulty a new form of hydraulic motor, belonging to the turbine class, the invention of Messrs. Purdon and Walters, has recently been brought into notice. It may be best described as a turbine in which the vanes, instead of being attached to a wheel, are fixed between two endless pitch chains. The two double chain-wheels, round which the chains pass, and which are driven by the latter, are supported by a pontoon or barge, through the lower part of which a channel is formed in which the vanes when under water move. In this channel are fixed guide-vanes, which impart to the water flowing between them the desired direction. The barge is moored across stream, so that the water flows through it at right angles to the motion of the chain. Half the moving vanes are immersed at any time, the axes of the chain-wheels being placed near the water-level. The machinery to be driven is erected on the barge and

PURDON-WALTERS' PATENT WATER MOTOR

Fig. 163.

Fig. 164.

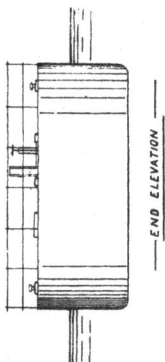

END ELEVATION

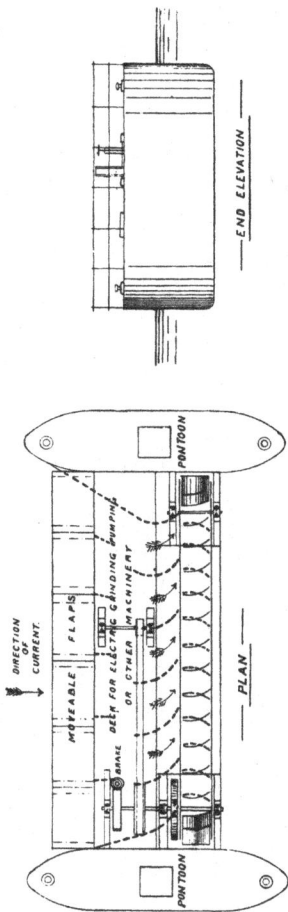

PLAN

Fig. 165.

ELEVATION

Fig. 166.

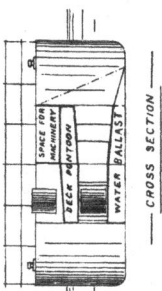

CROSS SECTION

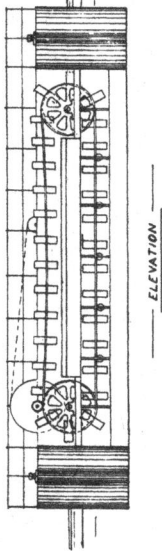

actuated through gearing from one of the chain-wheels, that namely which is situated on the *concave* side of the working vanes.

The chains are supported under water at regular distances by vertical rollers, and guided laterally by horizontal rollers, which take the pressure of the water on the vanes in the line of the stream. The construction will be easily understood from the accompanying illustrations (Figs. 163—166).

The first motor designed by Messrs. Purdon and Walters was constructed with plane floats (instead of curved vanes) and without guide-vanes, and the barge was placed so that the floats moved at an angle of 60° to the direction of the current. Trials carried out by Professor Unwin, with an experimental motor thus constructed, showed an actual efficiency—as measured by a brake—of about 11 per cent. with a current velocity of about 4·24 ft. per second. The theoretical hydraulic efficiency of such a motor would, according to Professor Unwin, be 32 per cent., and it is certain that a large proportion of the loss, represented by the difference between 32 and 11 per cent., was due to mechanical friction.

At Professor Unwin's suggestion the motor was re-designed and constructed in its present form. Preliminary trials carried out at Chepstow by Professor Unwin are said to have proved that the motor in its present shape can work with an efficiency of 25 per cent. at low velocities. This is, for a stream motor, a very satisfactory result. It is anticipated, that by modifying in some particulars the form of the vanes a still better efficiency can be obtained. Such a motor is of course not intended to compete with ordinary turbines, but would be applicable in very many cases in which the employment of the latter would be out of the question. It is expected that the motor may prove useful in tide-ways as well as in rivers having rapid currents,

and it is suggested that it could be advantageously employed
in connection with the pontoons of floating bridges.

The leading idea in the design of the motor is to obtain
a large sectional area of the stream entering the motor in
as convenient a manner as possible.

For a stream motor there is of course no fall, and the
available energy is only that due to the velocity of the
current. It seems however probable that the machine,
moored across the stream, would act to some extent as a
dam, and cause a slight head to be formed on the up-
stream side of the guide passages, so that in effect the
motor would work under a fall. This fall would be
influenced by the shape of the vanes and buckets themselves, but in every case the available energy remains the
same, and the maximum possible head could only be that
equivalent to the velocity of the current.

The theory of the motor is essentially that of a turbine,
but, in the absence of experimental data, the uncertainty
as to the pressure conditions obtaining at the inflow and
outflow render accurate calculations in advance impossible.

The following table gives the equivalent head and
total energy per square foot of sectional area of a stream
for velocities of from 1 to 6 miles per hour—

Velocity of Stream.		Equivalent Head.		Total Energy.
Miles per Hour.	Feet per Second.	Feet and Decimals.	Inches and Decimals.	H. P. per Square Foot Sectional Area.
1	1·467	·033	·43	·0055
2	2·933	·134	1·62	·0445
3	4·4	·300	3·60	·15
4	5·867	·534	6·42	·355
5	7·333	·834	10·07	·694
6	8·8	1·200	14·39	1·2

Fourneyron Turbine at St. Blaise.

As an instance of an outward-flow turbine of exceptional proportions and for unusual conditions, one of the first wheels constructed by Fourneyron, and erected at St. Blaise in the Black Forest, deserves notice.

The turbine in question was driven by a fall of 354 feet at a speed of no less than 2300 revolutions a minute, and developed 30 effective horse-power. The outer diameter of the wheel was 12·99 inches, the inner diameter 7·47 inches.

Two larger turbines for the same fall were afterwards designed by Fourneyron, each with an outer diameter of 21·66 inches, to run at 2300 revolutions per minute and develop 60 horse-power. As might be expected at the enormous speed, the bearings had to be renewed every ten to fourteen days.

These early Fourneyron turbines discharged above water and were in reality impulse wheels. They were subsequently replaced by tangent wheels, and at a still later date by Girard turbines.

The Best Speed of Turbines as proved by Experiment.

The Table A, in which the experimental best speeds of various turbines are compared with the results of calculation, shows that there is not in some of the cases analyzed a very close agreement between theory and practice but that in all of them the best speed as calculated does not deviate from the true best speed to a sufficient extent to cause any important difference in the efficiency—at the worst about 2 per cent.—it being

fortunately a peculiarity of turbines, that in the neighbourhood of the most favourable velocity of rotation the
efficiency does not alter very rapidly with the speed.
The discrepancies between theory and practice are due
to several causes which have already been incidentally
mentioned, but call for further consideration in some
respects.

In the first place, the assumption that the best speed
is that at which the water enters with least resistance
and leaves with the smallest possible residual velocity, is
not strictly correct, since the velocity of flow varies with
the speed, and consequently the losses from friction,
bends, abrupt changes of section, &c., also vary and do
not attain their minimum value simultaneously with the
minimum values of the losses by shock and unutilized
energy. In the second place, many of the turbines in
question are not designed so that inflow without shock
and vertical outflow can take place simultaneously. In
the third place, the *assumed* conditions under which
inflow without shock occurs are not always those which
really correspond to inflow with least resistance. The
last of these causes tending to produce discrepancies
between theory and practice, is probably the most important. Professor Albert Fliegner of Zürich carried out
about 1879 a number of experiments[1] to ascertain to
what extent the direction of inflow *with least resistance*
deviates from that theoretically requisite for inflow without shock. These experiments were made with stationary
guide passages and buckets so arranged that the direction
of the streams entering the buckets could be varied
relatively to the latter.

The streams issued from mouthpieces shaped like the

[1] *Zeitschrift des Vereins Deutscher Ingenieure*, 1879, vol. xxiii.,
p. 459.

TABLE A.

Experimental and Calculated Results at Best Speed.

No.	Description of turbine	h Total head (Feet)	Q Quantity of water per second (Cubic feet)	E_a Available energy (Horse-power)	a Angle of outflow from guides (Deg. Min.)	g Angle of inflow into buckets (Deg. Min.)	β Angle of outflow from buckets (Deg. Min.)	r_1 Radius of wheel where water enters / r_m Mean radius for axial turbines / r_2 Radius of wheel where water leaves (Feet)	c Width of guide passages at outflow (Feet)	b_1 Width of wheel-buckets at inflow (Feet)	b_2 Width of wheel-buckets at outflow (Feet)	n Number of guide-vanes or passages	n_1 Number of wheel-vanes or buckets	t Thickness of guide-vanes at outflow (Feet)	t_1 Thickness of wheel-vanes at inflow (Feet)
1	Rittinger's turbine	8·32	3·61	7·334	67 30	+61 0	64 0	4·782 / 2·5743	0·6436	0·6436	1·477	12	24	0·0208 (24) / 0·0312 (8)	0·0208
2	,, ,,	6·32	3·14	7·166	75 50	+65 0	74 0	6·150	,,	2·503	2·503	24	24	,,	,,
3	,, ,,	6·48	4·39	6·443	75 30	+54 0	75 0	5·775	,,	1·148	1·148	24	12	,,	,,
4	,, ,,	9·47	4·20	4·513	76 0	61 0	65 0	4·022	,,	1·361	1·361	24	24	,,	,,
5	,, ,,	9·54	7·12	5·115	78 0	61 40	63 50	6·150	,,	,,	,,	24	24	,,	,,
6	,, ,,	18·08	5·62	4·193	73 0	89 40	61 0	5·775	,,	,,	,,	24	20	,,	,,
7	,, ,,	17·94	6·82	4·133	78 0	96 0	61 0	4·782	,,	,,	,,	12	20	,,	,,
8	,, ,,	17·90	8·32	4·960	68 0	45 0	63 40	—	,,	,,	,,	24	24	,,	,,
9	Haenel's turbine	5·127	47·029	27·827	71 10	9 25	73 25 / 73 30 / 73 20 / 73 0	0·7301	0·838	0·838	1·477 / 2·503 / 1·148 / 1·361	32	32	0·491	0·0425
9A	Goeppingen turbine	5·989	82·301	41·656	72 15	12 15	71 0	3·375	0·970	0·9368	0·9314	34	44	0·00833	0·0117
10	Oelzing turbine	—	210·134	284·00	,,	,,	,,	—	,,	,,	,,	,,	,,	,,	,,
11	Outer ring	—	404·12	458·83	,,	,,	,,	2·065	,,	,,	,,	,,	,,	,,	,,
11A	Inner ring	13·58						4·146							
12	"Collins" turbine	16·55	64·40	130·47	—	—	—	3·6665	0·823	0·760	0·748	24	30		
12A	,,	16·57	65·30	132·900	,, ,,	,,	,, ,,	3·759	1·0066	1·0066	1·230	40	17	0·0156	0·015 / 0·0208
12B	,,	16·55	64·99	132·970	,, ,,	,,	,, ,,	4·689	1·330	1·310	2·800	24	55		
13	Tremont turbine	12·896	138·189	202·214	62 0	0	68 0	1·500	0·836	0·836	0·812	24	30	0·025	0·006* / 0·037
14	"Boyden" turbine	18·60	147·10	277·151	61 0	30 0	61 0	3·000	1·330	1·310	1·230	40	40	0·0156*	0·0208 / 0·015
15	Boott turbine	13·356	113·00	171·307	78 0	–8 0	70 0	0·815 (?) / 3·9365	1·0066	1·0066	0·748	24	17		
16	"Hercules" turbine	16·96	88·33	189·827	75 15	,,	73 0	2·3075	0·823	0·760	1·290	40	54		0·002* / 0·037
16A	,, ,,	,,	162·338	227·678	65 0	,,	65 0	4·05							
17	"Swain" turbine	16·92 / 12·372	78·00		,,	,,	,,			1·107	1·885	24	25	0·025	
18	"Humphrey" turbine	12·478	207·82	294·080	75 0	10	—	4·05	1·090	1·107	1·885	19	13	0·0125 / 0·0192*	0·087

* Thickness at extreme end.

† *Mean* of inflow without shock and vertical outflow.

N.B.—Where two values are given for the *calculated* velocity of circumference at inflow, the *upper* is for inflow without shock, the *lower* for radial or vertical outflow.

TABLE A.

Experimental and Calculated Results at Best Speed.

No.	Description of turbine.	t_2 Thickness of wheel-vanes at outflow. (Feet)	A_1 Measured area of guide passage orifices. (Square feet)	A_{41} Measured area of wheel-bucket outflow orifices. (Square feet)	A Effective area of guide passage orifices. (Square feet)	A_4 Effective area of wheel-bucket outflow orifices. (Square feet)	h_1 Height of lower edge of wheel above tail-water. (Feet)	d_2 Diameter of suction-tube. (Feet)	h_2 Clearance of wheel above tail-water when not drowned. (Feet)	$\frac{A}{A_1}$ Outflow areas.	$\frac{r}{r'}$ Radii or diameter.	$\frac{a}{a'}$ Widths of buckets at inflow and outflow.	$\frac{a}{b}$ Radius to width at inflow.	W Effective power developed (measured on brake). (Horse-power)
18	"Humphrey" turbine	—	13·317	—	11·985	—				1·000	1·30	$\frac{1}{1·703}$	1·94	240·863
17	"Swain" turbine	"	9·880	9·558	8·992	9·080				0·673	1·84	$\frac{2·105}{1}$	2·71	190·278
16a	"Swain" turbine		3·830		3·447									130·26
16	"Hercules" turbine	0·013	4·782	7·925	4·2768	6·3648				0·867	1·169	0·82	3·60	145·72
15	Boott turbine	0·0238	9·904	6·9082	5·3136	6·1383				1·166	0·817	1·016	2·49	136·623
14	"Boyden" turbine	0·000* 0·015	6·814	5·660	6·1283	5·094	7·13	5·44		0·850	0·814	1·006	3·50	160·287
13	Tremont turbine	0·0117	6·6371	7·6869	5·8834	6·9182	"	"		"	"	"	"	222·04
12	"Collins" turbine	0·017	7·9869	2·822	2·889	2·822				1·023	"	"	"	97·03
12a	Outer ring	"	2·912									0·5		102·70
12b	Inner ring	"										"		102·62
11	Olching turbine		25·435 13·853 11·682	21·254 19·201 9·253	24·280 13·247 11·033	21·254 12·001 9·253				1·142 1·104 1·192	1	1	2·72 2·53 5·38	28·748 289·46 365·62
10	Greggiogen turbine			9·950	3·0516	6·3883			0·3215	0·4784	"	"	4·0	18·887
9	Haenel's turbine	0·0425	—								"	"	"	4·904 4·510 5·067
1–8	Rittinger's turbine	0·0208	0·486 0·550 0·777 0·256 0·203 0·1362 0·1202 0·1202	0·438 0·388 0·421 0·421 0·437 0·250 0·250 0·250	0·1136 0·212 0·183 0·699 0·495 0·1082 0·1082 0·1082	0·438 0·388 0·421 0·437 0·210 0·250 0·250 0·250				0·433 0·433 0·504 0·540 0·418 1·061 1·274 0·522	1	1	2·5	3·395 3·262 2·881 3·507

* Thickness at extreme end.

† *Mean* of inflow without shock and vertical outflow.

N.B.—Where two values are given for the *calculated* velocity of circumference at inflow, the *upper* is for inflow without shock, the *lower* for radial or vertical outflow.

TABLE A.

Experimental and Calculated Results at Best Speed.

No.	Description of turbine.	Actual efficiency. r	Velocity of circumference at inflow or mean circumference. Experimental. Feet per second.	Velocity of circumference at inflow or mean circumference. Calculated. Feet per second.	Ratio of velocity of circumference to velocity due to head. Experimental. K_1	Experimental velocity of flow from guides. c Feet per second.	Ratio of velocity of flow to velocity due to head. Experimental. K_2	Ratio of velocity of flow to velocity due to head. Theoretical. K_2	Theoretical velocity of flow due to head. V	Number of revolutions per minute. n	Remarks.
18	"Humphrey" turbine	0·819	20·113	17·533	0·730	17·342	0·612		28·344	47·424	
17	"Swain" turbine	0·8357	21·708	16·376 / 21·289	0·770	18·075	0·644	0·638	28·224	69·006	At 0·806 gate.
	,,										
16A	,,										
16	Boots turbine	0·871	22·088	24·042	0·669	22·628	0·685	0·629	33·047	140·62	At full gate.
15	"Hercules" turbine	0·858	18·668	19·233	0·637	29·653	0·625	0·659	29·311	38·182	
14	"Boydan" turbine	0·7975	20·392	20·837 / 20·293	0·634	21·283	0·726	0·755	32·702	63·50	
13	Tremont turbine	0·8017	18·051	21·303 / 15·065	0·626	24·346	0·744	0·856	28·809	51·064	
12B	,,	0·7037	19·704	21·140 / 22·445	0·906	23·480	0·815	0·672	32·653	89·17	Draft tubes.
12A	,,	0·838	19·960	22·337	0·946	22·603	0·692	0·673	32·669	96·00	Draft tube with inverted cone.
12	"Collins" turbine	0·843	19·467	21·579	0·900	22·495	0·689	0·672	32·653	90·25	
11B	Outer ring	0·804	14·393	22·692	0·567	22·291	0·680	0·658	19·493	39	
11A	Inner ring										
11	Oishing turbine	0·832	18·593	16·640	0·733	16·640	0·656	0·646	29·569	45·52	
10	Goeginger turbine	0·690	16·863		0·666	20·419		0·933	25·367	28·87	
9A	,,	0·691	19·157		0·648	15·411	1·0475	0·937	18·168	33	
9	Haenel's turbine	0·695	10·515		0·539		0·8482				
8	,,	0·678	8·897	16·640	0·4897	33·418	0·988	0·991	33·994	99	
7	,,	0·700	15·660	11·061	0·461	33·150	0·944	0·981	34·122	135	
6	,,	0·645	15·921	8·347	0·468	32·150	0·800	0·882	33·994	151	
5	,,	0·715	11·896	13·518	0·475	27·771	0·791	0·879	24·692	164	
4	,,	0·632	12·365	14·780	0·482	19·750	0·970	1·009	24·789	159	
3	,,	0·697	14·800	11·888	0·499	10·175	0·503	0·476	20·190	308	
2	,,	0·673	13·235	17·561 / 15·451	0·733	11·362	0·555	0·561	20·431	302	
1	Rittinger's turbine 1		9·705	9·392	0·483	15·602	0·775	0·879	20·074	297	

* Thickness at extreme end.

† *Mean* of inflow without shock and vertical outflow.

N.B.—Where two values are given for the *calculated* velocity of circumference at inflow, the *upper* is for inflow without shock, the *lower* for radial or vertical outflow.

TABLE B.

Particulars of Turbines manufactured by Messrs. J. J. Rieter and Co., of Winterthur, Switzerland.

14	13	12	11	10	9	8	7	6	5	4	3	2	1	Description of turbine.
														Impulse Turbines (cols. 7–14); Reaction Turbines (cols. 1–6).
„ „	„ „	„ „	„ „	Radial outward-flow	„	„ „	Parallel-flow			Journal { Outer ring / Inner ring	„ „			—
695·500	1804·483	590·559	570·873	144·358	5·413	9·187 to 10·987	9·842	12·139	3·590	42·650	8·838	7·217 to	19·685 / 32·153	h Total head. (Feet.)
15·89	0·035	19·602	8·476	24·730	70·630	105·960 to 114·780	176·380	211·900	111·247	6·712	183·630	141·270 to	125·610 / 211·900	Q Quantity of water passing through turbine per second. (Cubic Feet.)
70°	82°	72° 30'	73°	62°	„	64°	64°	70°	59° 30'	71°	72°	„	68°	a Angle of outflow (vane angle) from guide-vanes.
54°	—	48°	57° 30'	56°	32°	40°	40°	0°	0°	0°	„	0°	0°	β Angle of inflow into wheel.
70°	73°	68°	76°	73°	71°	„	68°	66°	73°	71°	68°	72°	66°	θ Angle of outflow from wheel.
8·200	0·984	7·875	7·875	4·130	6·558	8·200	6·725	9·841	8·325	7·375	5·900	5·900	5·708 / 1·448	d_0 or d_m Diameter of wheel where water enters, or, for Jonval wheels, mean diameter. (Feet.)
9·417	1·230	9·116	9·016	5·118	—	—	—	—	—	—	—	—	—	d_1 Diameter of wheel where water leaves. (Feet.)
5·51	0·59	4·91	4·32	3·35	7·87	„	7·87	11·02	9·84	11·81	14·96	13·39	13·39 / 6·69 / 2·76	b Clear width of buckets at inflow. (Inches.)
16·14	2·36	15·75	15·75	9·45	21·46	„	15·75	27·36	10·23	19·29	14·96	13·39	13·39 / 6·69 / 2·76	b_1 Clear width of buckets at outflow from wheel. (Inches.)
4·91														c Clear width of outflow from guide passages. (Inches.)
2	1	3	1	2 × 10	70	„	54	64	36	30	36	28	36 / 12	n Number of guide-vanes.
100	85	90	110	54	60	„	56	60	40	36	38	28	36 / 14	N Number of wheel-vanes.
—	—	0·39	—	0·14	0·31	„	0·35 and 0·24	0·35 and 0·20	0·24	0·47	0·39	0·39	0·43 / 0·28	t Thickness of guide-vanes at outflow. (Inches.)
—	0·04	0·39	0·50	0·24	0·31	„	0·35	0·47	0·47	0·79	0·43	1·06	1·06 / 0·43	t_1 Thickness of wheel-vanes. (Inches.)
—	1·38	4·91	4·32	3·35	10·23	„	10·63	8·46	9·84	12·79	10·23	9·64	8·66 / 2·76	k Depth of guides.* (Inches.)
7·30	2·36	7·08	6·29	5·31	7·87	„	10·94	11·02	10·23	12·79	10·23	9·64	8·66 / 2·76	k_1 Depth of buckets.* (Inches.)
									1·3125				8·858 / 7·050	δ Diameter of suction-tube, if used. (Feet.)

* Measured parallel with axis for axial wheels and radially for radial wheels.

† Maximum thickness.

TABLE B.

Particulars of Turbines manufactured by Messrs. J. J. Rieter and Co., of Winterthur, Switzerland.

Description of turbine	Sym.	Units	14	13	12	11	10	9	8	7	6	5	4	3	2	1
			Impulse Turbines.								**Reaction Turbines.**					
			Radial outward-flow	,,	,,	,,	,,	Parallel-flow	,,	,,	Jonval { Outer ring / Inner ring }	,,	,,	,,	,,	,,
Measured outflow area from guides.	A	Sq. feet	—	0·000984	0·134	0·1565	0·289	5·102	5·167	9·062	11·550	12·387	9·308 / 9·453	3·532 / 3·532	4·522	9·308
Measured outflow area from buckets.	A'	Sq. feet	—	—	—	—	—	—	—	—	11·194	10·785	3·198 / 3·532	4·522	9·906	9·906
Clearance between guide apparatus and wheel.		Inches	—	0·04	0·12	0·12	0·20	0·20	0·20	0·20	0·12	0·14	0·08 to 0·12	0·08 to 0·12	0·08	0·08 to 0·12
Height of suspension above surface of tail-water.	h	Inches	—	—	—	—	3·94	3·94 to 0	3·94	3·94	—	—	—	—	—	—
Ratio of diameter at inflow to diameter at outflow.	d/d		1·149	1·55	1·159	1·16	1·257	—	—	—	—	—	—	—	—	—
Ratio of width of buckets at outflow to width at inflow.	a/a		2·36 / 1	4 / 1	3·20 / 1	3·638	2·628	2·725	—	—	2·5	1·94	1·633	—	—	—
Ratio of measured area of outflow from guides to measured area of outflow from buckets.	A/A'		—	—	—	—	—	—	—	—	1·082	1·145	1·00 / 1·00	1·00 / 1·00	1·00	1·011
Number of revolutions of wheel per minute.	n		200	2500	180	210	200	24·5	30·5	27	33	28	530	140	94 / 365	38
Velocity of flow theoretically due to head, per second.	v		165·636	240·314	194·079	191·982	96·396	18·964	24·317	25·170	27·951	32·391	14·330	21·553	43·579 / 35·662	35·662
Ratio of velocity of wheel where water enters to velocity of flow due to head.	k_s		0·438	0·877	0·290	0·430	0·467	0·480	0·314	0·590	0·8	0·836	0·8 / 0·63	0·64 / 0·63	0·64	0·684
Ratio of velocity of flow calculated from measured area to velocity due to head.	k_1		0·630	0·730	0·811	0·779	0·741	0·843	0·775	0·606	0·619	0·676	0·616	0·616	0·619	0·746
Ratio of velocity of flow due to *effective* area to velocity due to head.	k_2		0·930	0·900	0·946	0·946	—	0·987	0·941	0·606	0·675	0·675	0·673	0·673	0·716	0·723
Effective work done by wheel.	W	Horse-power	800	4·5	1000	400	300	33	100	150	222	30	20	100	340 / 380	350 / 100
Height of wheel in suction-tube above tail-water.		Feet	—	—	—	—	—	—	—	—	—	—	—	—	13·135	Very small
Remarks			Wheel with horizontal axis at Terni.	,,	,,	Regulated by segment slides.	Regulated by semi-circular horizontal slides.	Regulated by segment slides.	Turbine constructed to work both drowned and with free deviation; works drowned with the smaller head.	Regulated by hinged valves.	Regulated by hinged valves.	Regulated by vertical slides.	Regulated by throttle valve.	Two rings; regulated by covers to guide passages.	Regulated by cylindrical sluice.	Regulated by cylindrical sluice.

* Measured parallel with axis for axial wheels and radially for radial wheels.

† Maximum thickness.

guide passages of turbines, and entered channels made
to resemble as nearly as possible the buckets of turbines·
The experiments were made with nine different forms of
channel, and for each of these ten mouthpieces were tried,
each inclined at various angles to the channel, which may
for convenience be termed the *turbine* bucket.

In carrying out the experiments the pressure ξ immediately above the orifice of the mouthpiece was measured
by a pressure gauge and the quantity of water flowing out
observed; the area of the orifice being known, the velocity
of flow c_r could be calculated, and denoting by h_o the
height of the orifice above the tail-water level, or the
depth of immersion (as the case might be), the available
head h of the water leaving the mouthpiece would be
expressed by

$$h = \frac{c_r^2}{2g} + \xi + h_o$$

The whole of the energy represented by this was expended in producing the velocity of outflow c_2 from the
turbine bucket and in overcoming the resistances of the
latter, including shock, so that

$$\frac{c_r^2}{2g} + \xi + h_o = (1 + \zeta) \frac{c_2^2}{2g},$$

where ζ is an experimental co-efficient of resistance or loss.

Since c_2 could be determined from the observed quantity
of water and the known outflow area of the bucket, the
only unknown quantity in the above equation was ζ, and
this could be calculated in every case.   Over 1000
experiments were made, from the results of which the
values of ζ were calculated. Each different bucket was
tried with several heads of water ranging from 3·28 to
98·4 feet, and the values of ζ plotted as ordinates with the
heads as abscissæ.

Professor Fliegner found that the co-efficient ζ diminished as the head h increased, and that for any given head its minimum value did not occur for inflow without shock, but when the mouthpiece was placed at an angle a_o differing from the vane angle a_1 of the turbine bucket.

In the following table the corresponding values of a_2, a_1, and a_o as found by Fliegner are given :—

	a_2	a_1	a_o
I.	75°	30°	21°
II.	60°	30°	19°
III.	75°	0°	15°
IV.	60°	0°	12°
V.	75°	−30°	$8\frac{1}{2}°$
VI.	75°	−30°	− 5°
VII.	75°	−30°	−10°
VIII.	75°	−60°	− 4°
IX.	65°	−45°	−17°

In the case of buckets Nos. VI. and VII. the width was greater at the outlet than at the inlet, the form being otherwise the same as that of No. V.

In their *general tendency* these results agree with those recorded in Table A (which relate to actual turbines). From that table it will be seen that for nearly all the turbines with a *positive* angle a_1 the best velocity of rotation is *less* than that corresponding to inflow without shock, and consequently the best relative angle of inflow a_o *smaller* than the vane angle a_1. For turbines with a negative value of a_1 the best speed is *greater* than indicated by theory, and consequently the angle a_o—in absolute value apart from its sign, positive or negative— under these circumstances also *less* than a_1.

For Jonval wheels of ordinary proportions, as constructed in Europe, with $a_1 = 0°$ and a_2 from 64° to 75°,

the theoretical best speed agrees very well with experiment for practical purposes. For radial and mixed-
flow wheels the deviations are somewhat more considerable, but this is *in part* due to the fact that they have not
been designed accurately for simultaneous inflow without
shock and radial or vertical outflow.

A further explanation of disagreements between calculated and practical results is in some cases due to the
uncertainty as to the real value of the outflow areas A
and A_2 used in calculating the velocities of flow; this
uncertainty arises from ignorance of the true allowance to
be made for contraction and of that required for the
obstruction of the guide passages by the wheel-vanes, and
occasionally also is aggravated by causes already explained
in analyzing the performance of particular wheels, in
which there is a considerable space between the guide-
vanes and buckets, so that the absolute angle of outflow
a from the former is indefinite. In radial wheels, too, as
already pointed out, there may be some doubt as to the
proper way of estimating the measured area of outflow A_1.

Fliegner's investigations show that the loss by shock is
not so great as it should be according to Carnot's theory,
which was explained in connection with the losses in
reaction turbines, and this fact causes a turbine to be in
practice less sensitive to deviations from the best speed
than theory indicates.

For turbines which have *not* been designed for simultaneous inflow without shock and vertical or radial outflow,
the *mean* of the speeds corresponding to these two assumptions is generally a sufficiently good approximation to the
best speed for practical purposes. It is not difficult to see
how the method of calculation might be modified to agree
with the data furnished by Fliegner's experiments. If,
for instance, the relative vane angle of inflow were $a_1 = 0°$

and the relative vane angle of outflow $a_2 = 60°$; instead of introducing the velocity of rotation corresponding to inflow without shock for $a_1 = 0°$, the relative angle of inflow should be taken as $a_o = 12°$, for which the resistance is a minimum, and the speed determined accordingly; a_o simply takes the place of a_1 in the calculations. More experiments are required before this method can be generally adopted, as the data are insufficient; as far as they go Fliegner finds that the results may be expressed by the empirical formula,

$$a_o = \frac{a_1 + a_2}{5} - 30 \frac{e_2 - e_1}{e_1}.$$

For most of the radial or mixed-flow turbines referred to in Table A, the best speed agrees more closely with that for radial or vertical outflow than with that for inflow without shock, even when allowance is made for the difference between a_1 and a_o. This must occur when the increased loss by unutilized energy due to a given deviation from the best speed in one sense is greater than the loss from shock resulting from an equal deviation in the opposite sense. To take an extreme example, suppose the velocity of rotation for vertical outflow to be 20 feet per second, while that corresponding to inflow with least resistance is 16 feet. If over 16 and up to 20 feet the loss from shock is inappreciable, then it is clear the best velocity will be nearer 20 than 16.

In the following short table, for more convenient reference, the corresponding experimental and calculated velocities of rotation are given for all the turbines mentioned in Table A, with the exception of Rittinger's. The calculated velocity is in every case except one the mean of the velocities for inflow without shock and vertical or radial outflow.

With impulse turbines fewer experiments extending
over a considerable range of speed have been made than
with reaction wheels, but there is no doubt that practice
and theory are in substantial agreement in their case,
which is in some respects simpler than that of reaction
turbines, there being less uncertainty as to the true values
of the dimensions and quantities on which the velocities
of flow and rotation depend.

Description of turbine.	Experimental velocity.	Calculated velocity.
Haenel's Turbine. Parallel-flow, buckets widened towards outflow	10·515	9·906
Olching Turbine. Parallel-flow	16·863	17·422
"Collins" Turbine. Parallel-flow . . .	20·960	22·337
Tremont Turbine. Radial outward-flow .	18·051	18·102
"Boyden" Turbine. Radial outward-flow	20·392	20·967
Boott Turbine. Radial inward-flow . .	18·668	22·102
"Hercules" Turbine. Mixed inward-flow	22·088	21·637
"Swain" Turbine. Mixed inward-flow .	21·708	18·832
"Humphrey" Turbine. Mixed inward-flow	20·113	*17·533

* Velocity for inflow without shock.

CHAPTER XII.

AMERICAN TURBINES.

General remarks.—Distinctive features of American practice.—
Holyoke tests: probable accuracy, agreement of Francis's and
Braschmann's formulas.—Comparative analysis of mixed-flow and
axial turbines: balance of advantages.—Desirability of Comparative
tests by same methods and under similar conditions.—"Humphrey"
and "Hercules" turbines: free deviation not possible; evidence of
reaction.—Explanation of performance from construction.—Dimensions of American turbines.—Peculiarities in angles and form of
buckets, &c.—Diameters.—Number of vanes.—Comparison (numerical) of a mixed-flow with a parallel-flow turbine.—Object of
investigation.—Dimensions assumed.—Velocity of flow.—Velocities
of rotation.—Residual velocities at various points.—Loss from
unutilized energy.—Mean residual velocity.—Conclusions.—Result
of modified assumptions regarding outflow.—Estimation of efficiency
of mixed-flow wheels.—Calculations for "Hercules" turbine and
comparison with result of trials.—Calculation for Jonval turbine.
—Explanation of superior efficiency of "Hercules" wheel at part
gate.—Probable values of co-efficients of resistance.

AMERICAN turbines have already been incidentally referred
to in the course of this work, and as in general their
construction differs considerably from that usually adopted
in Europe, and energetic measures are taken to introduce
them on this side of the Atlantic by claiming in many
cases extraordinary efficiency, some more detailed consideration of their peculiarities, in so far as information on
the subject is available, is desirable.

American turbines are with few exceptions of the mixed-flow type, the direction of flow being generally inward and downward. Formerly outward-flow Fourneyron turbines appear to have been made, and many are still in use in the United States. Francis introduced the radial inward-flow wheel (invented, it is believed, by Professor J. Thomson), and this by an easy transition led to the mixed inward- and downward-flow turbine, of which the Leffel turbine seems to have been one of the first used in America.

As compared with European practice the following are the leading characteristics of American turbines: smaller diameter and greater number of revolutions for a given fall and power, greater relative width of buckets both at the inflow and outflow, smaller number of guide- and wheel-vanes, and regulation by means of pivoted guide-vanes or circular sluice between guide passages and wheel.

The majority, if not all American turbines, belong to the *reaction* system, and work drowned. There appears to be an idea on the part of some American makers of allowing the water to follow its own course, especially in *leaving* the wheel, thus permitting free deviation under water; as will be subsequently shown, however, in the case of two wheels in which this is supposed to occur, no such thing as free deviation in the ordinary sense of the term takes place, the phenomena observed being those incidental to reaction turbines.

In some American wheels the water leaves, not only in a downward (axial), but also partly in a radial outward direction, some of it being deflected through an angle approaching 180° in a path at right angles to that in which the driving force has to be exerted. In other wheels, again, the deflection of the fluid on its passage through the buckets is very abrupt, and it would seem

as if the designers of such motors had aimed at giving the
water the most tortuous course possible, with what object
it is difficult to imagine.

Probably more trials of turbines have been made in
America than in all the rest of the world put together,
and of these a large number have been carried out at
the celebrated testing "flume" at Holyoke, described in
another place (p. 391).

There can be little doubt of the relative accuracy of the
Holyoke tests, which bear all the evidence of careful
observation, and—it may be noted in passing—it would
not be in the interests of the Holyoke Water-power Company to credit the motors of their customers, who pay for
the water consumed by their turbines, with a greater
efficiency than that actually obtained.

It is, however, noticeable that the recorded efficiencies
of turbines tested in America are all remarkably high as
compared with the results obtained in Europe, and this
has not unnaturally given rise to the surmise that the
method of computing the quantity of water consumed is
answerable to some extent for this, and that the co-
efficients employed in the calculations, necessary when a
weir is used for measurement, are too small. The formula
applied at Holyoke is that of Francis, and the results
given by it agree very well with those obtained with, for
instance, Braschmann's formula, which is very generally
used in Germany, so that there appears to be little probability of errors arising from this cause. On the other
hand, it should be noted that in the best reliable trials
made in Europe with Jonval turbines, the quantity of
water consumed was measured by current-meters and not
over weirs, so that the details are not directly comparable
with those of American experiments.

American practice in the construction of turbines has

been rather savagely attacked by a late well-known continental writer on the subject, and it will be well to compare, carefully and dispassionately, the merits and faults of a wheel of the American type with those of the Jonval turbine generally accepted in Europe.

(1) In an inward- and downward-flow wheel the inflow of the water takes place without shock *over the whole width of the buckets*, while in a Jonval turbine this can only happen *at one diameter*, and consequently there is a loss from shock in the latter which is absent in the American type of motor.

(2) Owing to the relatively great width, in proportion to the diameter, of the buckets of American mixed-flow turbines at the outflow, *the deviations from the best speed*— the speed corresponding to vertical outflow—are greater than in a Jonval wheel of the usual proportions, and it has been inferred from this that the loss from unutilized energy must therefore also be greater; this, however, it will be shown, is not necessarily the case, the disadvantages arising from this cause being compensated by the superior efficiency due to inward-flow turbines, as previously explained in comparing the different types of wheel (p. 192).

Under similar conditions, for very narrow buckets the residual velocity of the water from an inward- and downward-flow wheel is less than from a parallel-flow wheel. As the width (of the outflow) is increased the average residual velocity becomes greater, but there is in many cases a sufficient margin between the values for the mixed-flow and for the parallel-flow turbines to compensate for a very considerable excess of width of the former over the latter at the outflow. This will be illustrated subsequently by a numerical example.

(3) It has been objected to the mixed-flow turbine that

H H

the water on entering the guides and passing through the buckets is twice uselessly deflected, whereby a certain loss is incurred. This is no doubt true, but there is reason to believe that this loss is unimportant in a well-designed wheel, although that portion of it due to downward deflection in the buckets themselves may be very appreciable where the curvature is abrupt, as in some American designs.

(4) As the course of the water while acting upon the vanes is partly radial, the downward pressure on the bearings is less in mixed- than in parallel-flow wheels, and is still further reduced by the lighter weight of the former, arising from their smaller diameter as generally made.

(5) For a given area-ratio $\dfrac{A}{A_2}$ the velocity of flow both in guide passages and buckets is lower in a radial or mixed-flow turbine than in an axial turbine, consequently the loss from hydraulic friction, &c., is also less, assuming the same experimental co-efficients to be applicable in both cases. This advantage may, however, be partly neutralized by the additional resistance due to sharp bends in a mixed-flow wheel.

This point will be more fully dealt with in the sequel by means of a numerical example.

Provided, therefore, it can be proved that the loss from residual velocity under given conditions is no greater in mixed-flow turbines, as proportioned in America, than in Jonval turbines, it is evident that the former have distinct theoretical advantages over the latter—first, in the absence of shock; secondly, in the diminution of friction due to less downward pressure.

In addition to these are the practical advantages of smaller diameter and consequent reduced cost.

It has been already noticed that the number of vanes in American wheels is generally less than in European turbines of the same diameter; provided the streams of water are sufficiently guided, this is an advantage, as it tends to reduce the friction. Whether the guidance afforded by the vanes is sufficient or not, depends mainly on the proportion between the length and depth of the passages; the passage in mixed-flow wheels is necessarily longer than in Jonval turbines, and hence the depth and pitch may safely be greater.

As the diameter of the shaft and bearings is inversely proportional to the cube root (according to some rules, the fourth root) of the speed for a given power, the work absorbed by friction with a certain pressure on the bearings is relatively greater as the diameter of the wheel is reduced. Thus the advantage due to the smaller pressure on the bearings of mixed-flow turbines is partially neutralized.

It will be seen on reference to the results of Mr. Bernhard-Lehmann's experiments on the shaft friction of various types of turbine, that while for inward-flow wheels this friction ranges from 0·9 to 1·7 per cent. only, for axial wheels the range is from 1·4 to 3·4 per cent. at full gate, and may be as high as 5·4 per cent. at half gate. At full gate therefore shaft friction alone may easily account for 2 per cent. efficiency.

Taking 83 per cent. efficiency as the best record of a Jonval turbine—somewhat higher efficiencies than this have been occasionally observed—an inward-flow wheel of *the same hydraulic efficiency* might thus show an actual efficiency of 85 per cent. From the causes already enumerated, however, the hydraulic efficiency may also be greater, so that an efficiency of 87 per cent. is by no means impossible.

Making every allowance for the merits of American hydraulic motors, the average results obtained at Holyoke still appear somewhat high as contrasted with trials made in Europe, and it would be satisfactory if some of the wheels reported to have given such admirable results at Holyoke could be compared under precisely the same conditions with Jonval turbines of the best European construction.

Apart from this, there is no doubt that *at least* as high efficiencies can be obtained with inward- and downward-flow wheels as with those having parallel-flow, while at the same time securing the practical advantages of greater lightness and lower cost.

The vanes of American wheels are generally thinner than those of European wheels, and at the ends are tapered almost to a knife-edge; frequently, also, they are constructed of brass or gun-metal.

Among the accounts of experiments with turbines will be found the results of trials with a "Humphrey" and a "Hercules" turbine, both of American design. In each of these there is reason to believe the water is intended, on leaving the wheel, to choose to some extent its own course, that is to say, to act with free deviation. On referring to the experimental value of K_1, it will be found that this is, for the "Humphrey" turbine, 0·621, and for the "Hercules" turbine, 0·625. These figures point to one of two conclusions: either the wheels work with so-called "reaction," or the efficiency must be exceedingly low, as the velocity of flow c could only be reduced, in a wheel with free deviation, to the stated proportion of the theoretical velocity by an excessive amount of useless resistance.

The results of the trials show that the latter is not the case, since the efficiencies in both instances are very good ; it is consequently clear that the motors are *reaction* turbines.

In the "Humphrey" wheel the vanes project below the revolving wheel-casing, and on the outside are practically unconfined, being connected merely by a narrow ring.

The stationary outer casing, however, entirely encloses the buckets, but there is a considerable space between the outer edges of the vanes and the interior surface of the stationary casing. It might naturally be supposed that this free space would interfere with the proper action of the water, which would escape outwards without doing work. The experiments prove that this does not happen to any great extent, and the explanation is, probably, that *the water itself*, between the outside of the wheel and the inside of the casing, acts as a wall, nearly stationary eddies being formed, which resist the outward pressure of the fluid in the wheel.

If free deviation really took place, the value of the coefficient $K_1 = 0\cdot621$ would indicate that the water on leaving the guide passages had already *lost* about **69** *per cent.* of its energy; with reaction the greater portion of this is still available as *pressure* to be converted into velocity during the passage of the fluid through the wheel-buckets.

Dimensions of American Turbines.

The angles of the vanes of American wheels have much the same range as in European practice; in many cases there is a tendency to make the angle a_2 as great as possible, with a view to reducing the residual velocity of the water leaving the buckets, and in some instances this seems to be carried so far, that the relative direction of the outflowing water is almost at right angles to the axis of the wheel. This, of course, is a mistake, as the absolute motion left in the fluid is not sufficient to carry it off, and a certain amount of pressure is then necessary for the purpose, which retards the flow and proportionately reduces the efficiency.

In many mixed-flow wheels the outflow orifices of the buckets do not lie in a plane, but each is of a semi-annular or crescent form, the edges of the vanes being curved. The vane-angle a_2 of outflow at any point must under such circumstances be measured in a plane defined by the tangent to the circumference through that point and the tangent to the vane-curve; the angle a_2 is then the angle formed by the latter, with a direction at right angles to the former.

The outer radius of American inward- and downward-flow turbines ranges from

$$r_1 = 0{\cdot}70 \sqrt{A} \text{ to } 1{\cdot}11 \sqrt{A};$$

from which it will be seen that the maximum in American design is about the same as the minimum in European practice for the *mean* radius of Jonval wheels—the outer radius being of course still greater, and at least $1{\cdot}25 \sqrt{A}$. For a given area of the passages, their width in American wheels exceeds that in Jonval turbines in proportion as the diameter is less.

As regards the number of guide-vanes and buckets the practice varies very considerably, but generally speaking these are fewer in number in American than in European turbines; the tendency in the former seems to be, to leave the water more at liberty to follow its own course, with the intention of thereby reducing the friction. This probably accounts for the loss from sudden bends in the passages not being so great as might be anticipated ; in the sharp corners stationary eddies are formed, over which the flowing water passes with less resistance and loss than if it were compelled to follow exactly the curve of the vanes.

Comparison of a Mixed-flow (inward and downward) with a Parallel-flow Turbine.

The object of this investigation is to show in how far —if at all—the disadvantages arising from great relative width of the buckets in an inward- and downward-flow turbine may be compensated by the more effective manner in which it utilizes the energy of the water as compared with a parallel-flow turbine. In the first instance the comparison will be confined entirely to the losses by unutilized energy.

As an example, a somewhat extreme case will be assumed, in which the mixed-flow turbine has the following proportions :—

Ratio of radius where water enters to greatest outflow radius . . . . $\dfrac{r_1}{r_2} = \dfrac{6}{5}$

Ratio of radius where water enters to mean outflow radius . . . . $\dfrac{r_1}{r_2^{m}} = \dfrac{12}{7}$

Ratio of radius where water enters to least outflow radius . . . . . $\dfrac{r_1}{r_2^{1}} = \dfrac{3}{1}$

Ratio of width at outflow to mean outflow radius . . . . . . $\dfrac{c_2^{m}}{r_2^{m}} = \dfrac{6}{7}$

The angle a_2 is (in the first instance) taken as constant for the whole width,

$$a_2 = 68°;$$

in consequence of this the distribution of the water is not uniform for the whole width, but the quantity flowing through a given portion is proportional to the diameter.

The relative loss from unutilized energy at any point —as elsewhere shown—is not represented simply by the square of the residual velocity, but by the product of that square and the corresponding proportional diameter. The quantity of water passing through at the mean radius $r_2{}^m$ may be taken as unity, and from the curve representing the losses at all points the *average loss* determined.

The angle a is assumed

$$a = 69°;$$

the ratio $\dfrac{A}{A_2} = 0\cdot 75.$

The Jonval turbine with which the mixed-flow wheel is to be compared has the same angles a and a_2 and the same ratio $\dfrac{A}{A_2}.$

For calculating the velocity of flow, the efficiency of the turbines may be taken as equal, and then from the well-known formula

$$c = K_1 \sqrt{2gh} = \sqrt{\frac{\epsilon}{2\, \frac{r_1}{r_2}\, \frac{A}{A_2}\, \sin a \, \sin a_2}}.$$

$$\frac{c_p}{c^m} = \frac{\text{Velocity of flow of parallel-flow wheel}}{\text{Velocity of flow of mixed-flow wheel}} = \sqrt{\frac{2\, \frac{r_1}{r_2{}^m}\, \frac{A}{A_2}\, \sin a \, \sin a_2}{2\, \frac{A}{A_2}\, \sin a \, \sin a_2}};$$

or since a, a_2 and $\dfrac{A}{A_2}$ are the same for both turbines,

$$\frac{c_p}{c_m} = \sqrt{\frac{r_1}{r_2{}^m}} = \sqrt{1 \cdot 714} = 1 \cdot 316.$$

For a distribution of the water *proportional to the diameters*, the correct *average* value for $\dfrac{r_1}{r_2}$ to be substituted in the formula is actually the ratio of the outer radius r^1 to the mean outflow radius $r_2{}^m$, as above. For an *even distribution* of the water, however, such as occurs with a helical vane surface, the average value of $\dfrac{r_1}{r_2}$ would not be that given, but the mean of all the values of $\dfrac{r_1}{r_2}$ between the greatest and least outflow radius; this can be found by constructing a curve representing successive values of $\dfrac{r_1}{r_2}$ and determining the mean ordinate; the average ratio in that case is $1 \cdot 835$.

As only proportions are in question, it may for convenience be assumed that the velocity of flow for the mixed-flow wheel

$$c_m = 10 ;$$

then the relative velocity of outflow from the buckets

$$c_{2m} = \frac{A}{A_2}\, c_m = 0 \cdot 75 \times 10 = 7 \cdot 5.$$

The velocity of rotation w_1, relative angle of inflow a_1, and residual velocities, at the various diameters, can now be determined by graphical construction.

In Fig. 167, let $O D = r_1$, $O A = r_2{}^m$, $O B = r_2{}^1$ and $O C = r_3$, to any scale, so that the proportions be those given. Let each of the angles $O A a$, $O B b$ and $O C c = a_2 = 68°$.

According to the usual principles of design it may be assumed, to begin with, that the absolute direction of outflow at the mean radius $r_2{}^m$ is perpendicular to the direction of rotation, or, for exclusively downward outflow, parallel with the axis.

$A\,a$, $B\,b$ and $C\,c$ are each equal to $c_2 = 7\cdot5$, and $E\,a$, perpendicular to $A\,E$ the momentary direction of rotation of A, then represents the absolute velocity of outflow u_m at the mean radius $r_2{}^m$, while $A\,E$ is the corresponding velocity of rotation. $O\,E$ prolonged intersects $B\,F$, $C\,G$ and $D\,H$ in F, G and H ($B\,F$, $C\,G$ and $D\,H$ being at right angles to $O\,D$), and $B\,F$, $C\,G$ and $D\,H$ are the velocities of rotation at the radii $r_2{}^1$, r_2 and r_1. $F\,b$ and $C\,G$ are the absolute residual velocities u^1 and u at B and C.

By measurement it will be found that

$$E\,a = u_m = 2\cdot875.$$
$$F\,b = u^1 = 4\cdot125.$$
$$G\,c = u = 4\cdot125.$$

The square of each of these values is proportional to the loss *per unit of weight* from unutilized energy, but, to take into account the distribution, must be multiplied with the corresponding ratio of the radius at the point in question to the mean radius $r_2{}^m$, thus :—

Loss from unutilized energy at A

proportional to $\qquad u_m{}^2 \times \dfrac{r_2{}^m}{r_2{}^m} = \left(2\cdot875\right)^2 \times 1.$

Loss from unutilized energy at B

proportional to $\qquad u^{12} \times \dfrac{r_2{}^1}{r_2{}^m} = \left(4\cdot125\right)^2 \times 0\cdot57.$

Loss from unutilized energy at C

proportional to $\qquad u^2 \times \dfrac{r_2}{r_2{}^m} = \left(4\cdot125\right)^2 \times 1\cdot43.$

These values can easily be ascertained for any desired number of points, and the curve XY constructed. The average value of the proportional loss, as determined from the mean ordinate of the curve, is

$$U^2 = 11 \text{ (nearly)},$$

whence the mean residual velocity

$$U = \sqrt{11} = 3\cdot32.$$

The relative velocity and direction of inflow is found as follows—

The angle IDZ is made equal to $a = 59°$, and $DI = c^m = 10$; DH is the velocity of rotation w_1 at the outer circumference, then $HI = c_1$ the relative velocity of inflow, and angle $HIK = a_1 = 37\frac{1}{2}°$.

With regard to the parallel-flow turbine, it is assumed that the proportion of the width of the buckets to the mean diameter is so small that practically the variations in the absolute residual velocity u at different points are negligible, and the latter may be taken as constant for the whole width. This assumption is extremely favourable to the parallel-flow wheel.

The ratio of the velocity of flow c_p in the parallel-flow turbine to that c_m in the mixed-flow turbine was found to be $1\cdot316$; hence $c_p = 10 \times 1\cdot316 = 13\cdot16$, and the relative velocity of outflow $c_{2p} = 0\cdot75 \times 13\cdot16 = 9\cdot87$.

By construction the residual velocity u is found to be—

$$u = 3\cdot75,$$

whence $$u^2 = 14\cdot06.$$

Notwithstanding, therefore, the relatively great width of the buckets of the mixed-flow wheel, it appears that the loss by unutilized energy is *less* than in the Jonval turbine, although the assumptions in this respect were all in favour of the latter. For a parallel-flow wheel of the given proportions the energy represented by the velocity of flow c is about *one-half* of that due to the head—K_1 having a value of about 0.7; hence the total available energy is proportional to $2 \times (13.16)^2 = 346.4$, and the ratio to this of the loss by unutilized energy is for the parallel-flow turbine—

$$\frac{14}{346.4} = 0.04, \text{ or } 4 \text{ per cent.},$$

for the mixed-flow turbine—

$$\frac{11}{346.4} = 0.032, \text{ or } 3.2 \text{ per cent.};$$

there being consequently a difference in favour of the latter of 0.8 per cent.

In addition to this trifling advantage, and apart from those before-mentioned, the mixed-flow turbine gains in efficiency from another source, namely, that owing to the velocity of flow being less in it than in the parallel-flow turbine working with the same head, the various losses by hydraulic friction and other resistances are reduced proportionally to the square of the velocity of flow.

Taking the same co-efficients as hitherto,
for *the parallel-flow turbine*—

the loss in the guide passages

$$= \frac{1}{2g} \, 0.125 \, (13.16)^2 = \frac{21.6}{2g}$$

the loss in the wheel-buckets

$$= \frac{1}{2g} \, 0.2 \, (9.87)^2 = \frac{19.5}{2g}$$

$$\text{Total} \quad . \quad . \quad \frac{41.1}{2g}$$

for *the mixed-flow turbine*—

the loss in the guide passages

$$= \frac{1}{2g} \, 0.125 \, (10)^2 = \frac{12.5}{2g}$$

the loss in the wheel-buckets

$$= \frac{1}{2g} \ 0 \cdot 2 \ (7 \cdot 5)^2 = \frac{11 \cdot 25}{2g}$$

$$\text{Total} \quad . \quad . \quad \frac{23 \cdot 75}{2g}$$

The difference in favour of the mixed-flow wheel is therefore—

$$\frac{41 \cdot 1 - 23 \cdot 75}{2g} = \frac{17 \cdot 35}{2g},$$

and this is about **5** *per cent.* of the total available energy.

From this it will be seen that there is theoretically a balance of 5·8 per cent. to the credit of the mixed-flow turbine, apart altogether from any advantages arising from absence of shock and reduced friction of the bearings; assuming that from the latter causes there is a further gain of 1 per cent.—a very moderate estimate—the total advantage would be **6·8** *per cent.* Supposing half of this to be neutralized by the increased loss due to the double deflection of the water previously referred to, there is still a nett gain of **3·4** *per cent.* as compared with the Jonval wheel.

If, instead of a *constant value* of the relative angle of outflow a_2, the vane surface at the lower end be *helical,* so that the distribution of the water is *nearly uniform,* the loss by unutilized energy remains almost unchanged, supposing the velocity of flow to be the same. The construction and curve of unutilized energy in this case are shown in the diagram Fig. 167 by the broken lines. The average loss is 10·95, corresponding to a velocity $u =$ 3·309.

As a matter of fact, for even distribution of the water, the velocity of flow must be rather less than with a

Fig. 167.

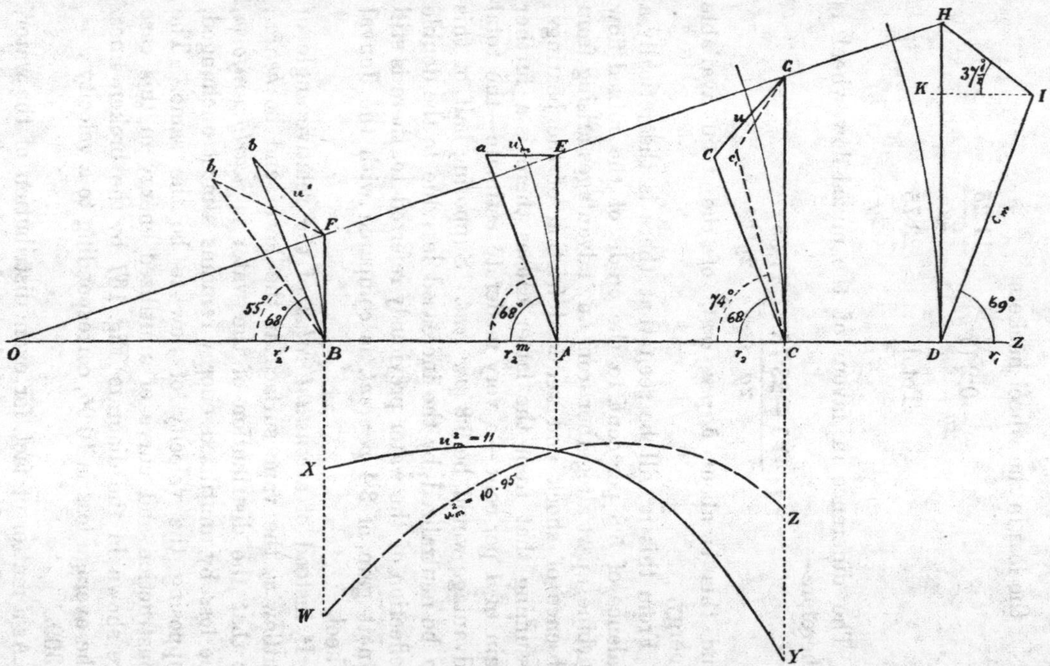

Diagram of Results for Mixed-flow Turbine.

constant angle a_2, since the mean ratio $\dfrac{r_1}{r_2}$ to be used in calculating K_1 is greater under the former conditions in the proportion of 1·835 to 1·714; with *even distribution*, therefore, the advantage of the mixed-flow over the parallel-flow turbine is even greater than that calculated, and the assumption of equal velocities of flow in both cases errs on the side of caution.

Experiments show a superiority of the best American mixed-flow wheels over the best Jonval turbines of about 4 per cent. (in one case 2 per cent. only), so that the estimate above made agrees very well with facts.

Under otherwise similar conditions, *the greater the ratio of the outer to the average inner diameter, the higher is the efficiency of a mixed-flow turbine.*

The theoretical superiority of the inward-flow wheel was pointed out in comparing the different systems of reaction turbines, but the investigation just completed was necessary to prove that this superiority is not neutralized by the great relative width of the buckets which characterizes so many mixed-flow wheels.

If the mixed-flow turbine be so designed, that instead of the water leaving vertically at the *mean* radius it does so at the *outer* radius r_2, it will be found that, with the assumed proportions, the effect on the loss by unutilized energy is comparatively slight, it being about 13 instead of 11, and the speed of the turbine is considerably reduced, while of course the angle a_1 must be modified to insure inflow without shock. A reduction of the speed with a very slight reduction in the efficiency is in itself a very desirable result, but this is accompanied by an increase of the velocity of flow which necessitates a smaller area (less width with the same diameter) of the passages and buckets. The increase in the velocity of flow is explained as follows—

The usual formula for determining the velocity is only strictly accurate on the assumption that the water leaves the turbine at right angles to the direction of rotation at every point. This condition is only fulfilled for *one* diameter, but when that is the *mean* diameter, the result is practically correct, as the deviations on either side nearly balance each other: when, on the other hand, outflow at right angles to the direction of rotation occurs at the outer diameter, it is clear that the deviations from the assumed conditions being all in one sense must result in the average velocity of flow being that due to a lower speed than the speed of the outer circumference. In inward-flow turbines the velocity of flow *increases* as the speed *decreases*, hence it will be greater for the wheel in question than in one designed for vertical outflow at the *mean* diameter.

Increased velocity of flow is accompanied by increased losses by hydraulic friction, .&c., and results in a further reduction of the efficiency. The best effect is obtained, therefore, by arranging the wheel for vertical outflow at the mean diameter, but as the velocity of flow rises only gradually with diminished speed, a very considerable departure from this principle may be permitted without very serious detriment to the performance of the wheel.

This peculiarity is only possessed to a very limited extent by parallel-flow turbines, and not at all by those with purely inward-flow.

It having been proved that in an inward- and downward-flow turbine with a—relatively to the diameter—very great width of buckets at the outflow, the loss by unutilized energy is not greater, and may even be less than in a Jonval turbine of good proportions, the efficiency of the former may be estimated by the usual formula for any given angles a and a_2 and ratio $\dfrac{A}{A_2}$.

It will be evident from the preceding investigation that the average loss by unutilized energy with the assumed proportions is to the minimum loss—where the outflow is vertical (or at right angles to the direction of rotation) —as 4 to 3, or in the ratio $1\frac{1}{3}$: 1. In calculating the efficiency by the well-known formula, that portion of it representing the loss by residual velocity, $\left(\dfrac{A}{A_2}\right)^2 cos\ ^2a_2$, must be multiplied by a co-efficient, which in the above case would be $1\frac{1}{3}$, and in others might be greater or less according to the dimensions of the wheel, in order to allow for the excess of actual loss over that due to vertical outflow.

Taking as an example the " Hercules " turbine of which the test results are given on page 405, all the necessary dimensions are given except the relative angle of outflow a_2, and if it be assumed that the average value of the latter is 70°, it will probably not be very far from the truth; besides which the co-efficient of velocity K_1 is not very sensitive to variations in the relative angle of outflow.

A general description of the " Hercules " turbine has already been given in connection with the trials made with it at Holyoke, but for purposes of calculation it will be necessary to examine its construction somewhat more closely.

Fig. 168 shows a section through the axis of the wheel and some of the principal dimensions kindly furnished to the author by Mr. Clemens Herschel, at the time engineer of the Holyoke Water-power Company.

The flow of the water on leaving the buckets is both inward and downward, but the inward component of the direction of motion preponderates. In entering the buckets the water already has a downward tendency, and

I I

for the greater bulk of it this is comparatively slightly increased in passing through the motor; a part of the fluid, however, at the lower end of the buckets leaves axially, or possibly even with a slightly outward radial tendency.

Fig. 168.

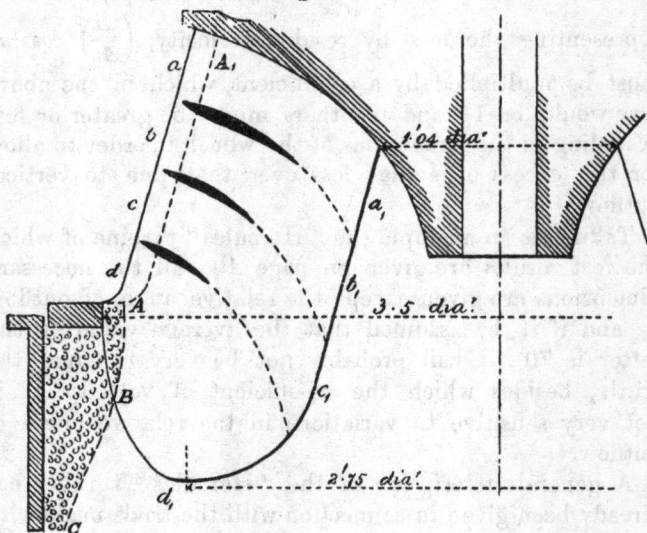

The curved form of the inner edges of the buckets is given with the object of securing a large area of outflow, which with the small diameter would otherwise be impossible.

Mr. Clemens Herschel expresses the opinion that the water " seldom if ever *fills* the bucket orifices." As was previously remarked, the co-efficient of outflow from the guide passages ($K_1 = 0.625$) proves that the turbine is the *reaction* type, and that therefore the water passing through the buckets is under a pressure greater than that of the atmosphere; if the bucket orifices are not entirely filled, then a portion of the water itself, not participating in the flow, forms itself into a wall, and the *effective* orifice is that bounded by the body of stationary eddying water. An action of this kind undoubtedly does often take place in the buckets of turbines, and it is extremely probable that it occurs in the "Hercules" wheel.

The water leaves by a short suction or draft-tube, and must take a course in leaving the buckets somewhat like that indicated in Fig. 168; the shaded area represents relatively stationary water in an eddying condition.

If the water *filled* the outflow orifices, a portion of it must leave in a nearly *radial outward* direction, and would immediately impinge nearly at right angles on the inner surface of the suction-tube, its flow being thus abruptly arrested and a great resistance offered to the flow.

The fluid will therefore not take this course, but follow the path of least resistance, and the line *A B C* will probably very nearly represent the outer boundary of a section through the axis of the body of water leaving the turbine.

If the edge of the vane at the *outflow* for the greatest diameter be supposed to lie in the plane of the diagram, then the true position of the projection of the *inflow* edge is represented by *A A₁*, as the line *a b c d* shows this edge brought round into the same plane as the outflow edge.

Assuming the course of the water to be as indicated, then only 0·89 of the full outflow area from the buckets, after allowing for contraction, would be occupied, and the ratio $\frac{A}{A_2} = 0.673$ instead of 0·509, as would be the value if the whole area were filled.

From the dimensions given the mean ratio of the diameters or radii $\frac{r_1}{r_2} = 1.84$; this is arrived at by supposing the turbine to be divided into four parts—by the transverse partitions—through each of which an equal quantity of water is assumed to flow, and taking the mean of the ratios $\frac{r_1}{r_2}$ of each division. The corresponding outer and inner diameters are measured at a and a_1, b and b_1, c and c_1, d and d_1 respectively.

Adding 3 per cent. for shaft friction to the actual efficiency makes the hydraulic efficiency $\epsilon = 0.888$, or say $0.89 = 89$ per cent.

The data are then as follows—

$$a = 75\tfrac{1}{4}°, = a_2 = 70°, \frac{A}{A_2} = 0.673, \frac{r_1}{r_2} \text{ (mean)} = 1.84,$$

$\epsilon = 0.89$, $sin\ a = 0.967$, $sin\ a_2 = 0.940$, r, mean $= 1.5$

Substituting these values in the formula for K_1^2

$$K_1 = \sqrt{\frac{0.89}{2 \times 1.84 \times 0.673 \times 0.967 \times 0.94}} = \mathbf{0.629};$$

the experimental value is **0.625**, so that the agreement is very good.

For the sake of further comparison the theoretical efficiency may be calculated, allowing **50** *per cent.* more loss by unutilized energy than that corresponding to radial outflow in a very narrow wheel. On this assumption the efficiency

$$\epsilon = \frac{2\,\frac{r_1}{r_2}\frac{A}{A_2}\,sin\ a\ sin\ a^2}{2\,\frac{r_1}{r_2}\frac{A}{A_2}\,sin\ a\ sin\ a_2 + \zeta_1 + \left(\frac{A}{A_2}\right)^2 (\zeta_2 + 1.5\ cos\ ^2 a_2)};$$

or substituting numerical values

$$\epsilon = 0.881.$$

as against $\epsilon = 0\cdot888$ for the experimental value with 3 per cent. added for shaft friction.

The ratio of the best speed as found by experiment to the velocity due to the head is

$$K_2 = 0\cdot669.$$

For inflow without shock calculated with the observed value of $K_1 = 0\cdot625$—

$$K_1 = 0\cdot582.$$

The value of K_1 which should strictly speaking be employed in calculating K_2 is probably not quite the same as that given above, since there is a clear space between the guide orifices and the inflow orifices of the buckets, so that the area of the stream on leaving the guides is not the same as at the moment of entering the buckets, the absolute direction of flow being also changed during the passage of the water through the clearance space.

In consequence of the diameter at which the water enters the buckets being considerably less than that at which it leaves the guides, the absolute area of the stream tends to diminish, but this tendency is partly counteracted by the absence of vanes to occupy a portion of the space; at the same time the angle a is very probably less than that at which the water leaves the guides, and this tends to increase the area A. It is impossible, owing to the uncertainty as to the direction taken by the water, to know exactly what would be the effective area of the stream, and consequently its absolute velocity at the moment of entering the buckets, even if the clearance between guides and wheel-buckets were given (which in this case it is not), but it is probable that the area is somewhat less than the value previously stated for A, and therefore the velocity of flow greater than that from the guide passages. Under these circumstances the velocity w_1 of the outer circumference, as calculated, would be also greater than the value corresponding to $K_2 = 0.582$.

The velocity of rotation at the inflow corresponding to radial outflow at the mean effective inner diameter $2r_{2m}$ is $0.7275 \sqrt{2gh}$, whence $\qquad K_2 = 0.7275$.

The mean of the values of K_2 resulting from the two assumptions of inflow without shock and radial outflow

is $\qquad K_2 = \dfrac{0.582 + 0.7275}{2} = 0.655$,

which corresponds very satisfactorily with the experimental value $\qquad K_2 = 6.669$.

In this case the correspondence between theory and practice is exceedingly good.

If the efficiency of a Jonval turbine with the same values of a, a^2 and $\dfrac{A}{A_2}$ be calculated in a similar manner, it will be found that $\quad \epsilon = 0.818$, the same co-efficients ζ_1 and ζ_2 being used in both instances, but no excess of loss by unutilized energy over that due to vertical outflow allowed for in the parallel-flow wheel (which of course tells in *favour* of the latter).

It will be seen that the theoretical advantage of the mixed-flow over the parallel-flow turbine in the case investigated is **5·4** per cent., apart from the question of any gain arising from absence of shock at the inflow. The actual superiority is not so great, as some peculiarities in the construction and circumstances before alluded to in connection with mixed-flow wheels tend to neutralize the advantages. The actual hydraulic efficiency of parallel-flow turbines is often greater than that obtained above; this is owing to the assumed values of the co-efficients of resistance ζ_1 and ζ_2, more especially the latter, being too high. Good Jonval wheels have in certain cases shown efficiencies of 85 per cent., or adding 3 per cent. for shaft friction, hydraulic efficiencies of 88 per cent.

A peculiar feature of the "Hercules" wheel under consideration is, that the efficiency at 0·806 gate was *higher* than at *full* gate.

This is easily accounted for by the construction of the turbine. The buckets are subdivided by partitions parallel with the direction of flow, so as to form practically several narrower wheels connected together and keyed to one shaft. At full gate, when all these divisions are full, the water leaving the uppermost division is, owing to the tapering form of the inside of the wheel and the obstruction of the boss, more abruptly deflected and impeded in its off-flow than the water from the lower divisions. When, therefore, the uppermost portion of the wheel is closed, the fluid meets with less resistance than at full gate, while the top division is filled up with nearly stationary water, which, owing to the presence of the boss, remains shut up between the latter and the sluice-gate regulator in the one direction and the top of the wheel and the flowing stream in the other. In addition to this, the area of the short draft-tube used with the motor, which is about 4 feet diameter, appears to be too small in proportion to the relative area of outflow A_2, so that the water is choked in leaving at full gate, while at part gate this does not occur. In calculating the theoretical efficiency the choking action was not taken into account, but when that is done, the efficiency becomes 0·871 instead of 0·881.

The author is of opinion that even better results than those actually achieved would be possible with the type of wheel in question if the inner and outer diameters were both proportionately greater, so that there would be more room for the water to escape, and the necessity might be avoided of giving the inside edge of the buckets the complicated curved form now required to secure the stated outflow area A_2, at the same time making the draft-tube of sufficient area to avoid all choking.

There is no reason to suppose that the co-efficients ζ_1 and ζ_2 would be higher in a turbine of the "Hercules" type than in a Jonval wheel; on the contrary, for the same angles a and a_2, the curvature of the vanes is less abrupt in an inward-flow than in a parallel-flow wheel, while owing to the relatively greater pitch and width of buckets compared with their length, the surface area in contact with a given volume of water is less in the former than in the latter motor. Both these circumstances would tend rather to reduce the co-efficients of resistance ζ_1 and ζ_2 than otherwise.

Against this may be set the double curvature (which, however, in the "Hercules" wheel is only slight for the greater proportion of the water passing through it), and the transverse divisions which offer additional resistance to the water.

In the "Hercules" turbine the transverse divisions, besides facilitating regulation, enable the guide passages and buckets to be made wider than would otherwise be practicable, since they transform a wheel of very great width in proportion to the diameter into several wheels of very moderate width. The drawback of making the total width of the buckets relatively great, in order to secure a small diameter, which is common to many American turbines, is illustrated in the "Hercules" wheel by the choking of the outflow. Through this choking, as has been mathematically shown, the efficiency is reduced below what it would otherwise be, and this is confirmed by the experiments, which prove the motor to work better with the gate partially closed than when it is fully open.

A more accurate method of analyzing the performance

of this motor would be to treat each division as a separate
turbine, but in that case a knowledge of the relative
outflow areas of the different parts would be necessary,
and this information the author does not possess. For
each division the ratio $\dfrac{A}{A_2}$ is probably somewhat different,
its value being least for the lowest division. Under these
circumstances the resultant velocity of flow should be
taken as the mean of the four different values obtained.

At the Centennial Exhibition in Philadelphia a large
number of turbine trials were carried out, and the motor
which on this occasion showed the best results was the
"Risdon" Turbine. This motor is illustrated on pages
27 and 28; it is of the inward- and downward-flow type,
the whole of the water leaving axially. It is reported to
have shown an efficiency of 87 per cent. when tested at
the Centennial Exhibition, and as much as **90** *per cent.*
when tested elsewhere.

The author is of opinion that even the efficiency of 87
per cent should be accepted with caution, while 90 per
cent. may be rejected as practically impossible with the
turbine in question. The ratio of the guide outflow area
to the bucket outflow area in this "Risdon" turbine was,
without allowing for contraction, as 1·25 to 1; allowing
for contraction at the guide orifices, and assuming there
was none at the bucket orifices, the ratio $\dfrac{A}{A_2}$ would be
nearly 1. As to the proportions between the diameter at
the inflow and the diameter at the outflow, no data are
published, but judging from the illustrations of this type
of wheel in the makers' catalogues, the average ratio $\dfrac{r_1}{r_2}$
would be about $\frac{4}{3}$. With low values of the co-efficients
ζ_1 and ζ_2 and the above ratios, a hydraulic efficiency of

about 90 per cent. or a little over is theoretically possible, which, allowing 3 per cent. for shaft friction, would leave an actual efficiency of about 87 per cent. The question is whether, with the abrupt deflection which a portion of the water is compelled to make from a radial to a vertical course, the assumption of such low values for the co-efficients is justified.

Comparison of Water Measurements.

As it might very naturally be suggested that the exceptionally good results obtained with the "Hercules" wheel are due to errors in the measurement of the water used by the motor, a comparison of the co-efficient used at Holyoke with that obtained from Braschmann's formula and by other experimenters appears desirable.

The mean length of the weir at Holyoke during the trials of the "Hercules" turbine was $L = 20 \cdot 035$ feet, being the same as that of the canal of approach.

For the best speed at full gate the depth on the weir was $H = 1 \cdot 2$ feet, and the quantity of water estimated to have passed through the turbine $Q = 88 \cdot 33$ cubic feet per second. If from these data the co-efficient C be calculated in the formula

$$Q = CL\sqrt{2gH^3},$$

it will be found that

$$C = 0 \cdot 4179.$$

By Braschmann's formula

$$C = \left(0 \cdot 3838 + 0 \cdot 0386 \, \frac{L}{B} + \frac{0 \cdot 00174}{H} \right);$$

since $\dfrac{L}{B} = 1$

$$C = 0 \cdot 3838 + 0 \cdot 0386 + 0 \cdot 0014 = 0 \cdot 4238.$$

This gives a difference of only $1 \cdot 4$ per cent. between the two values, which is well within the limits of probable error.

Fteley and Stearns found a co-efficient $C = 0.4106$ for a weir 19 feet wide without end contractions, with a depth of water on the weir of 1.166 feet; the difference between this and the Holyoke value is 1.75 per cent.

Before dismissing the subject of American turbines, it may be recalled to mind that with an inward-flow wheel the loss from unutilized energy with a *given outflow-angle* a_2 is not so great as in a parallel-flow turbine with the *same angle,* and this fact has been taken advantage of in some American motors, by making the outflow angle *less* than would be desirable in axial turbines in order to secure a greater outflow area without increasing the diameter. The smaller the angle a_2 can be made, the less sudden is the change of sectional area of the buckets for a given length of the latter. The loss in the buckets is reduced by making the change of shape more gradual, and this may in some instances compensate for the accompanying increase of loss from unutilized energy resulting from the smaller angle a_2. For this reason, of two turbines of the same class, but with different outflow angles, that with the smaller angle may give the higher efficiency, in spite of the greater residual velocity of the water leaving it.

The following Tables I. and II. (previously referred to) give respectively the values of the co-efficients of flow and efficiencies corresponding to various proportions of the outflow areas and radii, and to various angles of outflow. The figures contained in Table II. are not intended to be taken as having any absolute value, but merely illustrate the manner in which modifications in the proportions tend to influence the efficiency, on the assumption that the experimental co-efficients of friction, &c. remain unaltered.

TABLE I.

Co-efficients of Flow K_1 for various proportions of Turbines.

$a_2 = 80°$.

A/A_2	2					1·75				
n/r_2	1/2	3/4	1	5/4	3/2	1/2	3/4	1	5/4	3/2
a										
75°	0·528	0·494	0·454	0·415	0·385	0·573	0·535	0·490	0·448	0·414
70°	0·533	0·499	0·459	0·420	0·389	0·578	0·540	0·496	0·453	0·419
65°	0·540	0·506	0·466	0·426	0·395	0·586	0·548	0·503	0·460	0·426
60°	0·549	0·515	0·474	0·434	0·403	0·595	0·557	0·512	0·468	0·434
55°	0·559	0·525	0·489	0·444	0·412	0·607	0·569	0·522	0·479	0·445
50°	0·572	0·538	0·496	0·456	0·424	0·622	0·583	0·537	0·492	0·457

A/A_2	1·5					1·25				
n/r_2	1/2	3/4	1	5/4	3/2	1/2	3/4	1	5/4	3/2
a										
75°	0·627	0·585	0·535	0·487	0·451	0·695	0·647	0·591	0·538	0·497
70°	0·633	0·591	0·541	0·493	0·456	0·702	0·654	0·598	0·544	0·503
65°	0·642	0·599	0·549	0·501	0·463	0·712	0·664	0·607	0·553	0·511
60°	0·653	0·610	0·559	0·510	0·472	0·724	0·676	0·618	0·563	0·521
55°	0·666	0·623	0·571	0·522	0·484	0·739	0·690	0·632	0·577	0·534
50°	0·682	0·639	0·587	0·537	0·498	0·758	0·708	0·650	0·593	0·549

$a_2 = 80°.$

A/A_2 (top-left block)

r_1/r_2	$\frac{2}{3}$	$\frac{4}{5}$	1	$\frac{5}{4}$	$\frac{3}{2}$
a					
75°	0·785	0·730	0·665	0·605	0·558
70°	0·793	0·738	0·673	0·612	0·565
65°	0·805	0·749	0·684	0·622	0·574
60°	0·819	0·762	0·696	0·634	0·586
55°	0·836	0·779	0·713	0·649	0·600
50°	0·857	0·800	0·733	0·668	0·618

$A/A_2 = 0\cdot75$

r_1/r_2	$\frac{3}{2}$	$\frac{5}{4}$	1	$\frac{4}{5}$	$\frac{2}{3}$
a					
75°	0·646	0·700	0·771	0·847	0·912
70°	0·654	0·709	0·780	0·856	0·921
65°	0·665	0·720	0·792	0·869	0·934
60°	0·678	0·734	0·807	0·885	0·950
55°	0·695	0·753	0·827	0·904	0·971
50°	0·716	0·774	0·849	0·929	0·996

$A/A_2 = 0\cdot5$

r_1/r_2	$\frac{2}{3}$	$\frac{4}{5}$	1	$\frac{5}{4}$	$\frac{3}{2}$
a					
75°	1·107	1·029	0·939	0·854	0·788
70°	1·118	1·041	0·950	0·864	0·798
65°	1·134	1·056	0·966	0·879	0·811
60°	1·154	1·075	0·983	0·894	0·827
55°	1·178	1·099	1·005	0·917	0·847
50°	1·208	1·128	1·033	0·943	0·872

$A/A_2 = 0\cdot25$

r_1/r_2	$\frac{3}{2}$	$\frac{5}{4}$	1	$\frac{4}{5}$	$\frac{2}{3}$
a					
75°	1·082	1·164	1·274	1·386	1·480
70°	1·096	1·179	1·288	1·400	1·495
65°	1·112	1·198	1·307	1·419	1·514
60°	1·133	1·219	1·330	1·443	1·537
55°	1·159	1·247	1·358	1·483	1·565
50°	1·191	1·280	1·392	1·507	1·600

TABLE I.—continued.

$a_2 = 75°$.

$\dfrac{A}{A^2}$		1					1.75			
r^1/r_2	3/5	4/5	1	5/4	3/2	3/5	4/5	1	5/4	3/2
a										
75°	0·522	0·489	0·450	0·413	0·384	0·566	0·530	0·488	0·446	0·414
70°	0·526	0·494	0·455	0·418	0·388	0·569	0·536	0·493	0·451	0·418
65°	0·532	0·500	0·462	0·424	0·394	0·582	0·543	0·500	0·458	0·425
60°	0·540	0·508	0·469	0·432	0·401	0·588	0·552	0·508	0·466	0·433
55°	0·550	0·518	0·479	0·441	0·410	0·599	0·563	0·519	0·476	0·443
50°	0·562	0·530	0·491	0·452	0·421	0·612	0·576	0·532	0·489	0·455

$\dfrac{A}{A_2}$		1.5					1.25			
r_1/r_2	3/5	4/5	1	5/4	3/2	3/5	4/5	1	5/4	3/2
a										
75°	0·621	0·581	0·533	0·486	0·451	0·691	0·644	0·590	0·538	0·497
70°	0·627	0·587	0·539	0·492	0·456	0·697	0·651	0·596	0·544	0·503
65°	0·635	0·595	0·546	0·499	0·463	0·707	0·660	0·605	0·553	0·511
60°	0·646	0·605	0·556	0·509	0·472	0·719	0·672	0·616	0·563	0·521
55°	0·658	0·617	0·568	0·520	0·483	0·733	0·686	0·630	0·576	0·534
50°	0·673	0·632	0·583	0·534	0·496	0·750	0·703	0·647	0·592	0·549

$$a_2 = 75°.$$

$\frac{A}{A_2}$	0·75					1				
$\frac{r_1}{r_2}$	$\frac{3}{2}$	$\frac{5}{4}$	1	$\frac{4}{5}$	$\frac{2}{3}$	$\frac{3}{2}$	$\frac{5}{4}$	1	$\frac{4}{5}$	$\frac{2}{3}$
a										
75°	0·649	0·703	0·773	0·847	0·910	0·560	0·606	0·665	0·728	0·782
70°	0·657	0·711	0·781	0·856	0·919	0·566	0·613	0·673	0·736	0·790
65°	0·668	0·722	0·794	0·869	0·933	0·576	0·623	0·683	0·747	0·801
60°	0·681	0·737	0·809	0·884	0·949	0·587	0·635	0·696	0·760	0·815
55°	0·698	0·754	0·827	0·904	0·968	0·601	0·650	0·712	0·776	0·831
50°	0·718	0·775	0·850	0·927	0·992	0·619	0·668	0·731	0·796	0·852

$\frac{A}{A_2}$	0·25					0·5				
$\frac{r_2}{r_1}$	$\frac{3}{2}$	$\frac{5}{4}$	1	$\frac{4}{5}$	$\frac{2}{3}$	$\frac{3}{2}$	$\frac{5}{4}$	1	$\frac{4}{5}$	$\frac{2}{3}$
a										
75°	1·090	1·175	1·282	1·394	1·486	0·793	0·858	0·943	1·033	1·109
70°	1·102	1·187	1·295	1·407	1·499	0·802	0·868	0·953	1·043	1·119
65°	1·119	1·205	1·314	1·426	1·519	0·815	0·882	0·968	1·059	1·135
60°	1·140	1·227	1·337	1·449	1·542	0·831	0·899	0·988	1·080	1·156
55°	1·166	1·253	1·364	1·476	1·569	0·852	0·920	1·009	1·101	1·179
50°	1·198	1·285	1·397	1·510	1·604	0·876	0·946	1·035	1·130	1·208

TABLE I.—continued.

$$a_2 = 70°$$

A/A_2	2					1·75				
n/r_2	3/8	5/8	1	5/4	3/2	3/8	5/8	1	5/4	3/2
a										
75°	0·512	0·482	0·446	0·410	0·382	0·558	0·524	0·484	0·444	0·413
70°	0·517	0·487	0·451	0·415	0·386	0·562	0·529	0·489	0·449	0·417
65°	0·522	0·492	0·456	0·420	0·392	0·569	0·536	0·495	0·455	0·423
60°	0·530	0·500	0·464	0·428	0·399	0·577	0·544	0·503	0·463	0·431
55°	0·539	0·509	0·473	0·437	0·408	0·588	0·554	0·514	0·473	0·441
50°	0·549	0·520	0·484	0·448	0·418	0·600	0·567	0·526	0·485	0·453

A/A_2	1·5					1·25				
n/r_2	3/8	5/8	1	5/4	3/2	3/8	5/8	1	5/4	3/2
a										
75°	0·614	0·576	0·530	0·485	0·451	0·685	0·641	0·588	0·538	0·498
70°	0·620	0·581	0·535	0·491	0·456	0·691	0·647	0·594	0·544	0·504
65°	0·627	0·589	0·543	0·498	0·463	0·700	0·656	0·603	0·552	0·512
60°	0·637	0·598	0·552	0·507	0·471	0·711	0·667	0·614	0·562	0·522
55°	0·648	0·610	0·564	0·518	0·482	0·725	0·680	0·627	0·575	0·534
50°	0·663	0·625	0·577	0·532	0·495	0·741	0·697	0·643	0·591	0·549

$a_2 = 70°$

$\dfrac{A}{A_2}$			1						0·75		
$\dfrac{r}{r_2}$	$\tfrac{2}{3}$	$\tfrac{3}{4}$	1	$\tfrac{5}{4}$	$\tfrac{3}{2}$		$\tfrac{2}{3}$	$\tfrac{3}{4}$	1	$\tfrac{5}{4}$	$\tfrac{3}{2}$
a											
75°	0·778	0·726	0·665	0·607	0·562		0·909	0·847	0·775	0·706	0·653
70°	0·785	0·734	0·672	0·614	0·568		0·918	0·856	0·784	0·714	0·661
65°	0·796	0·744	0·683	0·625	0·578		0·931	0·869	0·796	0·726	0·671
60°	0·809	0·757	0·695	0·636	0·589		0·946	0·884	0·810	0·739	0·685
55°	0·825	0·773	0·710	0·650	0·603		0·965	0·903	0·830	0·756	0·701
50°	0·844	0·792	0·729	0·670	0·624		0·988	0·926	0·850	0·777	0·721

$\dfrac{A}{A_2}$			0·5						0·25		
$\dfrac{r}{r_2}$	$\tfrac{2}{3}$	$\tfrac{3}{4}$	1	$\tfrac{5}{4}$	$\tfrac{3}{2}$		$\tfrac{2}{3}$	$\tfrac{3}{4}$	1	$\tfrac{5}{4}$	$\tfrac{3}{2}$
a											
75°	1·112	1·037	0·948	0·864	0·799		1·493	1·403	1·292	1·184	1·100
70°	1·123	1·048	0·959	0·874	0·809		1·509	1·417	1·306	1·198	1·113
65°	1·138	1·063	0·973	0·888	0·822		1·527	1·435	1·324	1·216	1·129
60°	1·157	1·082	0·991	0·904	0·837		1·549	1·457	1·346	1·237	1·149
55°	1·181	1·104	1·013	0·926	0·858		1·578	1·486	1·374	1·264	1·175
50°	1·209	1·131	1·040	0·951	0·882		1·612	1·520	1·407	1·296	1·208

TABLE II.

Calculated Efficiencies with $a_2 = 70°$ for various proportions of Turbines.

$\dfrac{A}{A_2}$	2			1·75			1·5			1·25		
$\dfrac{r_1}{r_2}$	$\frac{3}{4}$	1	$\frac{3}{2}$	$\frac{3}{4}$	1	$\frac{3}{2}$	$\frac{3}{4}$	1	$\frac{3}{2}$	$\frac{3}{4}$	1	$\frac{3}{2}$
a												
75°	0·634	0·722	0·796	0·659	0·743	0·813	0·684	0·765	0·829	0·709	0·785	0·846
70°	0·628	0·717	0·792	0·653	0·738	0·809	0·678	0·760	0·826	0·703	0·781	0·842
65°	0·620	0·710	0·786	0·644	0·731	0·803	0·670	0·753	0·820	0·695	0·774	0·837
60°	0·609	0·700	0·778	0·634	0·722	0·796	0·660	0·744	0·813	0·686	0·766	0·831
55°	0·595	0·688	0·751	0·621	0·711	0·786	0·648	0·733	0·805	0·674	0·756	0·823
50°	0·579	0·674	0·756	0·605	0·697	0·775	0·632	0·720	0·794	0·659	0·743	0·813

$\dfrac{A}{A_2}$	1			0·75			0·5			0·25		
$\dfrac{r_1}{r_2}$	$\frac{3}{4}$	1	$\frac{3}{2}$	$\frac{3}{4}$	1	$\frac{3}{2}$	$\frac{3}{4}$	1	$\frac{3}{2}$	$\frac{3}{4}$	1	$\frac{3}{2}$
a												
75°	0·732	0·804	0·860	0·749	0·818	0·871	0·747	0·816	0·869	0·677	0·758	0·824
70°	0·727	0·800	0·857	0·744	0·813	0·867	0·742	0·812	0·866	0·670	0·753	0·820
65°	0·719	0·794	0·852	0·737	0·808	0·863	0·735	0·806	0·862	0·662	0·746	0·815
60°	0·710	0·786	0·847	0·728	0·801	0·858	0·727	0·799	0·856	0·652	0·737	0·808
55°	0·699	0·777	0·839	0·717	0·792	0·851	0·715	0·790	0·850	0·639	0·727	0·799
50°	0·684	0·764	0·830	0·703	0·780	0·842	0·701	0·779	0·840	0·623	0·713	0·788

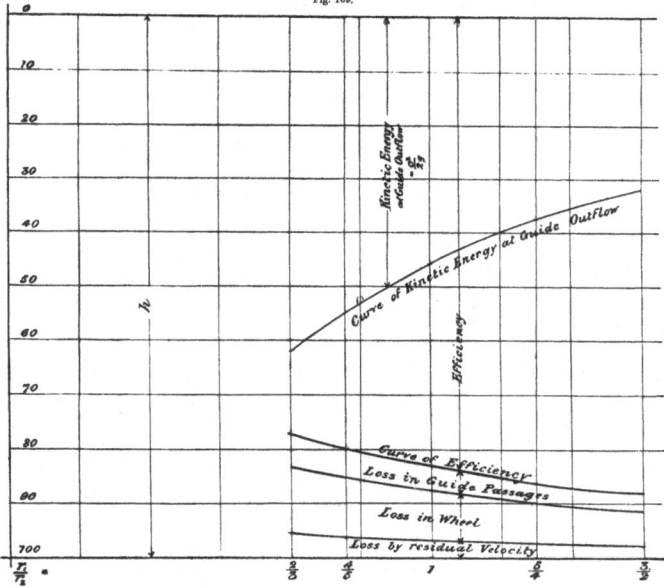

Fig. 169.

The curves shown in Fig. 169 illustrate clearly the effect of the ratio of inflow to outflow radius on the energy available in the kinetic form—as velocity—at the guide outflow, and on the various losses.

It will be seen that the loss from residual velocity decreases as the ratio $\dfrac{r_1}{r_2}$ increases.

Neglecting the loss in the guide passages, the ordinates of the curve of kinetic energy measured from the lower horizontal line upwards give the proportion of energy available as pressure at the guide outflow $h - \dfrac{c_2}{2g}$.

CHAPTER XIII.

HYDRAULIC PRESSURE ENGINES.

Definition.—Mode of action.—Motion of piston.—Influence of reciprocating masses.—Pressure on piston.—Available energy.—Useful work.—Losses.—Efficiency.—Motors of Armstrong, Rigg, Schmid, Wyss and Studer, Haag, Meyer, and Hastie.—General remarks.

A HYDRAULIC pressure engine is a motor in which water is made to do work by means of its hydrostatic pressure only acting on a piston or plunger reciprocating in a cylinder, or, in some cases, on a revolving piston similar to those employed in rotary steam-engines.

Reciprocating motors of this class have been made both single and double acting, and for pressure up to one ton per square inch and even greater.

In the case of turbines the pressure of a column of water is in the first instance employed in giving motion to the fluid itself, and after this motion has been produced the energy due to it is utilized in doing work. In hydraulic pressure engines the pressure of the water, instead of first being expended on accelerating the motion of the latter, is applied direct to the performance of work, by overcoming through the medium of a piston or plunger an approximately equal and contrary force or resistance. A portion of the available energy due to the pressure is, it is true, unavoidably expended in giving the fluid the velocity necessary to follow the piston and escape from the motor, but this portion should be as small as practicable, and an ideal pressure motor would be one in which the energy represented by the velocity is a vanishing quantity compared with the work done on the piston; this will be approximately the case when the pressure is very high and the speed of the engine low.

For pressure engines, as for other hydraulic motors, the available energy is expended in (1) doing useful work, (2) overcoming hydraulic resistances, and (3) producing the residual velocity of the fluid.

If the motion of the piston or plunger were uniform throughout the whole of its stroke, the pressure exerted on it would be constant, provided the pressure in the main did not vary; as a matter of fact, however, in reciprocating engines the speed of the piston varies very greatly at different points of the stroke, being *nil* at the commencement, a maximum near the middle, and again *nil* at the end. The water in the cylinder and in the pipes connected with it, as far back as the air vessel or its equivalent which is generally used, has to follow this varying motion of the piston, and consequently alternate acceleration and retardation of the whole of this mass of fluid, as well as the reciprocating masses of the engine, must occur. At the beginning of the stroke, when the piston is for a moment at rest, the whole available pressure will be transmitted through it to the crank-pin, but as the speed increases towards the middle of the stroke, a part of this pressure will be required to cause the acceleration of the moving masses, and the force acting on the crank-pin will be diminished by a corresponding amount. During the latter part of the stroke the energy thus absorbed will be again restored as the masses are brought to rest.

The pressure exerted on the piston is, in consequence of the action described, not constant, but varies according to the amount required to produce the velocity of the water necessary to follow the piston. The indicator diagram would have a form similar to that shown in Fig. 170.

The pressure, as recorded by an indicator, is not affected by the absorption of energy through the reciprocating masses of the engine, which only has the effect of altering the distribution of effort on the crank-pin.

Fig. 170.

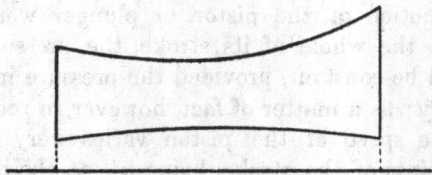

During the exhaust, in addition to the power required to overcome the atmospheric pressure, a certain amount of work has to be done in expelling the water from the cylinder, all of which represents loss of energy, and has the immediate effect of increasing the back pressure. The greater the piston speed the greater, with a given exhaust orifice, must be the back pressure, and the latter increases as the outflow area is reduced; it is therefore obviously desirable to keep the speed of the motor low relatively to the velocity due to the available pressure or head, and at the same time to make the area of the exhaust passages as great as possible.

An exact mathematical investigation of the effect of the acceleration and retardation of the masses of water and metal on the distribution of effort on the piston or crank-pin leads to very complicated expressions and has very little practical use; it may therefore be dispensed with.

The well-known formula

$$E = W + L,$$

where (as before) E denotes the available energy, W the useful work done, and L the losses, applies to water-pressure as to other motors; it remains to be seen of what components these quantities are made up.

Available Energy.

In the case of small water-pressure motors driven from the water-supply in the town mains, the available energy at the disposal of the motor is the product of the pressure in the pipes, measured as *head*, multiplied with the weight of water supplied. The pressure must be measured at a point immediately behind the orifice through which the water enters the motor.

The amount of the available energy for any given case is by no means invariable, but depends upon the position of the valve or cock through which the water leaves the main. If F denote the sectional area of the pipe, f the (variable) valve area, a the co-efficient of contraction, then the velocity of outflow

$$(1) \qquad c = \sqrt{\dfrac{2gh}{1 + \left(\dfrac{F}{af} - 1\right)^2}}$$

where a also varies under different conditions. The quantity of water passing through the valve-area f is $Q = f c$, so that Q is dependent on f in a double sense, and is not merely proportioned to the opening of the valve. If p denote the pressure near the valve-orifice, then the available energy

$$(2) \qquad E = c s f p = Q s p,$$

where s is the weight of the unit of volume of the fluid.

The quantity p is, however, dependent to a high degree on f (the valve-outlet area), and attains its minimum value when the valve is fully open, and its maximum (the full pressure in the main) when the valve is closed. The volume, on the other hand, increases with f, and hence it follows that E must reach its maximum with some intermediate position of the valve.

Fig. 171.

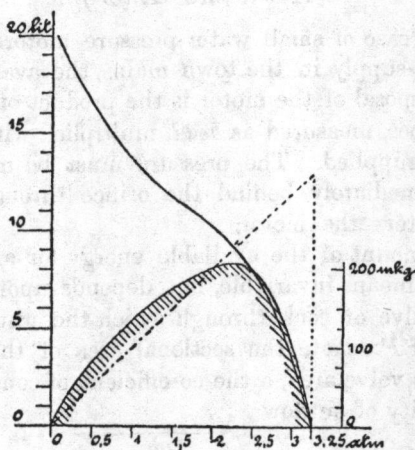

To follow the variations of E mathematically would be a somewhat tedious and unnecessary process, but experiments made on this subject by Schaltenbrand, the results of which were published in the *Zeitschrift des Vereins Deutscher Ingenieure* in 1881, p. 655, give valuable information on this point. The experiments were made with an 8 mm. valve in the Berlin mains, to which a pressure-gauge was connected.

When the valve was closed, the gauge indicated a

pressure of 3·25 atmospheres, and with fully open valve only 0·05 atmosphere.

The diagram (Fig. 171) shows graphically the results of this investigation. The pressures in atmospheres plotted as abscissæ and the quantities of water flowing out as ordinates give the curve shown by the full plain line. The product of the pressure and corresponding quantity of water is proportional to the available energy in each case, and is also plotted as a curve represented by the full line with shaded margin. The efficiency is shown by the dotted line. The maximum value of the available energy occurs at a pressure of 2·3 atmospheres, and with a quantity of water equal to 8·9 litres, and amounts to 205 metre-kilograms per minute, or 0·045 horse-power.

The amount of energy available for every cubic foot of water is evidently proportional to the pressure behind the valve-orifice.

The mean velocity of flow c through the main valve with a given motor is obviously directly proportional to the mean speed of the latter, so that with a given head in the main, there is some speed of the motor at which a maximum quantity of energy *is supplied to it* for conversion into work. It does not follow that the maximum work done by the motor is also coincident with this, since, as previously stated, the efficiency is greatest when the speed is least.

If h denote the total head of water available in the main, then for the relation between velocity of flow c and pressure p, the hydrodynamic equation applies—

$$(A) \qquad h = \frac{c^2}{2g} + p + \text{Losses} = (1 + \zeta)\frac{c^2}{2g} + p,$$

where ζ is co-efficient including all losses.

By (2) $\qquad E = c f s p,$

and since from (A)

$$c = \sqrt{\frac{(h - p) 2g}{1 + \zeta}}$$

(2a) $\qquad E = f s p \sqrt{\frac{h - p}{1 + \zeta} 2g},$

showing very clearly that the *total* available energy is not simply directly proportional to the pressure, although this is the case with regard to the energy available per unit of weight—

$$E_u = p.$$

Expressed in terms of c (instead of p as above)

(2b) $\qquad E = f s c \left\{ h - \frac{c^2}{2g} (1 + \zeta) \right\}$

Useful Work.

Let A denote the effective area of the piston,

$\quad S$ the stroke,

$\quad P$ the mean effective pressure on the piston, per square inch,

$\quad p_m$ the mean working pressure on the piston, per square inch,

$\quad p_o$ the mean back pressure above the atmosphere,

$\quad a$ the atmospheric pressure,

$\quad n$ the number of strokes per second of the piston,

then $P = p_m - p_o$, assuming the area of the piston to be the same on both sides, and the useful work per stroke $= P A S = (p_m - p_o) A S.$

The useful work *per second*

$$W = n\,P\,A\,S = n\,(p_m - p_o)\,A\,S,$$

or since the mean piston velocity per second

$$v = n\,S,$$
$$W = v\,(p_m - p_o)\,A = v\,P\,A.$$

The pressure p in the main, measured in the air vessel or its equivalent, would, if the velocity were uniform, equal the working pressure p_m, neglecting the losses due to friction and hydraulic resistances of various kinds. In practice, p and p_m are only equal at the commencement of a stroke, when the piston or plunger has the velocity *nil;* as the velocity increases, the pressure p (which for the sake of simplicity is assumed to be constant, as would be approximately the case with a very large air vessel), instead of being wholly expended in doing work on the piston, is partially employed in accelerating the mass of water between the air vessel and the piston, as well as the masses of the reciprocating portions of the engine. This has the effect of reducing the pressure on the piston below that in the air vessel, apart from hydraulic resistances which tend in the same direction, and causes the depression of the indicator diagram, shown in Fig. 170, previously referred to. As the piston is brought to rest during the latter part of the stroke, the pressure rises, and the kinetic energy due to the velocity of the moving fluid is reconverted into pressure, so that at the termination of the stroke, when the piston velocity is again *nil,* not only does the full pressure p in the air vessel make itself felt, but also an excess of pressure over this due to the restoration of the energy previously absorbed in acceleration. Hence the ordinate of the indicator diagram corresponding to the end of the stroke is higher than that at the beginning.

For practical purposes it is sufficient to assume a mean constant pressure p_m throughout each stroke, which is less than the mean pressure p in the air vessel, by the amount due to hydraulic resistances in the pipes, passages and cylinder.

With a small air vessel, as generally used with hydraulic reciprocating motors, the pressure p is not constant, since at the commencement and termination of every stroke (where only a single motor is in use) the body of fluid between the air vessel and piston is brought to rest while water continues to flow into the air vessel from the main, thus producing compression of the enclosed air and a corresponding increase of pressure. The actual conditions occurring are, it will be seen, of a very complicated character, owing mainly to the periodically varying piston speed.

Losses of Energy.

The hydraulic losses of energy taking place in water-pressure motors are similar to those occurring in the case of turbines, and are due to the following causes : fluid friction in pipes and passages, flow through orifices with sharp or imperfectly rounded edges, sudden changes of stream section, sharp bends in pipes and passages, and unutilized energy in outflowing water.

Some of these losses occur during the working stroke, others during the exhaust stroke.

Generally speaking, the water enters the cylinder through ports which are considerably less in area than the cylinder, so that in passing into the latter a sudden enlargement of the stream section takes place. If, as before, v be the piston speed and v_1 the velocity of flow through the ports, the loss for the unit of weight will be

$$L_1 = \frac{1}{2g} (v_1 - v)^2,$$

or, if R_1 denote the ratio of the area of the cylinder to

that of the ports, $L_1 = \dfrac{(R_1 - 1)^2 \, v^2}{2g}$,

since $\qquad\qquad v_1 = R_1 \, v.$

In passing through the ports a loss is sustained repre-

sented by $\qquad L_2 = \dfrac{\zeta_1 \, v_1^2}{2g} = \dfrac{1}{2g} \, \zeta_1 \, R_1^2 \, v^2.$

Previous to entering the cylinder a certain quantity of energy is absorbed by fluid friction in the pipe leading from the air vessel to the motor. A similar loss occurs in the cylinder both during the working and exhaust stroke.

Owing to the relatively small length of the cylinder, the loss in it by hydraulic friction may be neglected; that occurring in the supply pipe is expressed by the

well-known formula $\qquad L_o = \dfrac{l}{d_o} \, \zeta_r \dfrac{v_o^2}{2g},$

where l is the length of the pipe, d_o its diameter, v_o the velocity of the water in it, and ζ_r a co-efficient. If the exhaust takes place through a pipe of any considerable length, the loss by friction in this must also be taken into account. Denoting by R_o the ratio of the cylinder area to that of the pipe with the diameter d_o,

$$L_o = \frac{l}{d_o} \, \zeta_r \, R_o^2 \, \frac{v^2}{2g}.$$

The loss from unutilized energy has next to be considered. At every point of the exhaust stroke the velocity of the water expelled from the motor is different, varying with the piston speed, and the loss of energy, for a given quantity of water expelled is of course proportional to the square of the piston speed. If the squares of the piston velocities be plotted as ordinates, and the corresponding portions of the stroke traversed as abscissæ, a curve will result similar to that shown in Fig. 172. The mean ordinate of this curve—the height of a rectangle of the length of the stroke and of the area comprised between the curve and the axis of abscissæ— represents the *average loss* per unit of weight of water expelled, and the square root of this value the equivalent velocity of flow, which is in excess of the mean piston

Fig. 172.

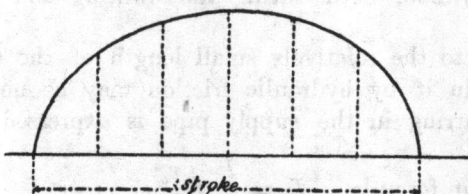

velocity. In a motor with an oscillating cylinder, and the distance of the centre of oscillation from the crank centre equal to three times the throw of the crank, the average loss by unutilized energy is to that due to the mean piston velocity as about 6·5 to 4, and the ratio of the equivalent velocity to the mean piston velocity as about 2·54 to 2, or say 1¼ times the latter.

In calculating the various hydraulic losses, this value should be taken for *v* instead of the actual mean piston velocity—it may be denoted by $C\,v$. The loss from unutilized energy may then be expressed by

$$L_3 = R_3{}^2 \frac{C^2 v^2}{2g},$$

where R_3 is the ratio of the cylinder area to that of the exhaust ports.

There will finally be a further loss during the passage of the water through the exhaust ports and passages, depending on the length, shape and surface of these, and whether the edges are well rounded or sharp.

All these losses may be summed up together in the form

$$L = \frac{v^2}{2g} \left\{ R_o{}^2 \frac{l}{d_o} \zeta_r + (R_1 - 1)^2 + R_1{}^2 \zeta_1 + R_3{}^2 C^2 + \cdots \right\};$$

let the quantities within the brackets be for brevity

denoted by ζ, then $\qquad L = \zeta \frac{v_2}{2g}.$

The mean effective pressure can now be expressed by

$$P = \left(\frac{p}{s} - \zeta \frac{v^2}{2g} \right) s.$$

Loss by leakage has not been included in the losses thus far considered, but it should in a well-constructed motor be very trifling.

Efficiency.

The efficiency of a hydraulic pressure engine

$$\epsilon = \frac{W}{E} = \frac{Q\,P}{Q\,p} = \frac{P}{p},$$

where both P and p are measured in pounds per square inch.

Substituting for P its value as previously found

$$\epsilon = \frac{p - \zeta \dfrac{v^2}{2g} s}{p}.$$

This equation shows very plainly how with a given velocity v the efficiency must increase with the pressure, or with given pressure increase as the velocity decreases.

Sir William Armstrong's Motor.

To Sir William Armstrong appears to be due the credit of having designed and constructed the first hydraulic pressure engine. His original idea in connection with the subject was to adopt the rotary type of engine, and having arrived at a design which he considered fulfilled the desired conditions, he sent it to the *Mechanic's Magazine*, in which journal it was fully described and illustrated under date December 29th, 1838. Fig. 173 shows this design, which, in the inventor's own words,[1] "may be briefly described as consisting of a wheel with a flat rim, containing four equidistant pistons folding into circular apertures, and intersecting longitudinally a curved tube open at the lower end, and communicating at the upper end with the supply pipe. The pistons open out as they enter the tube and fold up on leaving it, and each piston takes the pressure of the column before the preceding one loses it. The opening and closing of the pistons in the order required is effected by external cams and slides giving motion to the pistons through the axles on which they turn."

[1] "Minutes of Proceedings of the Institution of Civil Engineers," vol. i. p. 65.

Subsequently a working model of this motor was con-structed and tried at Newcastle and Gateshead, in con-nection with the town water-pipes.

Its efficiency is stated to have been 95 per cent. of the theoretic power of the fall as represented by the pressure. The pressure used was equivalent to a head of 131 feet, and at a speed of 30 revolutions per minute the motor developed about 5 horse-power. The rotary

Fig. 173.

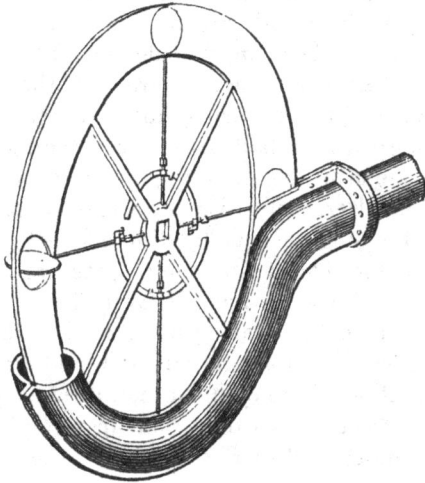

Sir William Armstrong's Rotary Motor.

type of engine seems never to have been actually applied in practice, and the first of Sir William Armstrong's hydraulic pressure engines commercially used was con-structed on the reciprocating principle. "Each engine was composed of two cylinders placed at an angle of 90° to each other and working upon the same crank-pin." "Balanced cylindrical valves were used, and the passages were made very large to keep down the velocity

L L

of the water. Relief valves were also applied to prevent
shock at the end of the stroke." The engines in question
replaced overshot water-wheels at Allenhead, and were
used for crushing and washing lead ore and for raising
material from the shafts. The head of water available
was about 200 feet.

Subsequently this type of motor was adapted to
working with accumulators under very high pressure.
The pattern now generally used by Sir William Armstrong
has two oscillating cylinders with differential pistons
working on over-end cranks placed at right angles to
each other. "The upper side of the piston presents
only one-half the effective area of the lower side, in
consequence of the displacement of the ram, and the
pressure upon that half area is constant, a passage being
always open to the supply pipe. The other side communicates alternately with the pressure and the exhaust,
so that the engine acts by difference of pressure in one
direction and by an unopposed pressure in the opposite
direction." "The action is governed by a two-port slide
valve of a cylindrical form, placed either within the
trunnion or in a prolongation of it, and worked direct by
the oscillation of the cylinder, and a relief valve is applied
to prevent concussion from shut-in water." This form of
motor is illustrated in Fig. 174.[1]

A. Rigg's Hydraulic Motor.

In this motor the inventor has sought to combine the
advantages of a rotary with those of a reciprocating
engine while avoiding their drawbacks. A perspective
view of it is shown in Fig. 175. Its construction

[1] For this and the preceding illustration the author is indebted
to the courtesy of the Institution of Civil Engineers.

Fig. 174.

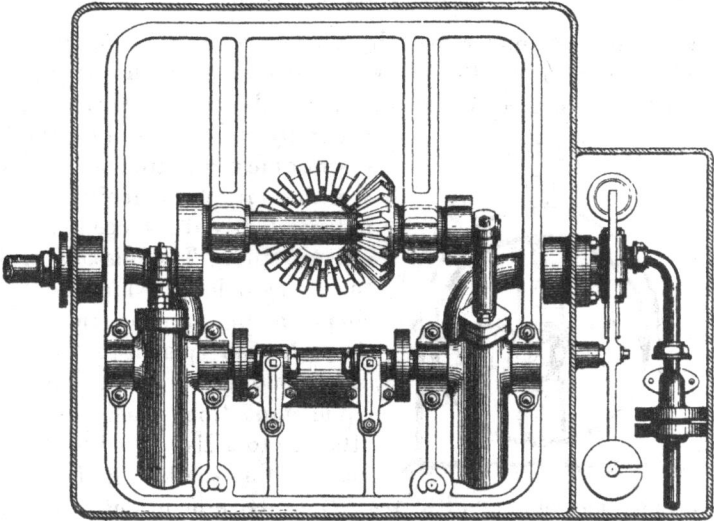

Sir William Armstrong's Motor.

Fig. 173.

Rigg's Motor.

will be best understood by reference to the diagram, Fig. 176;[1] from this it will be seen that the engine consists of four cylinders—in hydraulic motors there are

Fig. 176.

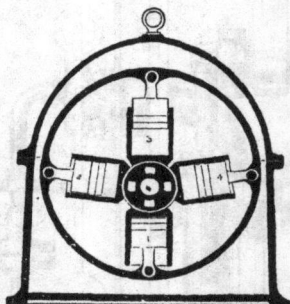

generally only *three*, but the essential features are the same —which revolve freely and independently about a common central stud. The pistons of these cylinders, which are single acting, are connected through the piston-rods to corresponding crank - pins, equidistant from each other, attached to a disc keyed to the main engine-shaft. The latter is not concentric with the stud above-mentioned, but the axes of both are parallel; if, therefore, the disc revolves, so must also the cylinders, but about another centre, and it is clear that to allow this movement every piston must reciprocate relatively to its cylinder, although both have absolutely only a simple rotary motion. In principle the result is the same as though the engines worked in the usual manner on a crank of a radius equal to the eccentricity of the two centres (that of the shaft and that of the stud), the stud taking the place of the crank-pin, but as the cylinders and pistons are respectively balanced among themselves, the effects resulting from the unbalanced dynamic action of reciprocating masses in the ordinary type of engine are avoided.

The stroke of each piston is equal to twice the eccentricity of the two axes, and the power developed may be calculated in the usual way.

[1] Reproduced, with the kind permission of Mr. A. Rigg and the editor, from *The Engineer.*

The turning effort on the disc to which the connecting-rod ends are attached is very constant, being in fact exactly equivalent to that exerted by four (or three) single-acting engines with the cranks evenly distributed, only in this case all act in one plane on a single crank, so that alternating stresses on the shaft are avoided. There are of course no dead centres with this arrangement, so that the engine can be started in any position.

Assuming the rotary motion of the shaft to be uniform, there must necessarily be a variation in the angular velocity with which the cylinders revolve; in moving from the position 3 to 1 in the diagram, there will be an acceleration, while from 1 to 3, during the other half of the revolution, a retardation must take place, but the power absorbed in effecting the acceleration is utilized in the subsequent retardation.

To proportion the expenditure of water to the power required, the hydraulic motors are arranged in such a manner that the centre about which the cylinders revolve can be moved so as to alter its distance from the crank-shaft centre, and thereby the stroke of the pistons. This movement is effected by a hydraulic cylinder and plunger acting on a cross-head carrying the stud. The admission and exhaust of the water under pressure to and from this is controlled by a centrifugal governor, and practically absorbs no power.

Each cylinder has at its lower end a boss with flat faces, through which the central stud referred to passes, and each boss is so arranged that its faces fit accurately against those of the neighbouring bosses, and contain ports for the passage of the water on its way to the cylinders.

The outside boss, at the end at which the inflow takes place, contains the admission and exhaust ports to all the cylinders; as the latter revolve, these ports are opened and closed in succession by a stationary valve, nearly balanced, which works against the outer face of the boss.

These motors run at a high speed. One of them, driving a dynamo in connection with glow-lamps, works regularly at 250 revolutions a minute, and can be made to revolve at a much higher speed if required.

It may not at first sight be obvious in what way the effects produced by rapidly reciprocating masses in ordinary engines are avoided in the case of Mr. Rigg's motor, since a relative reciprocation and reversal of the motion of the plungers actually occurs. Consideration will however make it clear, that the variation in the angular velocity with which the cylinders revolve takes the place of the acceleration and retardation of the reciprocating masses in an ordinary motor; the smaller the eccentricity of the central stud relatively to the crank centre, the less is the amount of relative reciprocation and variation in the angular velocity.

A. Schmid's Hydraulic Motor.

This was one of the first motors of the class under consideration designed to meet the wants of small manufacturers and for domestic purposes, and is in many respects the best. Its construction is shown in Figs. 177, 178, and 179.[1]

The cast-iron cylinder is closed at the end next the crank by a cover furnished with a long and massive stuffing-box, which serves as a guide for the piston-rod.

---

[1] For these and many of the subsequent illustrations of hydraulic pressure engines the author is indebted to Knoke's work, *Kraftmaschinen des Kleingewerbes.*

Fig. 177.

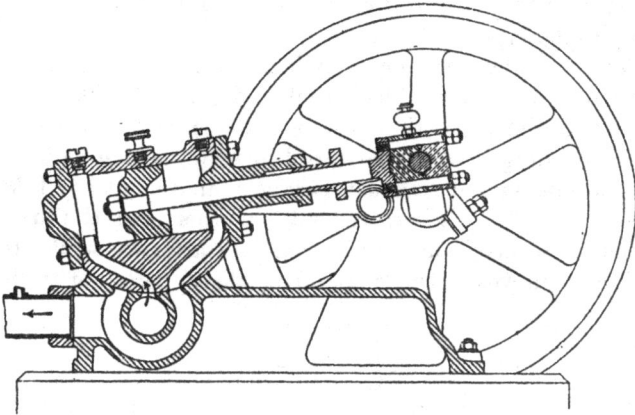

Fig. 178.

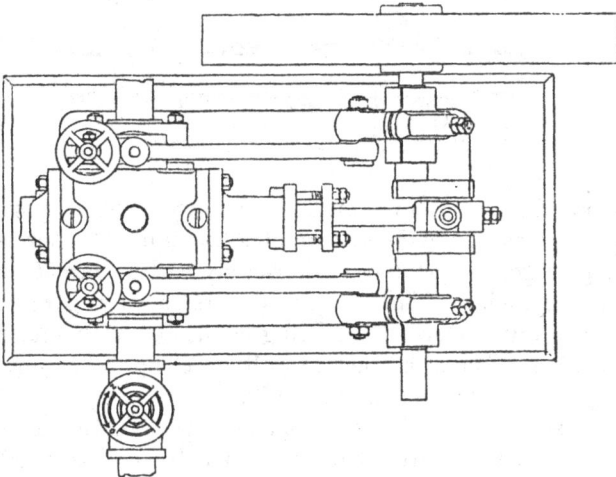

Schmid's Hydraulic Motor.

This is necessary, owing to the engine being of the oscillating type.

The cylinder rests on a cylindrical surface, about the axis of which the oscillation takes place, and this surface forms at the same time the valve-face and works in a corresponding hollow in the bed-plate, which contains the admission and exhaust passages, and in reality takes the place of a slide-valve, only in this case the usual order is reversed, and the valve is stationary while the cylinder moves. The valve has three ports, of which

Fig. 179.

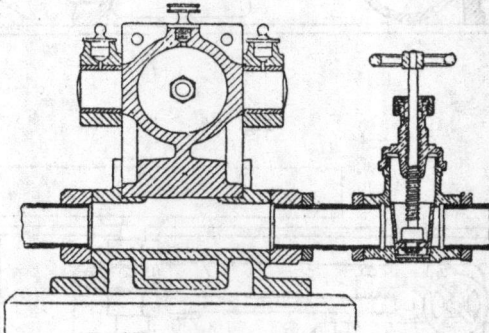

the middle one is in communication with the admission pipe, and as the cylinder oscillates communicates through the passages in the latter alternately with opposite ends of the cylinder. Through the two remaining ports the exhaust takes place, both being connected with one off-flow pipe. The construction will be clear from the accompanying illustration, Fig. 179.

On either side of the cylinder are necks, the axis of which coincides with that about which oscillation takes place. These necks have their bearings in two levers pivoted at one end to the frame near the crank-axle, and

by means of a hand-wheel and screw acting on the other
end of each lever the cylinder can be pressed down upon
its bearing surface. In order to lubricate or inspect
the valve-face the cylinder can be lifted quite off the
bed by unscrewing the hand-wheel.

The piston is solid. In small machines it is accurately
turned and ground in; in larger, a leather collar is used
to make it water-tight.

A small copper air vessel is used in connection with
the motor, and for larger sizes a small air-pump to
replace loss by leakage and absorption of the air by the
water.

The capacity of the air vessel is from 2 to $2\frac{1}{2}$ times
the volume swept through by the piston.

Experiments made by Professors Zeuner and Kronauer
show that an efficiency as high as 90 per cent. can be
obtained with Schmid's motors; the maker himself
guarantees 80 per cent.

Hydraulic Motor of Wyss and Studer.

This engine, illustrated in Figs. 180, 181, 182, and 183,
is also of the oscillating type, and differs from Schmid's
chiefly in the fact that the valve surface is flat instead of
curved.

The cylinder rocks on two necks, which have their
bearings in the main frame. On the under side of the
cylinder a chamber is formed, in the vertical sides of
which are the ports for the distribution of the water.
The fluid enters simultaneously from both sides into this
chamber, and thus any one-sided pressure is avoided.
A vertical partition (the prolongation of which would pass
through the axis of oscillation) divides the chamber into
two parts, each of which communicates through one

Fig. 180.

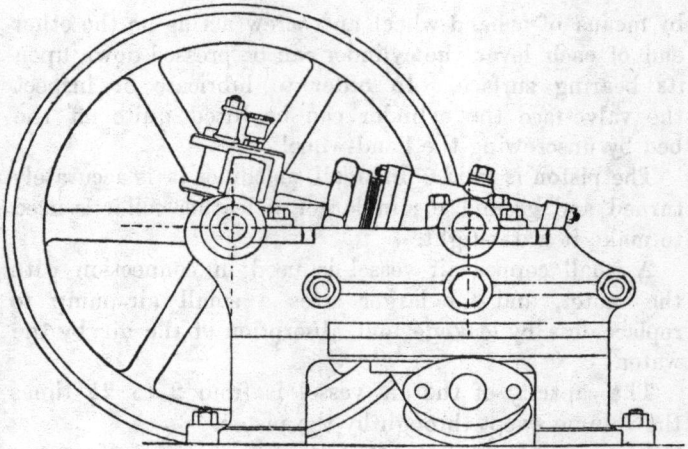

Fig. 181.

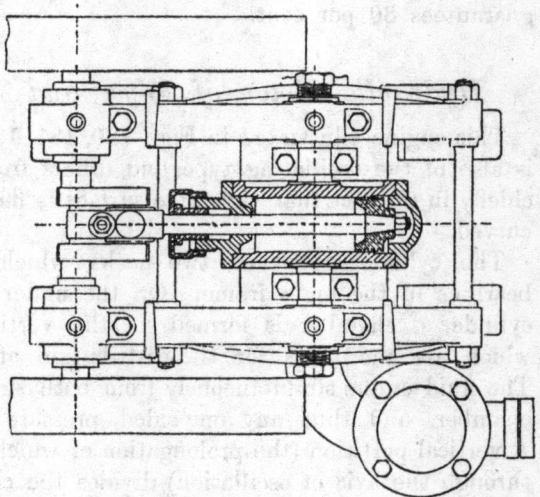

Motor of Wyss and Studer.

opening with one end of the cylinder, and through two lateral ports with the admission and exhaust passages alternately. The vertical surfaces, on the outside of the cylinder chamber containing the ports, work against corresponding faces on the sides of two valve-chests, attached to the main frame, through which the water enters and leaves the cylinder. Fig. 183 shows a section through one of these chests, which has three divisions. Through the middle the water is admitted, and by the two side passages exhausted.

Fig. 182.   Fig. 183.

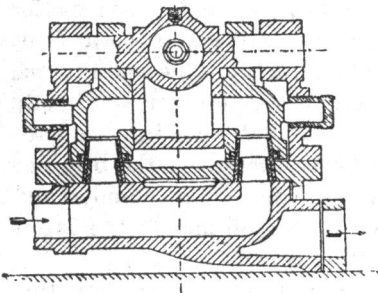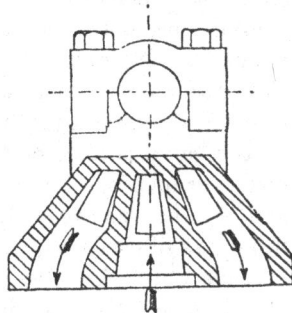

The area of each port in the valve-face is about 0·3 of the sectional area of the cylinder.

The valve-chests above referred to can be pressed by means of screws on both sides against the bearing surfaces on the cylinder chamber, so as to make a water-tight joint. Below the main frame is a casting containing the admission and exhaust passages to which the inlet and off-flow pipes are connected. This casting is so arranged that it can be turned round through 180°, so that, according to circumstances, the connections can be made on either side of the motor. The joint between the frame casting and each valve-chest is made tight by

means of two conical gun-metal tubes, one within the
other, which pass easily through the frame-plate, and
are forced by the pressure of the water against the sides
of the openings through which they project. The valve-
chests have bearings at the upper ends through which
the cylinder necks pass, to insure their being kept in the
proper relative position.

Haag's Hydraulic Motor.

In this hydraulic engine—of which Figs. 184 and 185
show the construction — the necks about which the
cylinder oscillates contain the ports and passages for
admission and exhaust, which takes place on both sides.
The water enters and flows off through the main frame,
as in Schmid's motor, but the necks do not rest directly
on the valve surface of the frame, but upon intermediate
bushes, which serve as the lower brasses of the bearings
and can be renewed when worn out. The cylinder and
frame are of cast-iron, the piston of gun-metal; the latter
is made water-tight by a leather collar.

The connecting-rod head is guided in a continuation
of the cylinder casting so as to relieve the stuffing-box
of all lateral pressure, and thus serves the purpose of a
cross-head.

It has been claimed as an advantage of this motor over
that of Schmid, that the surfaces for friction are smaller,
but as the friction depends chiefly on the total pressure
and very little on the area of the bearing surface, it is
desirable to make the latter as large as practicable in
order to reduce wear.

The following are the results of experiments made at
Erfurt in 1878 with one of Haag's hydraulic motors,
having a diameter of 50 millimètres (2 inches) and 75

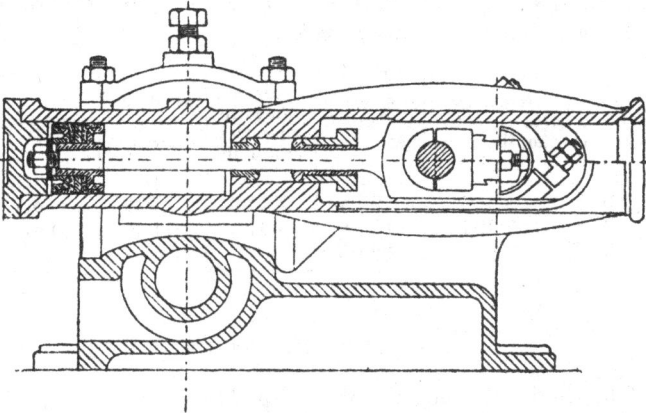

Fig. 184.

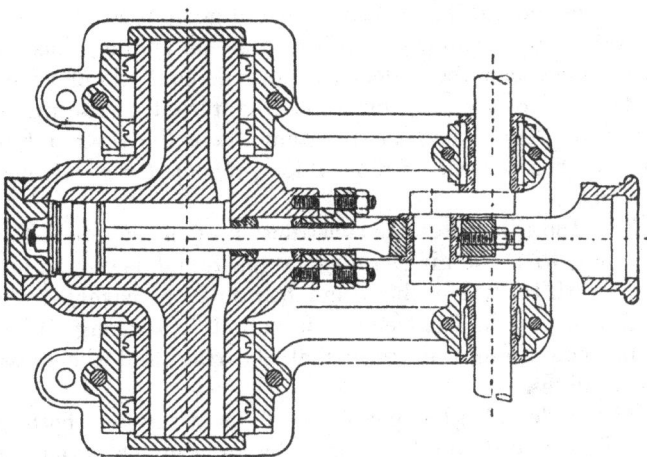

Fig. 185.

Haag's Hydraulic Engine.

millimètres (3 inches) stroke, as given in Knoke's work on *Kraftmaschinen des Kleingewerbes.*

Number of revolutions per minute.	Horse-power as measured on the brake.	Pressure in equivalent head, feet.	Water consumed in cubic feet per minute.	Available horse-power.	Efficiency.
351	0·406	98·4	3·920	0·728	0·56
372	0·451	114·8	4·132	0·900	0·50
335	0·406	123·0	—	—	—

Philip Meyer's Engine.

The hydraulic engine of Philip Meyer is of a special type, deserving notice chiefly on account of the fact that the admission of water can be cut off before the full stroke is completed.

The cylinder of the motor, which in general construction resembles an ordinary stationary steam-engine, has at either end a chamber filled with air, the contents of which at the end of a stroke must be compressed to the initial pressure. The water as it enters the cylinder therefore encounters an elastic cushion which prevents the occurrence of shocks.

Fig. 186 illustrates the construction of the engine. The cylinder, with the air chambers at each end, is cast in one piece with the valve chest, and has only one cover, at the back end; at the front end it is bolted to a strong forked frame containing the crank-shaft bearings and the cross-head guides.

The slide valve has parallel surfaces at top and bottom, and is balanced and kept tight by an adjustable plate on the back. This plate has recesses corresponding to the ports in the cylinder, which have the effect of making the

area of admission for a given opening of the valve twice as
great as it would otherwise be, and at the same time serve
to relieve more completely the pressure on the valve.

Fig. 186.

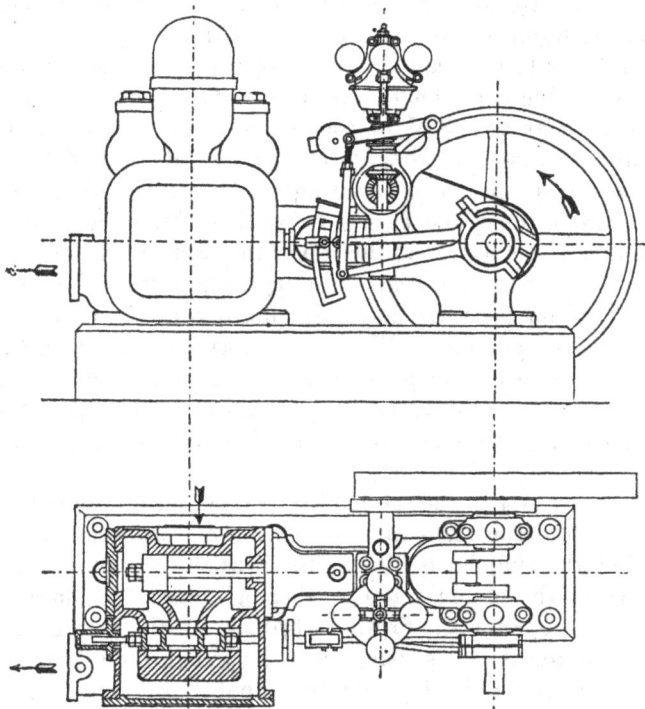

Meyer's Engine.

This arrangement, although it has not been found to
answer with steam, gives very satisfactory results with
water where the latter carries no sand or grit.

The valve is worked by an ordinary link motion, with open eccentrics (Stephenson's).

The governor, which is of a powerful type, acts on the link and regulates the cut-off, the peculiar feature of this motor which has yet to be explained.

The air vessel is placed on the top of the valve chest, and the water enters the latter from below.

In order to replace any air lost by leakage or absorption, the following arrangement is employed : Over each end of the air chambers, and in connection with the latter, are small expansion air vessels; at the side of each of these vessels is a suction or snifting valve, which, when the pressure in the cylinder sinks below that of the atmosphere, allows air to enter. At the top of the same air vessels are check valves communicating through a pipe with the main air vessel on the valve chest, which are kept down on their seats by the pressure. When the pressure of the air in the cylinder exceeds the pressure in the main, the air escapes through the check valves, but, instead of being lost in the atmosphere, passes into the main air vessel to compensate for loss.

Coming now to the action of the motor, let it be assumed that before the slide valve opens the air chamber at one end of the cylinder is filled with air at the pressure of the water in the main; water is then admitted, and does its work throughout the forward stroke. At the commencement of the return stroke, when the exhaust takes place, the pressure sinks very rapidly to that of the atmosphere. At a certain point before the end of the return stroke is reached, the exhaust passage is closed, and the contents of the cylinder are compressed during the remainder of the stroke, until the volume is reduced again to that of the air chamber. The work required for this compression, when water is admitted throughout the whole stroke, is entirely lost. The matter is however different when the cut-off takes place before the stroke is completed; in that case, up to the point of cut-off, the water does its work at —approximately—constant pressure, and during the rest of the stroke the compressed air expands, giving out again —when the proportions are correct—the energy required for compression. During the latter period the water in the cylinder merely acts as an extension, so to speak, of the piston. Under these conditions no loss is involved in the use of the air cushion, but it is necessary that for a given maximum pressure and air-chamber volume the latter should bear a definite ratio to the volume behind the piston at the moment the cut-off takes place. A deviation from this ratio entails loss. It will be seen that Meyer's motor is in reality a combined air and water engine, in which the air is compressed and expanded.

Theory of Meyer's Engine.

As in the case of this motor not merely the action of water under pressure, but the expansion of air, is concerned, it is necessary to undertake a separate theoretical investigation.

Let p_1 be the initial pressure, v_1 the volume of each air chamber, V_1 the volume swept through by the piston up to the point of cut-off, V the total volume swept through, and V_2 that at which the exhaust closes (not including v_1); further, p_o the pressure of the atmosphere.

As a result of the contact with a large body of water, the expansion or compression of the air in the cylinder may be assumed to take place at constant temperature, according to the law expressed by

$$p\,v = \text{constant,}$$

M M

where p and v are respectively the corresponding pressure and volume at any given moment.

At the time the exhaust closes all water has been expelled from the cylinder, and there remains only air of atmospheric pressure p_o and the volume $v_1 + V_2$; this is compressed to the volume v_1 of the air chamber and pressure p_1 of the main; hence there follows

$$p_1 v_1 = p_o (v_1 + V_2)$$

and

$$v_1 = \frac{p_o}{p_1 - p_o} V_2 \, ;$$

from this it is clear that for a given pressure p_1 the exhaust must always close at the same volume V_2 in order to obtain the necessary amount of compression.

The work required for this compression is

$$W = (p_1 - p_o)\, v_1 \, log.\ hyp. \ \frac{V_2 + v_1}{v_1},$$

$$W = (p_7 - p_o)\, v_1 \, log.\ hyp. \frac{p_1}{p_o}$$

and when the admission takes place throughout the whole stroke, Fig. 187, this work is lost; in order to minimize this loss v_1 should be as small as possible. Under these conditions it will be seen that to avoid shocks efficiency has to be sacrificed to some extent. The efficiency ϵ is expressed by

Fig. 187.

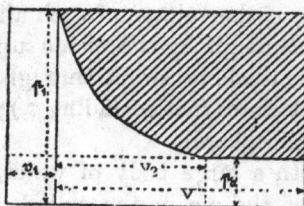

$$\epsilon = \frac{(p_1 - p_o)\, V - W}{(p_1 - p_o)\, V} = 1 - \frac{v_1}{V}\ log.\ hyp.\ \frac{V_2 + v_1}{v_1}.$$

When the cut-off takes place at the volume V_1, Fig. 188, before the end of the stroke, the following relations exist between the pressures and volumes:—

$$p_1 v_1 = p_o (v_1 + V - V_1),$$

and
$$p_o (V_2 + v_1) = p_1 v_1,$$

whence
$$V - V_1 = V_2; \qquad \text{consequently}$$

$$v_1 = \frac{V - V_1}{p_1 - p_o} p_o = \frac{p_o}{p_1 - p_o} (1 - E) V,$$

where E is the cut-off ratio $\dfrac{V_1}{V}$. It will be seen from this that v_1 the volume of the air chamber is dependent on the expansion ratio, and if this is changed with a constant value of v_1, the action of the engine ceases to be quite correct.

Fig. 188.

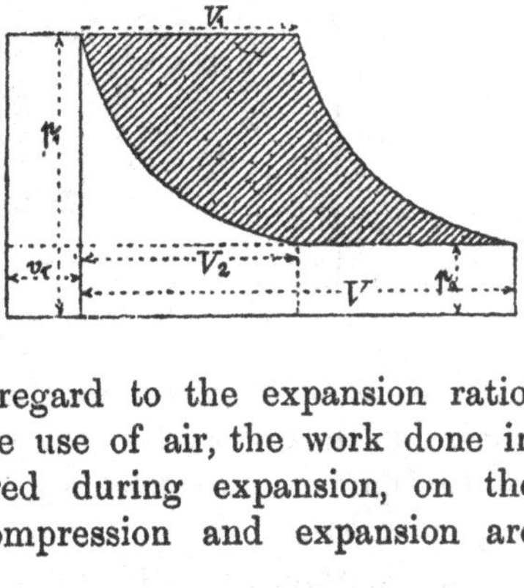

In the case just considered, when v_1 is correctly proportioned with regard to the expansion ratio, no loss is incurred by the use of air, the work done in compression being restored during expansion, on the assumption that both compression and expansion are isothermal.

When the ratio of expansion is changed, while v_1 remains unaltered, the process occurring is as follows: Supposing 0·5 to be the correct point of cut-off for a motor working with a pressure of say 60 lbs. per square inch absolute, and that this is suddenly changed to 0·8, the air, instead of expanding from v_1 to $v_1 + 0·5\ V$, can only expand to the volume $v_1 + 0·2\ V$, and consequently the corresponding pressure is above that of the atmosphere at which compression commenced. When the exhaust is opened this excess of pressure is blown off, and compression commences again at a volume of $v_1 + 0\cdot2\ V$, *less* than that of the previous stroke owing to the change in the cut-off. When the volume of the charge has been reduced to v_1 the pressure will only be increased in the ratio of $\dfrac{v_1 + 0\cdot2\ V}{v_1}$ (instead of as before $\dfrac{v_1 + 0\cdot5\ V}{v_1}$), and will be *less* than 60 lbs., so that on

Fig. 189.

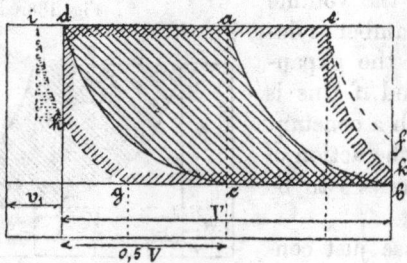

the admission of water at the latter pressure from the main the air has to be compressed by the water before equilibrium is established, and the work required for this purpose is *lost*.

The diagram, Fig. 189, illustrates the process; *a b c d* is the original indicator diagram with a cut-off at $0\cdot5\ V$; *e* is the new point of cut-off, and *e f* the expansion curve for the latter, with which the terminal pressure is in excess of that of the atmosphere by *b f*.

At *g* compression commences under the new conditions; *g h* is the compression curve up to the end of the back stroke, and *h i* represents the further compression by the

inflowing water. The area shaded with dotted lines is
equivalent to the work done by the water in compressing;
e i is proportional to the total quantity of water used,
while *e d* represents that usefully employed, the difference
d i being wasted. *d e k b g h* is the effective diagram for
the new point of cut-off when regular working has been
established.

When the cut-off takes place at an earlier point than
the correct one, the pressure at the end of compression
on the return stroke rises above that in the main, and
to meet this the check valves already described as
communicating with the main air vessel are provided.

Fig. 190.

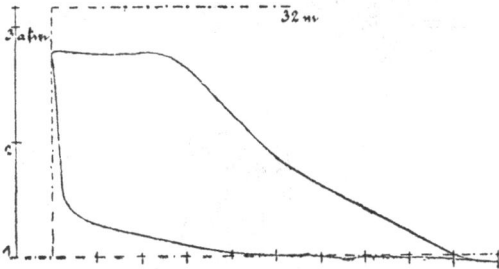

Fig. 190 shows an indicator diagram from Meyer's
engine.

Hastie's Motor, with Variable Stroke.

About the year 1879 Mr. John Hastie of Greenock
brought out a water engine designed to remove the chief
objection to hydraulic pressure motors, viz. their incapacity
for economic regulation. In Hastie's machine the stroke
is variable, and adapts itself automatically to the power
required. Figs. 191, 192, 193, 194, 195, and 196 show

Fig. 191.

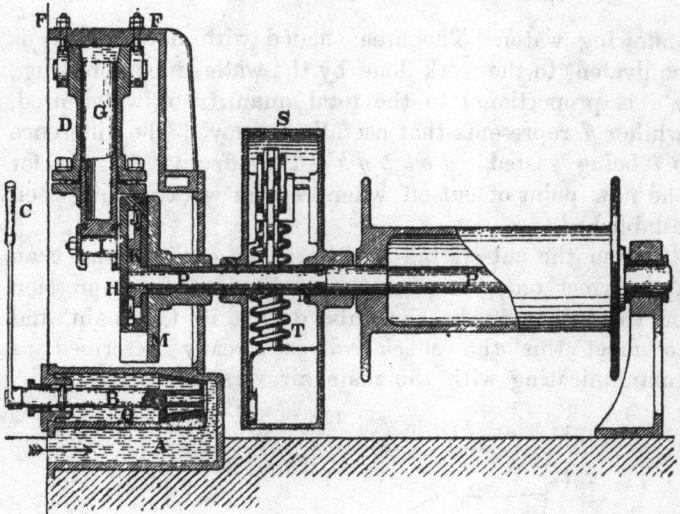

Fig. 192.

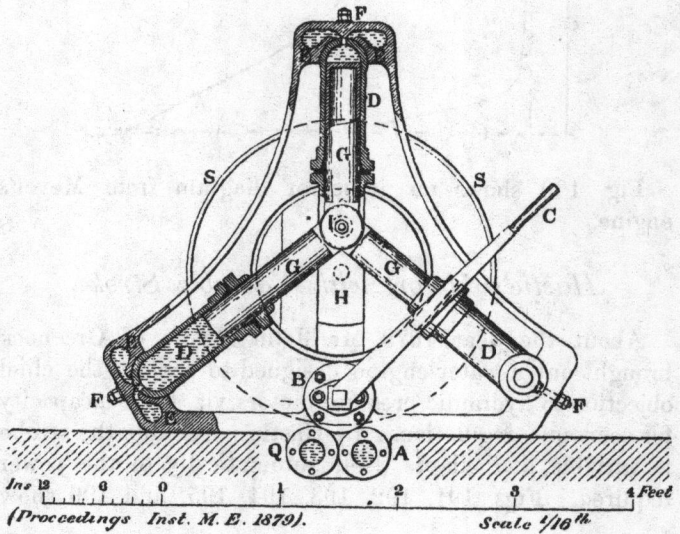

Scale 1/16th

the construction as applied to a hoist. There are three oscillating cylinders *D*, to which water is admitted through passages at *E*, acting alternately as admission and exhaust ports; the oscillating ends of the cylinders take the place of valves in distributing the water. The supply pipe *A* communicates with the cock *B*, worked by the handle *C*, which controls the action of the hoist, and can be used

Fig. 193.

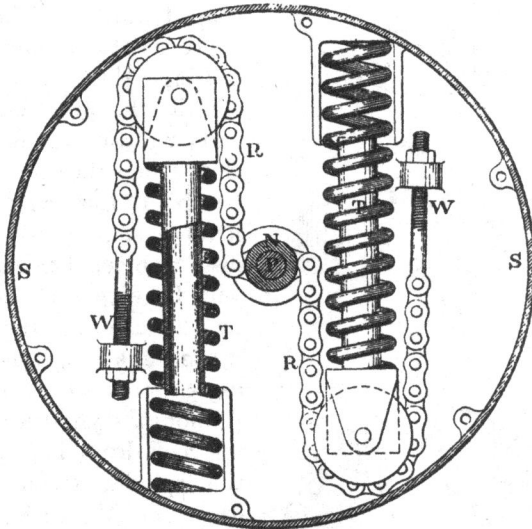

as a reversing valve at the extreme positions and a brake in the central position. In the latter case both ports are placed in connection with the exhaust passage *Q*, which is made with a bend so as to contain at least as much water as will fill the three cylinders.

All the plungers of the three cylinders are connected to the same crank-pin *I*, which, however, is not rigidly

secured to the crank disc *M*, but is attached to a block
H, sliding in a radial
groove in the latter. At
the back of this block,
revolving on pins projecting from it at either end,
are two small rollers *J* and
L, which run on the circumference of a peculiarly
shaped cam *K*. The cam
is keyed to a spindle *P*
passing through a con-
centric hollow shaft *N*
carrying the crank disc.
Supposing the spindle to
be held fast while the
hollow shaft and crank
disc revolve, the block *H*
will be displaced radially
by the cam *K*, and with
it the crank-pin, thus altering the stroke of the pistons.
A hollow cylindrical cas-
ing *S* or drum is keyed to
the spindle *P* and runs
loose on *N*; within this
are two rollers carried in
forks at the ends of hollow
rods. These rods with the
rollers are pressed out-
wards towards the circum-
ference by helical springs
T, while over each roller
passes a chain *R*, one end of which is attached to a snug

Fig. 194.

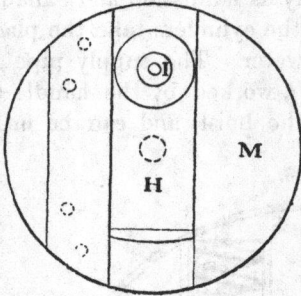

Fig. 195.

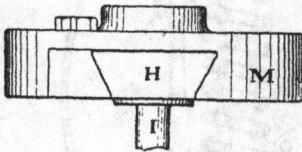

Fig. 196.

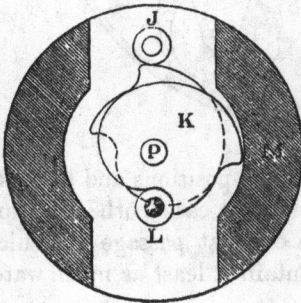

W projecting from the side of the drum, the other to the hollow shaft *N*.

When water is admitted to the cylinder under pressure the crank disc begins to revolve, and with it the hollow shaft *N*, while the spindle *P* carrying the cam *K* is held fast by the resistance of the load applied through the hoisting chain to the circumference of the chain barrel. The result of this is that the chains *R* are wound up on *N*; at the same time, owing to the motion of the crank disc relatively to the cam, the block *H* with the crank-pin is pushed outwards and the stroke increased. The winding up of the chains compresses the springs, and this compression and the simultaneous increase of the stroke go on until the resistance of the springs balances that of the pulley, and then the latter is driven round and a state of equilibrium established, which lasts as long as no change occurs in the load or pressure. If the load is increased, a motion of the crank disc relatively to the cam again takes place, until the turning moment on the crank equals that of the load on the pulley by the alteration of the crank radius or stroke.

In engines constructed to work with very high pressures the springs are not employed, but the arrangement shown in Figs. 197 and 198 is used instead. In this case the springs are replaced by two water rams *U*, which are in connection with the supply pipe through the centre of the shaft *P*, and are consequently under the same pressure as that used to work the engine. The chains *R* act in the same way as with springs, but instead of being wound directly on the hollow shaft *N*, they are wound on cams *V*; in this way greater power is required to force back the rams in proportion as the chains *R* act at an increasing distance from the centre of the shaft.

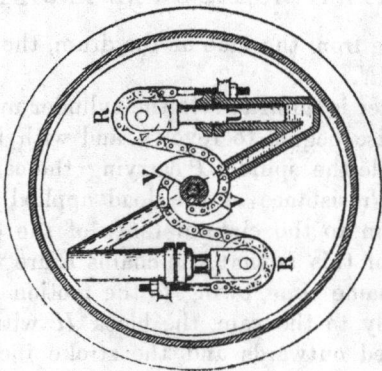

Fig. 198.

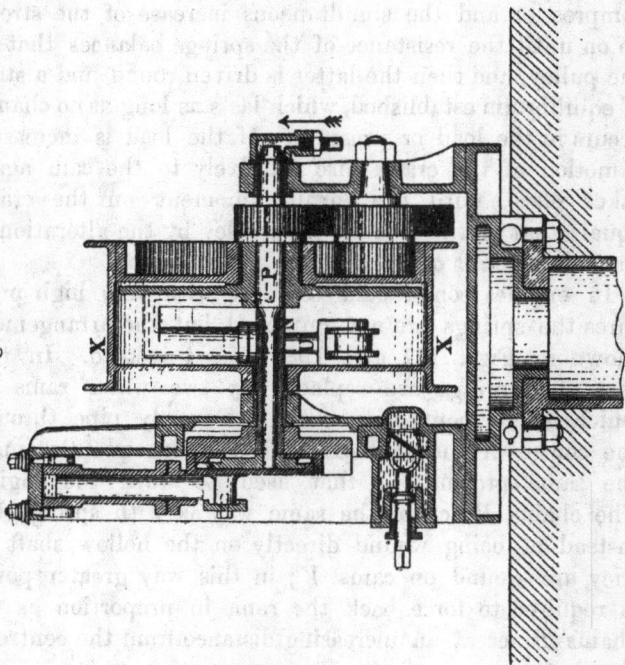

Fig. 197.

The following are the results of experiments made by Mr. Kinniple, chief engineer to the Greenock Harbour Trustees, on a small hoist for the Greenock Infirmary ; the lift was 22 feet and the pressure in the cylinder 80 lbs. per square inch :—

Weight lifted $\left\{ \begin{matrix} \text{Chain} \\ \text{alone} \end{matrix} \right\}$ 427 633 745 857 969 1081 1193

Water used, $\left. \right\}$ $7\frac{1}{2}$ 10 14 16 17 20 21 22
  gallons.

From this it will be seen that the water consumed is approximately proportional to the work done.

General Remarks.

Hydraulic pressure engines are suited for utilizing comparatively small quantities of water at higher pressures than are generally advantageous with turbines. Whereas the speed of a turbine is determined by the velocity of flow, which in its turn depends on the head, irrespective of the quantity of water supplied, in a hydraulic pressure engine with a given supply and head the speed may, within reasonable limits, be made high or low as desirable, by giving the piston a smaller or greater diameter and inversely corresponding stroke. Efficiency is promoted by a low piston velocity, but of course with high pressures the speed may, for a given efficiency, be greater than with lower pressures. Probably no other class of motor, except an electric motor, can be so efficient as a hydraulic pressure engine, but the latter suffers under the disadvantage, as generally constructed, that any regulation of the power developed is accompanied by a serious reduction in the efficiency. Such regulation can only be effected, where a constant speed is required, through a reduction of the pressure on the piston by throttling the supply of water. This means purposely wasting the available energy with the object of reducing the pressure. If, for example, the resistance to the piston decreases to one-half of its maximum value, and no change in the speed is allowed, the stop-valve orifice is reduced so that the velocity of flow through it is increased, and the sudden change from this velocity to that of the piston results in a greater loss of energy than occurs with the valve orifice fully open. The effect shows itself in a diminution of pressure, which must be adjusted to the resistance to be overcome, that is, in the assumed instance, reduced to one-half the pressure employed at full power. As the loss of energy is proportional to the reduction of pressure the disadvantages of this method of regulation are very evident.

The hydraulic motors of Rigg and Hastie, previously described, have been designed to overcome the defect in question by varying the *stroke* of the engines according to the power required. In this way the pressure on the piston is maintained practically constant, while the quantity of water used is varied, and no loss of efficiency is necessarily entailed by the variation.

In some places on the continent of Europe small hydraulic pressure engines are much used in connection with the water from the town mains. In mines hydraulic engines of a larger kind are often employed, on account of the comparative ease and safety with which the water power can be transmitted through pipes to different distant parts of the workings.

Water under high pressure is in many respects a good medium for the transmission of power to considerable distances, but its application to that purpose does not come within the scope of this work.

There are, of course, many varieties of hydraulic pressure engines besides those described, but the latter, it is believed, include the leading types with the exception of the hydraulic ram, which is used exclusively for pumping, and does not come within the category of general motors.

INDEX.

A.

ABSOLUTE path of stream, 19, 42, 160, 242.
velocity, 10, 14.
Air vessel, 501.
capacity of, 521.
American turbines, 192, 197, 198, 266, 462.
dimensions of, 469.
America, water power in, 2, 198. *See* American Turbines.
Analysis of results. *See* Descriptions of and experiments with turbines, *also* Table A *and* American turbines.
Angles, effect of, on flow, 69.
of flow, construction of, 134, 142, 147, 227.
of flow, 13, 61, 118, 134, 142, 227, 258, 387.
of outflow, influence on efficiency, 132, 167.
of outflow, mean, 124.
positive and negative, 69.
Areas, calculation of, 135.
effective, 120, 136.
measured, 117, 122, 225, 257, 375.
of ports in hydraulic pressure engines, 523.
of stream, 116.
ratio of, 107, 118.
Armstrong's hydraulic motor, 512.
Available energy, 3, 50, 503.
for wheel, 54, 105, 214.

Available head, 48.
horse-power, 131.
Axial-flow Girard turbine, 43, 224.
impulse turbines, 224.
impulse turbines, description of, 43.
outflow, 103.
turbines, 25, 36.
turbines, application of formulas to, 128.
turbines, comparison with radial and mixed-flow turbines, 192, 471.

B.

Bearings, collar, 272.
for turbine shafts, 271.
subaqueous step, 272.
overhead (Fontaine), 272.
Best speed of turbines, 73, 228, 455.
as proved by experiment, 455.
Bidone, experiments on flow over weirs, 322.
Blackwell, experiments on flow over weirs, 324.
Boileau, experiments on flow over weirs, 324.
Boott, experiments with turbines at, 384.
Boyden diffusor, action of, 264.
loss in, 203.
on the use of, 195, 198, 200.
Boyden turbine, experiments with, 400.

Brake, measurement of power by. *See* Descriptions of and experiments with turbines.

Braschmann's formula for flow over weirs, 318, 464, 490.

Buckets, depth of, 124.
loss in, 84, 92, 217, 241.
number of, 126.
of axial Girard turbine, 45, 229.
of impulse turbines, 43, 211, 227, 236.
of turbines, 24.
ventilation of, 43, 211.
width of, 115, 137, 225, 259.

C.

Callon, 39.

Canada, turbines in, 266.

Castel, experiments on flow over weirs, 320, 322, 324.

Centrifugal action in parallel-flow turbines, 247, 250.
corrections for, 252.

Channel, conducting, 48.

Characteristics of American turbines, 463.

Classification of turbines, 24.

Clearance space, leakage through, 84, 91.

Co-efficients, experimental, 64, 82.
of contraction, 120, 136.
of flow, 107, 109, 225 ; also Tables A and B.
of friction, 87.
of speed, 112.

"Collins" turbine, experiments with, 396.

Combined-flow turbines, 24, 26, 195, 463.

Comparison of axial with radial and mixed-flow turbines, 192, 465, 471.
of experimental and calculated results, 336, 356, 455.
of mixed flow with radial flow, 195.
of reaction with impulse turbines, 265.

Compound turbines, 270.

Concentric subdivision, regulation by, 184.

Construction of vanes of turbines. *See* Vanes.

Continuity of flow, 9, 49.

Contraction, co-efficients of, 120.

Correction of centrifugal action in impulse turbines, 252.
of vane angles for different diameters, 145, 239.

Current meters, 313.

Curve of efficiency, 76.
of kinetic energy, 499.
of loss in guides, 499.
of loss in wheel, 499.
of pressure difference, 76, 499.
of unutilized energy, 190, 499.
of velocity of flow, 76.
of work, 76.

Curved vanes, action of water on, 12.

Curve for Haenel's turbine, 361.
of vanes, 138, 237. *See* Vanes.

D.

Dam, 5.

Data for designing turbine, 131.

Deacon, experiments on flow over weirs, 326.

Definitions, 24.

Deflection of stream, 12.

Depth of buckets, 121, 124, 229.
of immersion, effect of, 418.
of passages, 121, 124, 229.
of stream, 115.

Descriptions of and experiments with turbines, 335.

Design of reaction turbines, 99.

Diameter of bearing spindles, 275.
of turbines. *See* Radii.
small, conditions for, 199.

Diffusor, action of, 264.
loss in, 205.
on the use of, 200, 203.

Dimensions, connection between various, 115.

Dimensions, effect of on performance of turbines, 164, 261.
independently selected, 131.
method of calculating, 135.
of American turbines, 469
of impulse turbines, 225.
Direction of inflow. *See* Angles of flow.
of outflow, observation of, 378.
Distribution of loss, 190.
of water in turbines, 187.
Donkin and Salter, on measurement of water over weirs, 318, 320.
Draft-tube. *See* Suction-tube.
Drowned, turbine working, 25.

E.

Effect of dimensions and proportions on performance of turbines, 164, 261.
of increasing width of buckets on performance of axial impulse turbines, 236.
Efficiency, 51, 72.
curve of, 76.
effect of regulation on, 74, 182.
effect of speed on, 72, 76.
hydraulic, 79.
influence of dimensions on, 132, 167, 261.
influence of reaction on, 134.
maximum, 73.
of hydraulic pressure engines, 511.
of impulse turbines, 221.
of properly designed turbines, 113.
Energy, available, 3, 48, 50.
available for wheel, 54, 105, 214.
curve of unutilized, 190.
due to residual velocity, 50.
kinetic, 3, 499.
losses of. *See* Loss.
potential, 3.

Energy, transformation of, 9.
unutilized, 50, 74, 85, 94, 190, 223.
ways in which applicable, 3.
Engines, hydraulic. *See* Hydraulic pressure engines.
Entry without shock. *See* Inflow.
Equations for velocity of flow, 63, 225.
general, 50.
Escher, Wyss & Co., design of, for Niagara turbines, 445.
turbines of, at Schaffhausen, 428.
Examples, numerical, 283.
Exhibition at Philadelphia, turbine trials at, 464.
Experimental and calculated results, comparison of, 455.
co-efficients, 64, 82.
turbines, 67, 77, 78, 337, 351.
Experiments on inflow with least resistance, 456.
with Boyden turbine, 400.
" Collins " turbine, 396.
" Hercules " turbine, 401.
" Humphrey " turbine, 406.
" Little Giant " turbine, 410.
"Swain " turbine, 446.
turbine at Immenstadt, 424.
turbines, 67, 77, 78, 335.
turbines at Boott cotton mills, 384.
turbines at Goeggingen, 365.
turbines at Olching, 368.
turbines at Varzin, 413.
turbines of Haenel, 351.
turbines of Rittinger, 67, 77, 337.

F.

Faesch and Piccard, turbines of, for Niagara Installation, 443.
Fliegner, experiments by, 207, 456.
Floats, measurement of water by, 315.

Flow, angles of, 13, 61, 118, 124, 132, 149, 237.
   co-efficient of, 107, 109, 225.
   of water in pipes, &c., 7.
Fontaine bearing, 272.
Forces acting on vane, 12.
Form of vanes for impulse turbines, 237.
   of vanes for reaction turbines, 140.
Formulas, summary of, 126, 235.
Fourneyron turbine, 29, 183.
   at St. Blaise, 455.
   triple, 31.
Francis, experimental co-efficient of, 89.
   experiment with diffusor, 206.
   experiments on flow over weirs, 319, 326, 328.
   experiments with turbines, 372.
   test by, 197.
   turbine, 32, 48, 463.
Friction, co-efficients of, 87.
   of shafts, 275.
   resistance due to, 50.
Fteley and Stearns, experiments on flow over weirs, 319, 326.

G.

Girard, 39.
   turbines, 39, 211. *See* Impulse turbines.
   turbines, formulas for, 235.
Goeggingen, turbine at, 182, 365.
Governors for turbines, 279.
Guide apparatus, 27, 257.
   passages, loss in, 82, 88, 215.
   passages, obstruction of, 119, 225.
   passages, regulation by closing, 173.
   passages, width of, 115, 230, 232, 259.
   vanes, 25. *See* Vanes.
   vanes, movable, 157, 180.

H.

Haag, hydraulic motor of, 524.
Haenel, experimental co-efficients of, 89.
   experiments with turbine, 351.
Hastie, hydraulic motor of, 533.
Height, maximum possible in suction-tube, 55.
Helical vane surfaces, 187.
   unutilized energy with, 189.
"Hercules" turbine, analysis of results, 481.
   experiments with, 401, 468.
Holyoke testing station, 391.
   Water-power Company, 330.
Horse-power, available, 131.
Humphrey turbine, 183, 197, 406, 468.
"Hurdy-gurdy," the, 435.
Hydraulic efficiency, 79.
   motors, general principles, 6.
   power, sources of, 4.
Hydraulic pressure engines, 500.
   available energy of, 503.
   efficiency of, 511.
   general remarks, 539.
   general theory, 501.
   indicator diagram of, 502.
   losses of energy in, 508.
   of A. Rigg, 514.
   of A. Schmid, 518.
   of Haag, 524.
   of Hastie, 533.
   of Ph. Meyer, 529.
   of Sir William Armstrong, 512.
   of Wyss and Studer, 521.
   rotary, 512.
   useful work of, 506.
Hydrodynamic equation, 8.
Hydropneumatization, 278.

I.

Immenstadt, experiments with turbine at, 424.
Immersion, effect of, 418.
Impact. *See* Shock.
Impulse turbines, 25, 39, 211.
   axial, 43, 224.

Impulse Turbines, centrifugal action in, 250.
dimensions of, 226.
essential principle of, 211.
on the difference between impulse and reaction turbines, 262.
radial, 256.
relative velocity of flow, 218.
summary of formulas for, 235.
theory of, 213.
with partial admission, 25, 234.
with suction-tube, 279.
Indicator diagram of hydraulic pressure engines, 502, 530, 533.
Inflow, direction of. *See* Flow, angles of.
with least resistance, experiments on, 456.
without shock, 14, 101, 456.
Introduction, 1.
Inward-flow combined turbines, 26.
radial turbines, 26.
turbines, difference between their action and that of outward-flow turbines, 195.

J.
Jonval turbine, 27, 36, 119.
construction of vanes, 140.
double, 37.
See also Axial turbines.

K.
K_1 co-efficient of flow, 107, 109, 225.
K_2 co-efficient of speed, 112.
Kennedy, experiments on flow over weirs, 320.
Kinetic energy, 3, 499.

L.
Leakage, loss by, 84, 91.
Lehmann, summary of losses by, 95.

Lesbros, experiments on flow over weirs, 319, 320, 324.
" Little Giant " turbine, experiments with, 410.
Loss by friction in suction-tube, 86.
by leakage, 84, 91.
by residual velocity, 50, 85, 94, 190, 472.
by shock, distribution of, 191.
by shock or impact, 82.
by unutilized energy, distribution of, 190, 472.
from sluice gate, 86, 94.
in guide passages, 82, 88, 215.
in supply pipe, 81, 87.
in wheel, 84, 92, 217, 241.
Losses due to width of wheel, 129, 145, 471.
in impulse turbines, 214, 217, 223.
of energy in hydraulic pressure engines, 508.
summary of, 95.

M.
Measurement of areas of outflow, 117, 122, 375.
Measurements of water, 312.
by current meters, 313.
by floats, 315.
by meters, 328.
by weirs, 316.
direct, 313.
levels, 329.
comparison of, 490.
Meissner, experiments of, 78.
on correction of centrifugal action, 254.
summary of losses by, 95.
Meter, " Venturi," 330.
current, 313.
Meters, measurement of water by, 328.
Meyer's hydraulic motor, 526.
Mixed-flow turbines, 25, 34.
comparison with radial-flow, 195.

Mixed-flow turbines, development of, 33.
  expression for useful work, 58, 506.
Momentum of water, 16.
Morris Co., turbines of, at Niagara Falls, 443.
Motion, relative and absolute, 10.
Mouthpieces, conical, 207.

N.

Nagel and Kämp, turbine of, 184.
Niagara Falls Installation, 442.
Notation, 47.
Number of buckets. *See* Number of vanes.
  of vanes, 126, 136, 231, 258, 260.
Numerical examples, 283.

O.

Olching, experiments with turbines at, 368.
Outflow areas, ratio of, 107.
  effect of ratio on co-efficient of flow, 108.
  effect of ratio on efficiency, 134.
Outflow, mean angles of, 124.
  observation of direction of, 378.
  radial, 103.
Outward-flow Girard turbine, 43.
  radial turbines, 26, 39.
  turbines, difference between their action and that of inward-flow turbines, 195.

P.

Parallel-flow turbines, 25, 36. *See also* Axial turbines.
  centrifugal action in, 247, 250.
Partial admission, 25, 234.
Passages, depth of, 121, 229.
  width of, 136, 230, 259.
Path, absolute, of water, 19, 160.
  relative, of water, 20.
"Pelton" wheel or turbine, 435.

Philadelphia, turbine trials at, 489.
Pipe, supply, loss in, 81, 87.
  system compared with turbine, 49.
Pipes, motion of water in, 7.
Pitch of vanes, 136, 235, 258.
Poncelet and Lesbros, experiments on flow over weirs, 319.
Poncelet water-wheel, 267.
Power, measurement of. *See* Descriptions of and experiments with turbines.
Pressure difference, 104, 263.
  difference, curve of, 76, 499.
  effect of excess of, 62.
  engines, hydraulic, 500.
Pressures at orifices, 51, 104, 262.
Principles, general, of hydraulic motors, 6.
Proportions, effect of on performance of turbines, 164, 261.
Purdon-Walters stream motor, 452.

Q.

Quantity of water, 118.

R.

Radial-flow turbines, comparison with mixed-flow, 195.
Radial impulse turbines, 256.
  inward-flow turbines, 32, 195, 256, 463.
  outflow, 103.
  outward-flow turbines, 28, 39, 195, 256, 455.
  turbines, comparison with axial turbines, 192.
Radii of turbines, 18, 60, 125, 136, 226, 234.
Ratio of areas, 107, 118.
  of areas, effect on co-efficient of flow, 108.
  of areas, effect on efficiency, 168.
  of radii, effect on efficiency, 479.

Reaction turbines, 25.
  compared with impulse turbines, 265.
  compared with pipe system, 49.
  differences between reaction and impulse turbines, 262.
  essential parts of, 45.
  turbines, general theory, 47.
  losses in, 81.
Regulating reaction turbines, methods of, 171.
Regulation by closing guide passages, 173.
  by head-race sluice, 171.
  by subdivision, 184.
  by tail-race sluice, 172.
  by throttle-valve, 173.
  correct system of, 184.
Relative velocity of flow, 10, 14, 104, 218, 259.
  effect of speed on, 219.
  for impulse turbines, 218, 259.
Residual velocity, 16, 50, 66.
  loss from, 85, 94, 190, 217, 472, 509.
Resistance of sluice, 86, 94.
Resistances, classification of, 50.
Rieter, J. J., turbines of, at Schaffhausen, 428.
Rigg's hydraulic motor, 514.
"Risdon" turbine, 27, 489.
Rittinger, experimental turbines of, 67, 77, 337.
Rotary pressure motor, 512, 514.

S.

Schaffhausen, Jonval turbines at, 428.
Schmid's hydraulic motor, 518.
Shaft friction of turbines, 275.
Shock, avoidance of, 14, 101.
  effect on velocity of flow, 22, 71.
  loss by, 82, 129.
Sluice gate, loss from, 86, 94.
  in head-race, regulation by, 171.

Sluice gate, in tail-race, regulation by, 172.
Sources of hydraulic power, 4.
South Germany, water power in, 2.
Speed, best, 73, 228, 455.
  co-efficient of, 112.
  effect on efficiency, 73, 77, 221.
  effect of on velocity of flow, 63, 69, 221.
  effect of variations on performance of turbines, 75.
  of turbine without load, 73.
Stream, deflection of, 12, 61.
  depth of, 115.
  motor of Purdon and Walters, 452.
Suction-tube, 25, 33, 36.
  best form of, 202.
  diameter of, 138, 210.
  on the use of, 200.
  position of wheel in, 55, 209.
  theory of, 52.
Suction-tubes, loss in, 205.
Summary of facts about reaction turbines, 96.
  of losses, 95.
  of rules and formulas, 281.
"Swain" turbine, analysis of results, 450.
  experiments with, 446.
Switzerland, water power in, 2.

T.

Table A, comparison of experimental and calculated results, 455.
  B, dimensions of turbines as constructed, 456.
  I., co-efficients of flow, 492.
  II., efficiencies, 498.
Tangent wheel, Zuppinger's, 39, 256.
Terni, turbines at, 421.
Testing station, Holyoke, 391.
Tests of turbines in America, 196. *See also* Experiments, &c.

Theory, general, of hydraulic pressure engines, 501.
  general, of reaction turbines, 47.
  of impulse turbines, 213.
  of suction-tube, 52.
Thickness of vanes, 136, 260.
Thomson's turbine, 156, 463.
Throttle-valve, regulation by, 173.
Tremont turbines, experiments with, 372.
Trials. *See* Experiments.
Tumbling bays, 316. *See also* Weirs.
Turbine, at Goeggingen, experiments with, 182, 365.
  at Immenstadt, experiments with, 427.
  experiments with "Boyden," 400.
  "Hercules," 401.
  "Humphrey," 406.
  "Little Giant," 410.
  "Swain," 416.
  of Professor James Thomson, 463.
  the "Pelton," 435.
Turbines, 4.
  at Olching, experiments with, 368.
  at Terni, 421.
  at Varzin, experiments with, 413.
  compound, 270.
  classification, 24.
  description of and experiments with, 335.
  description of typical, 26.
  essential construction, 24.
  experiments with, 335.
  experiments with at Boott cotton mills, 384.
  experiments with at Tremont, 372.
  experiments with "Collins," 396.
  impulse, 25, 39, 211.

Turbines, in general, 24.
  radial impulse, 256.
  to act both as impulse and reaction wheels, 242.
  "Boyden," 400.
  "Callon," 39.
  "Collins," 396.
  "Fourneyron," 28, 183, 455.
  "Francis," 32, 48, 384.
  "Girard," 39. *See* Turbines, impulse.
  "Henschel," 144.
  "Hercules," 401, 468, 481.
  "Humphrey," 183, 197, 406, 468.
  "Jonval," 27, 36.
  "Little Giant," 410.
  "Nagel and Kämp," 184.
  "Pelton," 435.
  "Risdon," 27, 489.
  "Swain," 450.
  "Vortex," 156.
  "Zuppinger," 39.
Types of motors, 4.

U.

Unutilized energy, loss from, 50, 74, 85, 94, 190, 223, 472.
  relative losses by, in parallel and mixed-flow turbines, 472.
Use of water for work, 1.
Useful work, 50, 57, 215, 506.

V.

Vane, angles, choice of, 149.
  correction of, 145, 239.
  curve, volute, 124, 156.
  curves, 140, 237.
  surface, helical, 187, 251.
Vanes, action of water on, 12.
  angles of, 13, 61, 124, 144.
  back, 154.
  construction of, 140, 237, 239, 252.
  general remarks on construction of, 159.
  guide, 25.

Vanes, number of, 126, 136, 236, 258.
of Jonval turbine, 36.
thickness of, 136, 260.
wheel, obstruction by, 119, 225.
Varzin, experiments with turbine, at, 413.
Velocity, absolute, 10, 14.
of flow, 15, 62, 67, 106, 213.
of flow, curve of, 76.
of flow, effect of angles on, 69.
of flow, effect of speed on, 63, 69.
of flow, formulas for, 64, 106, 213.
of flow, relative, for impulse turbines, 218.
of rotation, 18, 22. *See* Speed.
of rotation, mean, 59.
of vane, 13.
relative, 10, 14.
residual, 16, 50, 66.
residual, loss from, 85, 94, 190, 217, 472, 509.
theoretical, 64.
Ventilation of buckets, 43, 211.
Venturi, experiments of, 206.
"Venturi" meter, 330.
Volute vane curve, 124, 156.
"Vortex" turbine, 156.

W.

Water, capacity of for work, 3.
distribution of in turbines, 187.
flow of, in closed pipes, &c., 7.
levels, measurement of, 329.
measurement of, 312.
measurements, comparison of, 490.
meter, the "Venturi," 330.

Water, meters, 328.
power in Europe and United States, 2.
power in United States, 198.
power, real agent in, 3.
pressure engine, 4, 500.
quantity of, 118.
Water-wheel, 4.
Poncelet, 267.
primitive form of, 1.
Weight of water flowing through turbine, 58.
Weir, 5.
measurement of water by, 316.
Weisbach, experimental co-efficients of, 89.
Weisbach's formula for flow over weirs, 318.
Wheel, Poncelet, 267.
the "Pelton," 435.
vanes, construction of, 143.
Wheels of various turbines, 27.
Width of guides and buckets, 115, 125, 136, 231, 259.
of turbines, effect on performance, 129, 145, 236, 471.
Wood and Co., turbines of, at Niagara Falls, 445.
Work, conditions under which water can do, 3.
curve of, 76.
effective, 57.
lost on friction, &c., 50.
performed on curved vane, 17, 57.
useful, 50, 57, 215, 506.
Wyss and Studer, hydraulic motor of, 521.

Z.

Zuppinger's tangent wheel, 39.

www.ingramcontent.com/pod-product-compliance
Lightning Source LLC
Chambersburg PA
CBHW011301210326
41599CB00035B/7082